Practice in Physics

fourth edition

Tim Akrill
George Bennet
Chris Millar

HODDER EDUCATION
AN HACHETTE UK COMPANY

Hachette UK's policy is to use papers that are natural, renewable and recyclable products and made from wood grown in sustainable forests. The logging and manufacturing processes are expected to conform to the environmental regulations of the country of origin.

Orders: please contact Bookpoint Ltd, 130 Milton Park, Abingdon, Oxon OX14 4SB. Telephone: (44) 01235 827720. Fax: (44) 01235 400454. Lines are open 9.00–17.00, Monday to Saturday, with a 24-hour message answering service. Visit out website at www.hoddereducation.co.uk

First published in 2011 by
Hodder Education
An Hachette UK Company,
338 Euston Road
London NW1 3BH

Impression number 6
Year 2015 2014

Typeset in 10.5/12pt Minion Pro by Fakenham Prepress Solutions, Fakenham, Norfolk NR21 8NN
Printed and bound by CPI Group (UK) Ltd, Croydon, CR0 4YY

A catalogue record for this title is available from the British Library

ISBN 978 1 444 121254

Contents

About this book ...

This is a book of questions to help you understand Physics during the two years before you go to University or College, perhaps while following an AS or A Level Physics course or preparing for other examinations such as the Pre-U or the IB. None of the questions are from previous examination papers; questions in examination papers are meant to test you at the end of your course or module. What you need during the course is to do questions which will help you to check whether you have understood what you are being taught. That is why we have called this book *Practice in Physics*.

The first edition of this book was originally published in 1979 and it has been in print ever since. We have revised it to take account of the new specifications (syllabuses) for Advanced Level courses that started in September 2008. Some of the questions test whether you have understood the principles, and a few should make you think quite hard. Questions indicated by an asterisk (*) are designed to give you practice at answering questions about *How Science Works*; for example, appreciating the tentative nature of scientific knowledge or interpreting data presented in a variety of forms.

We have included a chapter of questions (Chapter 10) which will help you elsewhere in the book when you need certain mathematical techniques. Some of these are very general – others will help you with particular chapters. There are also a few questions suggesting that you search the web for information to learn more about an up-to-date application of physics or to illustrate an historical breakthrough. Specific references are not given as these tend to change rapidly year by year.

At the end of the book there are answers to nearly all the questions. You will not need to do them all! But we hope you enjoy doing most of them because part of the pleasure of doing Physics is to discover that you can get the right answers, showing that you understand the ideas. Throughout the book answers are given to the same number (usually 2) of significant figures as the data in the question, but when answers to the later parts of a question depend on the answers to earlier parts, you should use the unrounded figures for the later parts.

Tim Akrill and Chris Millar
June 2011

Photo Credits

p.7 TRL LTD./SCIENCE PHOTO LIBRARY; **p.8** © Houghton Mifflin Harcourt Publishing Company, Printed with permission; **p.14** Neil Tingle/Action Plus; **p.21** Neil Tingle/Action Plus; **p.25** Imagestate Media; **p.37** © manfredxy - Fotolia.com; **p.46** Sascha Wilsrecht – Fotolia; **p.67** HATTIE YOUNG/SCIENCE PHOTO LIBRARY; **p.75** Kallista Images/Getty Images; **p.83** *r* Darren Brode – Fotolia; **p.87** *l* © Houghton Mifflin Harcourt Publishing Company, Printed with permission; *r* © Houghton Mifflin Harcourt Publishing Company, Printed with permission; **p.88** © Houghton Mifflin Harcourt Publishing Company, Printed with permission; **p.89** © IOP Publishing Ltd , Printed with permission; **p.96** *l* Reprinted from *Advances in Biological and Medical Physics,* Vol. 5 no. 211, A. Rose, "Quantum effects in human vision", pp.211–242, Copyright 1957/ Reprinted with permission by Elsevier, *r* Reprinted from *Advances in Biological and Medical Physics,* Vol. 5 no. 211, A. Rose, "Quantum effects in human vision", pp.211–242, Copyright 1957/Reprinted with permission by Elsevier; **p.103** *l* © Houghton Mifflin Harcourt Publishing Company, Printed with permission, *r* © Houghton Mifflin Harcourt Publishing Company, Printed with permission; **p.129** Betsie Van der Meer/Stone/Getty Images; **p.151** NASA; **p.153** NASA; **p.158** NASA; **p.159** © MBI/Alamy; **p.168** © Science Museum/Science & Society Picture Library; **p.180** LAWRENCE BERKELEY NATIONAL LABORATORY/ SCIENCE PHOTO LIBRARY; **p.190** LAWRENCE BERKELEY LABORATORY/ SCIENCE PHOTO LIBRARY; **p.201** NASA/JPL/California Institute of Technology, **p. 233** © Science Museum/Science & Society Picture Library.

Mechanics: linear motion

Data: free fall acceleration at the Earth's surface $g = 9.81\,\text{m}\,\text{s}^{-2}$

1 mile $\equiv$ 1.6 km

speed of sound in air = $340\,\text{m}\,\text{s}^{-1}$

speed of light in a vacuum $c = 3.00 \times 10^{8}\,\text{m}\,\text{s}^{-1}$

1.1 Speed and velocity

In this section you will need to

- use the equation average velocity = $\Delta s/\Delta t$
- understand that displacement and velocity are vector quantities
- remember how to measure speed
- understand that the gradient of a displacement–time graph is the velocity
- understand that the area between a velocity–time graph and the time axis represents the displacement.

1.1 Starting from home, a jogger runs 4.0 km (about 2.5 miles). She returns home after 20 minutes. What is **(a)** her average speed **(b)** her average velocity?

1.2 The diagram shows the oil spots left on a road by a motorbike with a leaky sump as the bike travels from A to G. Describe the journey, assuming that the drips come at regular time intervals.

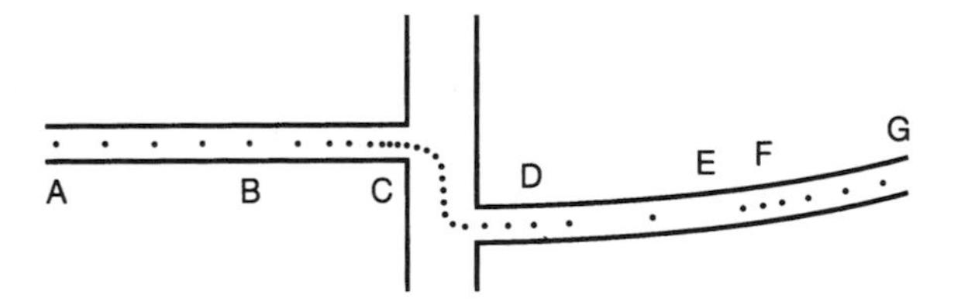

1.3 The diagram shows the observed movement of a smoke particle in a Brownian motion experiment.

(a) Use a ruler to find **(i)** the total distance moved by the smoke particle in going from A to B **(ii)** the displacement AB.

(b) If it took 1.20 s to travel from A to B, calculate **(i)** the average speed **(ii)** the average velocity of the smoke particle.

A
B
0.01 mm

1.4 What is the *change* of velocity when

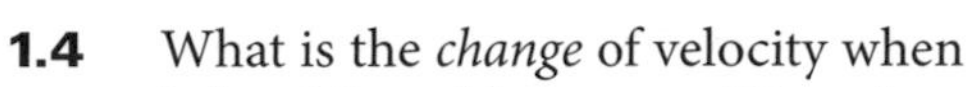

(a) $+6.0\,\text{m}\,\text{s}^{-1}$ becomes $+15\,\text{m}\,\text{s}^{-1}$
(b) $+6.0\,\text{m}\,\text{s}^{-1}$ becomes $-15\,\text{m}\,\text{s}^{-1}$
(c) $+6.0\,\text{m}\,\text{s}^{-1}$ becomes $-6.0\,\text{m}\,\text{s}^{-1}$
(d) $5.0\,\text{m}\,\text{s}^{-1}$ east becomes $15\,\text{m}\,\text{s}^{-1}$ west?

1.5 A skier moves at 11.0 m s^{-1} down a 16° slope. What is the skier's **(a)** vertical velocity **(b)** horizontal velocity?

1.6 The European Space Agency's *Giotto* spacecraft encountered Halley's comet in March 1986, when the comet was 93 million kilometres from Earth. How long did it take radio signals (travelling at the speed of light) to reach Earth from *Giotto*? [Use data.]

1.7 The graph shows the forward motion of a swimmer who dives in and swims a length in a 50-m pool.

(a) What is the swimmer's average speed during **(i)** the first 10 s **(ii)** from 20 s to 35 s **(iii)** the last 10 s?
Suggest why the average speeds are different.

(b) At 20 s, at what speed is the swimmer moving?

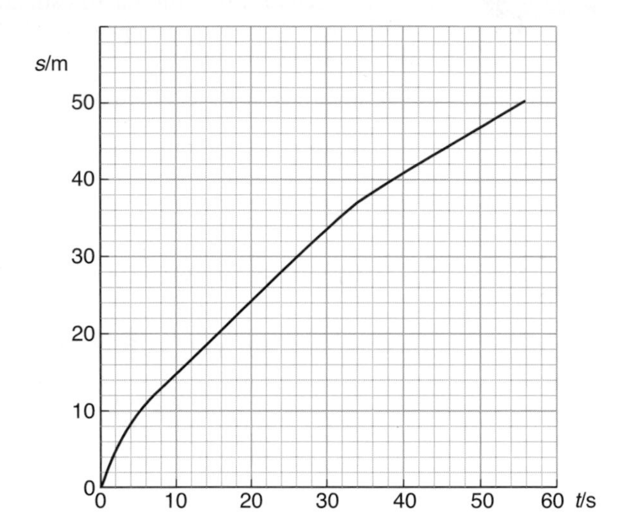

1.8* When races are timed manually, timers start their watches when they see the smoke from the starting gun rather than when they hear the gun. How much error is introduced in timing a 100-m race if the watch is started on the sound rather than on the smoke? [Use data.]

1.9* **(a)** Show that 25 m.p.h. is equivalent to about 11 m s^{-1}. [Use data.]
(b) In an emergency, you have a reaction time of 0.60 s. Calculate how far you would travel in this time on a bicycle moving at 25 m.p.h.

1.10 The diagram has been drawn after studying a stroboscopic photograph of a golf swing. The stroboscope was flashing 50 times per second.

(a) Sketch a graph showing how the speed of the head of the golf club varies with time from A to D.

(b) Using the scale from the 1.00 m ruler at the bottom of the diagram, estimate the speed of the club-head from **(i)** B to C **(ii)** P to Q.

(c) Estimate the speed of the golf ball between X and Y.

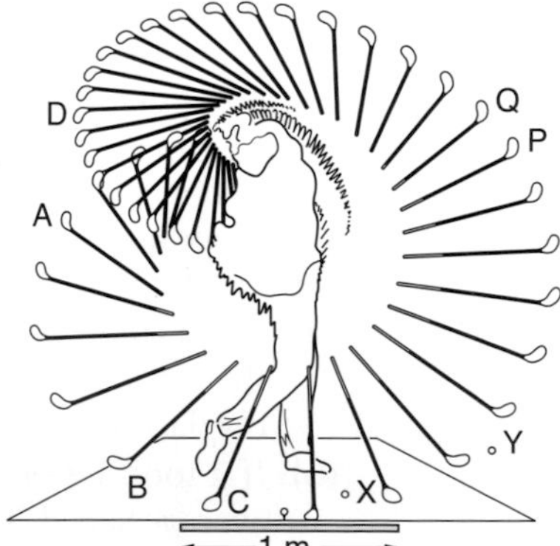

1.11 Approaching Terminal 3 at Heathrow Airport, passenger P uses the walkway and, having heavy luggage, allows it to take him along. Passenger Q walks alongside the walkway and passenger R walks on it, both walking briskly at 1.2 m s^{-1}. The walkway is 40 m long and moves at 0.80 m s^{-1}.

(a) Calculate how long it takes P, Q and R to reach the other end of the walkway.
Suppose a small boy on the walkway moved across it from one side to the other at a speed of $0.6\,\text{m s}^{-1}$.
(b) Draw a vector diagram illustrating the boy's motion and calculate his resultant velocity.

1.12 Show that 50 m.p.h. is only 1% larger than $80\,\text{km h}^{-1}$. [Use data.]

1.13 A speed skier registers an average speed of $233.7\,\text{km h}^{-1}$ over a distance of exactly one kilometre.
(a) How long did he take to cover the kilometre?
(b) Express his average speed in m s^{-1}.

1.14 The graph shows the motion of a stone thrown vertically upward. Calculate the maximum height reached by the stone
(a) by first finding the average velocity of the stone
(b) by finding the area under the graph.

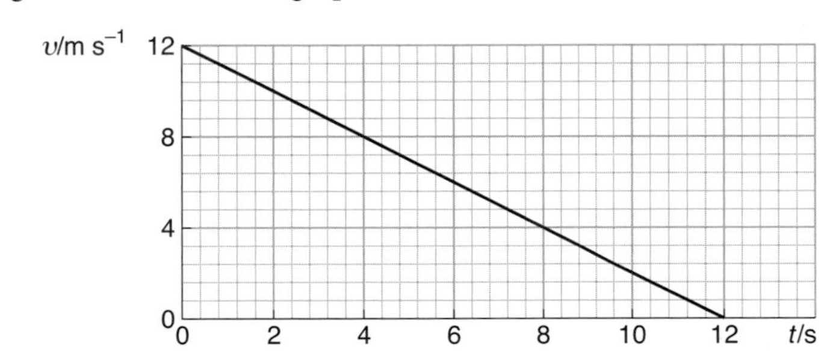

1.15 A snooker ball is rolled to strike the side cushion of a full-sized snooker table at right angles, so that it bounces across the table several times.
The graph shows how its velocity changes during its first three crossings (assuming it loses speed only when it bounces).
(a) Carefully describe the motion of the snooker ball.
(b) Do two *separate* calculations to determine the width of the snooker table.

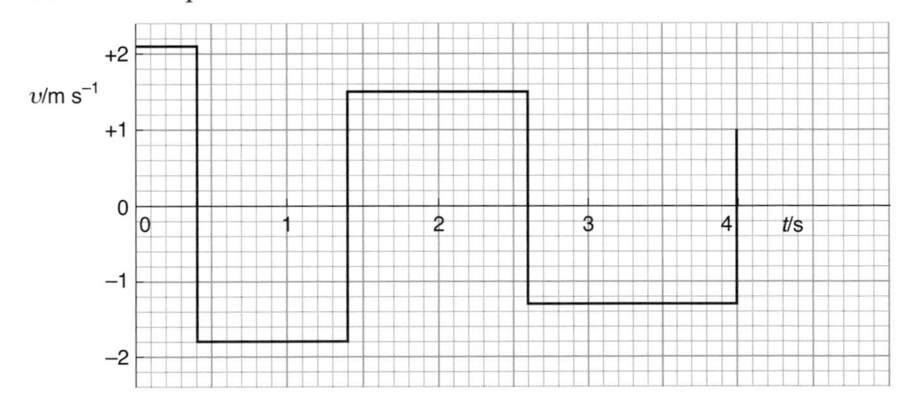

1.16 Draw a velocity–time graph for a tennis ball which is being volleyed backwards and forwards by two players close to the net. Assume that the ball travels horizontally and perpendicular to the net but that the players hit it so that it travels at a variety of speeds.

1.17 An ultrasonic displacement sensor is used to study the motion of a trolley sliding down a ramp in the laboratory. The displacement against time data is presented on-screen as a graph and then converted to a velocity–time graph as shown.

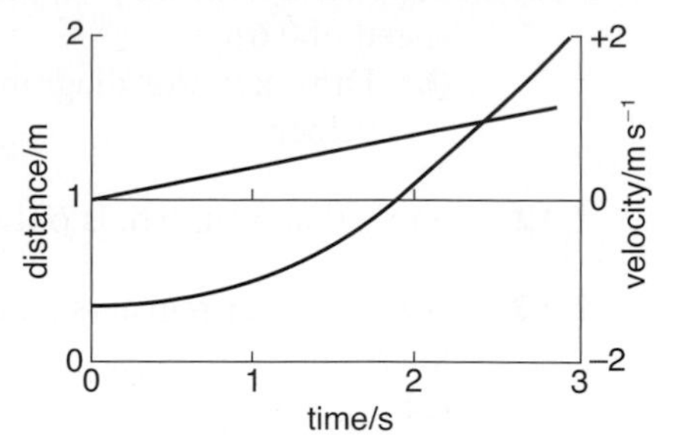

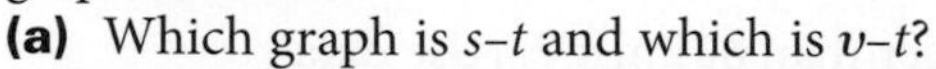

(a) Which graph is s–t and which is v–t?

(b) Explain the relationship between the two graphs.

(c) Verify that the computer software has drawn the v–t graph correctly at $t = 2$ s.

1.18 Use a scale diagram

(a) to add a displacement of 5.6 km north to a displacement of 3.5 km north-west

(b) to add a velocity of 32 m s^{-1} south-west to a velocity of 18 m s^{-1} 20° west of north.

1.2 Acceleration along a line

In this section you will need to

- use the equation: average acceleration = $\Delta v/\Delta t$
- understand that the gradient of a velocity–time graph is the acceleration
- remember how to measure acceleration
- understand how to draw graphs for displacement, velocity or acceleration against time when given only one of them
- use the following equations for uniform acceleration: $v = u + at$, $s = \frac{1}{2}(u + v)/t$ and (when $u = 0$) $s = \frac{1}{2}at^2$

1.19 A baby buggy rolls down a ramp which is 15 m long. It starts from rest, accelerates uniformly, and takes 5.0 s to reach the bottom.

(a) Calculate its average velocity as it moves down the ramp.

(b) What is its velocity at the bottom of the ramp?

(c) What is its acceleration down the ramp?

1.20 A man, John L. Stapp, travelling in a rocket-powered sledge, accelerated from 0 to 284 m s^{-1} (about 630 m.p.h.) in 5.0 s and then came to a stop in only 1.5 s. Calculate his acceleration

(a) while he is speeding up

(b) while he is slowing down.

1.21 The graph shows, in idealised form, a velocity–time graph for a typical short journey.

(a) Calculate the acceleration at each stage of the journey and display your answers on an acceleration–time graph.

(b) Sketch a displacement–time graph for this journey.

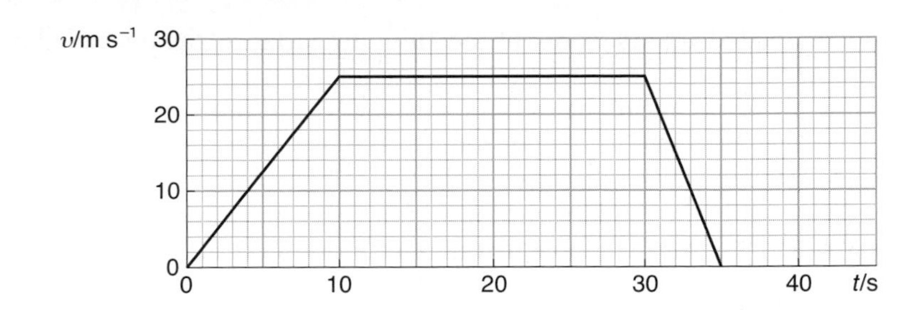

1.22 One type of aeroplane has a maximum acceleration on the ground of $3.5\,\mathrm{m\,s^{-2}}$.

(a) For how many seconds must it accelerate along a runway at this value in order to reach its take-off speed of $115\,\mathrm{m\,s^{-1}}$?

(b) What is the minimum length of runway needed to reach this speed?

1.23 Sketch a displacement–time curve, a velocity–time curve and an acceleration–time curve for an electrically powered milk trolley moving from one house to another on a straight road. Use the same time axes for all three graphs. [Hint: Start with the v–t graph.]

1.24 The graph shows the horizontal speed of a long jumper from the start of his run-up to the moment when he takes off.

(a) What is his maximum acceleration?

(b) Estimate the distance he runs before he takes off.

(c) Sketch the general shape of his acceleration against time.

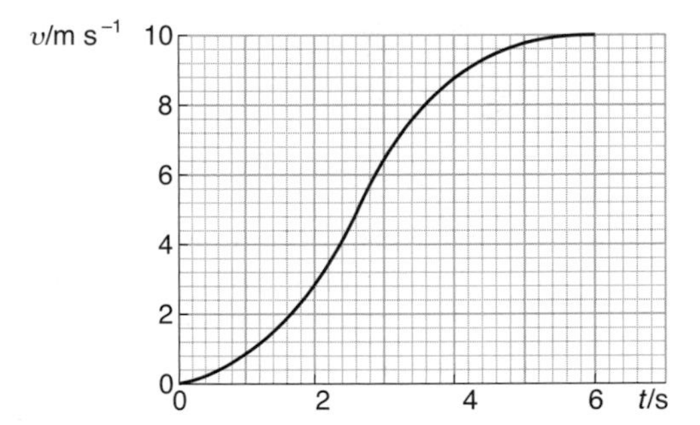

1.25 A particle moves in a straight line. Its motion can be described as follows:

at $t = 0$, $v = 0$

$0 < t < 10\,\mathrm{s}$, $a = 4.0\,\mathrm{m\,s^{-2}}$

$10\,\mathrm{s} < t < 20\,\mathrm{s}$, $a = -4.0\,\mathrm{m\,s^{-2}}$.

Sketch the velocity–time graph and use it to find the change of displacement of the particle between $t = 0$ and $t = 20\,\mathrm{s}$.

1.26* The UK *Highway Code* has a table of 'Typical Stopping Distances' on straight roads in dry conditions. The diagram is based on this information.

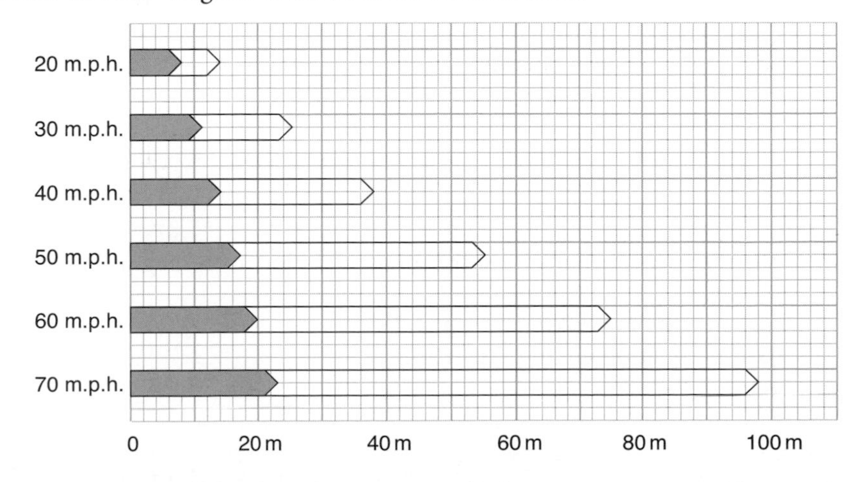

The shaded arrows represent the 'thinking distance', the unshaded part the 'braking distance' and the whole arrow the 'stopping distance'.

(a) Make a table of the thinking distances d_{think} and sketch a graph of d_{think} (y-axis) against speed v (x-axis) from $v = 0$ to $v = 70$ m.p.h.

(b) Deduce a relation between these two variables.

(c) Predict the thinking distance for a police car travelling at 90 m.p.h.

(d) Do you think that the driver's consumption of alcohol would affect the stopping distances? Explain your answer.

1.27* **(a)** Using the *Highway Code* data from the previous question, make a table of the braking distances d_{brake} and sketch a graph of d_{brake} (in metres on the y-axis) against speed v (in m.p.h. on the x-axis), from $v = 0$ to $v = 70$ m.p.h.

(b) The relationship here is that $d_{brake} = kv^2$, where k is a constant. Test this statement by calculating k for three numerical values of d_{brake} and v.

(c) Use the data to calculate the acceleration when braking from **(i)** 50 m.p.h. **(ii)** 70 m.p.h.

1.28* In France the motorways have different speed limits depending on the road condition. They are:

when dry $130\,\text{km}\,\text{h}^{-1}$

when wet $110\,\text{km}\,\text{h}^{-1}$

Using the data translate these into m.p.h. and comment on the French system.

1.29 **(a)** Slow motion photography shows that a jumping flea pushes against the ground for about 0.001 s during which time it accelerates upwards to a maximum speed of $0.8\,\text{m}\,\text{s}^{-1}$. What is its upward acceleration during this 'take-off'?

(b) It then moves upwards with an acceleration of $-12\,\text{m}\,\text{s}^{-2}$. (This is assumed to be constant and includes the effect of air resistance.) Calculate **(i)** how long it takes from leaving the ground to the top of its jump **(ii)** how high it jumps.

1.30 Electrons in a particle accelerator are moving at $8.0 \times 10^5\,\text{m}\,\text{s}^{-1}$ and are then accelerated to $6.5 \times 10^6\,\text{m}\,\text{s}^{-1}$ in 6.3×10^{-7} s.

(a) What is their acceleration in the tube?

(b) How far do they move during this time?

1.31 The graph shows the result of studying a sprint start.

(a) What was the maximum velocity reached? [Use data.]

(b) Estimate the acceleration of the sprinter **(i)** as she leaves her blocks **(ii)** after 2.0 s.

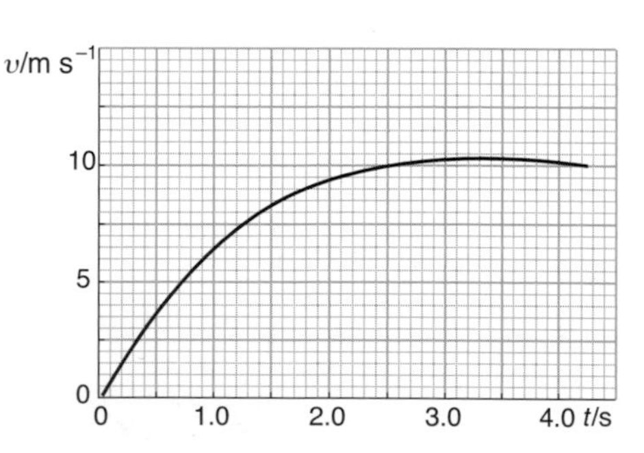

1.32 Use the two equations for uniform acceleration at the beginning of this chapter to produce

(a) an equation linking u, v, a and s

(b) an equation linking s, u, a and t.

1.33 A gazelle accelerates at $4.1\,\mathrm{m\,s^{-2}}$ from rest for a distance of 55 m in order to outrun a predator. What is its final speed?

1.34* A person who is properly held by a seat belt has a good chance of surviving a car collision if the deceleration does not exceed $30g$. Assuming uniform deceleration at this rate, calculate the distance that the front section of the car must 'crumple' if a crash occurs at $65\,\mathrm{km\,h^{-1}}$ (40 m.p.h.).

1.35 The best throwers in the world are baseball pitchers. They can release a ball travelling at $40\,\mathrm{m\,s^{-1}}$. In so doing they accelerate the baseball through a distance of 3.6 m. Calculate the average acceleration of the ball.

1.36 The graph describes the motion of a train moving in a speed-restricted area and then accelerating as it clears the area. You are to calculate the total distance travelled by the train in the 40 s shown in three different ways.

(a) Use the average velocity of the train during each 20 s interval to calculate two separate distances and add them together.

(b) Use equation **(b)** from question 1.32.

(c) Find the number of squares under the graph and the distance represented by one square.

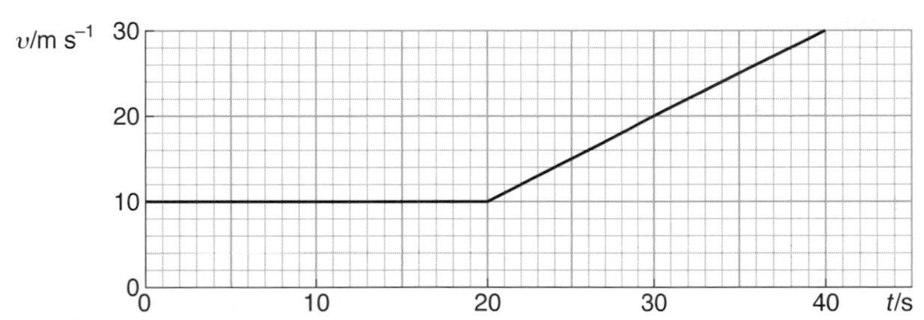

1.3 Free fall and projectile motion

In this section you will need to

- remember that the free fall acceleration at the Earth's surface is $9.8\,\mathrm{m\,s^{-2}}$
- remember how to measure the free fall acceleration in the laboratory
- use the equations $v^2 = 2gs$ and $s = \frac{1}{2}gt^2$ for free fall from rest
- understand that when an object is falling freely its vertical motion is independent of its horizontal motion
- remember that velocity vectors can be resolved into two perpendicular components, $v_x = v\cos\theta$ and $v_y = v\sin\theta$, where θ is the angle between v and the x-axis.

1.37 A ball is thrown vertically upwards at $19.6\,\mathrm{m\,s^{-1}}$.

(a) Make a table showing its velocity after 1.0 s, 2.0 s, 3.0 s and 4.0 s.

(b) What is its displacement after 2.0 s and 4.0 s?
(c) How far does it travel in the first 4.0 s?

1.38 **(a)** Ignoring air resistance, how long does an object take to fall from rest a distance of **(i)** 1.0 m **(ii)** 2.0 m?
(b) Why is the answer to **(ii)** not twice the answer to **(i)**?

1.39* **(a)** Explain how the apparatus shown can be used to measure *g*.
(b) What sources of error are there likely to be in such an experiment?

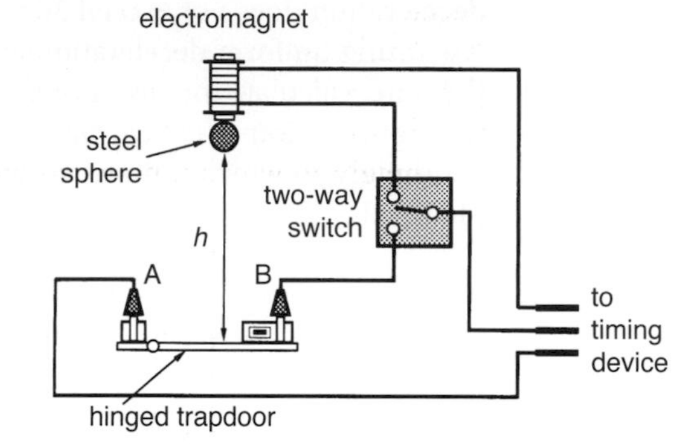

1.40 In an experiment with the above apparatus a steel sphere is found to fall a distance 456 mm in 301 ms. Calculate
(a) the average velocity of the sphere as it falls
(b) the velocity with which the steel sphere hits the trap door
(c) the acceleration of the steel sphere.

1.41 Parachutists hit the ground at about 6 m s^{-1}. How high a platform is needed for them to jump off in order to give them practice at hitting the ground at this speed?

1.42 In a cartoon two characters are standing by a well. One drops a stone down the well and starts to count. He stops counting when he hears the stone hit the water. He then announces proudly, 'Your well is exactly three seconds deep'. How deep is the well really?

1.43 A salmon moving upstream to its breeding grounds jumps a waterfall 2.5 m high. With what minimum speed must it leave the water below to reach the top level?

1.44 The photograph shows two golf balls, one released from rest and the other projected horizontally at the same moment.
(a) How does the photograph confirm that vertical motion in free fall is independent of horizontal motion?
(b) How would you use the diagram to confirm that the *horizontal* velocity of the projected ball remains constant as it falls?

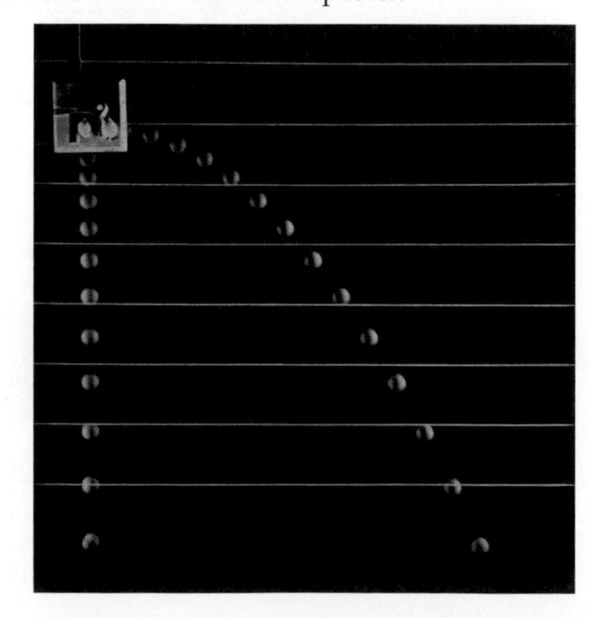

1.45 A bullet is fired horizontally at a speed of 200 m s^{-1} at a target which is 100 m away.

(a) Ignoring air resistance, calculate **(i)** how far the bullet has fallen when it hits the target **(ii)** the bullet's vertical velocity as it hits the target.

(b) What is the angle it then makes with the horizontal?

1.46 The diagram shows a velocity–time graph for a ball bouncing vertically on a hard surface.

(a) Explain the shape of the graph.

(b) Use the graph to calculate three separate values for the acceleration of free fall.

(c) Use the graph to calculate the height from which the ball was dropped and the height to which it bounced on **(i)** its first bounce **(ii)** its second bounce.

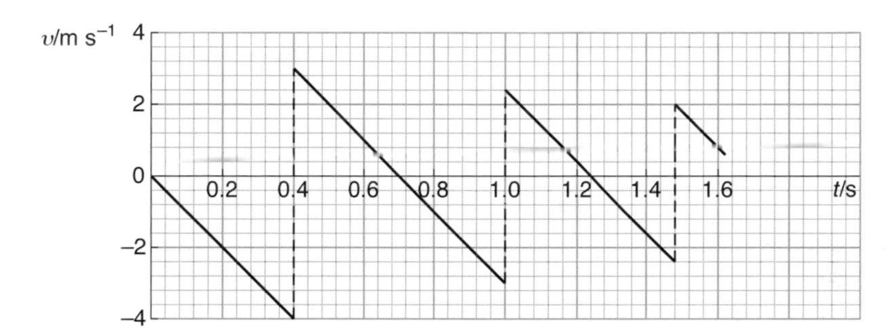

1.47 The Olympic flame at the 1992 Barcelona Olympics was lit by a flaming arrow that followed the trajectory shown in the diagram.

(a) The arrow was at the peak of its trajectory when it lit the flame. Show that it took 2.2 s to rise 24 m after being fired. Ignore air resistance.

(b) Hence calculate **(i)** the horizontal speed and **(ii)** the initial vertical speed of the arrow and confirm that it was fired at about 40° to the horizontal.

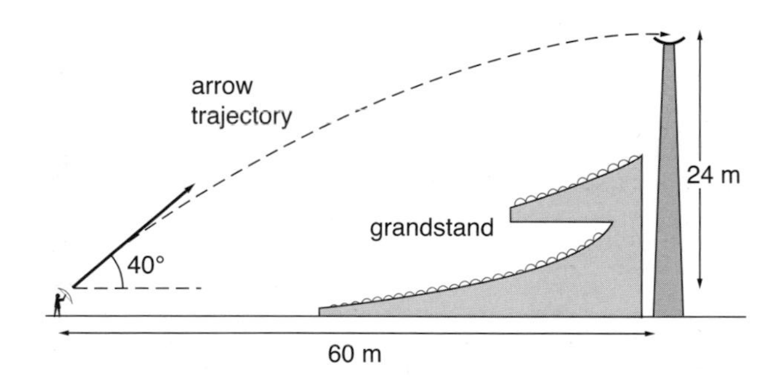

1.48 You drop a heavy stone from a high suspension bridge and one second later you drop a second stone.

(a) Draw two *v*–*t* graphs for the two stones on the same graph axes.

(b) Explain how the distance between the two stones changes as they fall towards the water.

1.49 You throw a stone vertically upwards and catch it as it comes down.
How could you best describe the motion of the stone to another person who is also studying physics?

1.50 The long jumper in the diagram is shown at the instant he leaves the ground, at three positions during his flight and at the instant he first touches the sand. His long jump measures 7.5 m and he is recorded as being in the air for 0.80 s. His centre of gravity falls 0.95 m between his take-off and landing.

(a) Calculate his horizontal velocity at take-off.

(b) Use the equation $s = ut + \frac{1}{2}at^2$, developed in question 1.32, to show that the vertical velocity u of the long jumper at take-off is 2.7(4) m s^{-1}. (Be careful with the signs of quantities in this equation: if upwards is positive, both s and g will have negative values.)

(c) Hence calculate the angle at which he projects himself at take-off.

1.51* The men's long jump world record went up from 7.61 m in 1901 (Peter O'Connor, GB) to 8.13 m in 1935 (Jesse Owens, US) to 8.95 m in 1991 (Mike Powell, US).

(a) Show that these roughly represent a steady rise during the 20th century.

(b) Bob Beamon (US) broke the world record in 1968. Use the data to predict what his record might have been.

(c) In fact Beamon jumped 8.90 m in Mexico City at the 1968 Olympic Games. Suggest why his record was much more than your prediction.

2 Balanced and unbalanced forces

Data: gravitational field strength $g = 9.81\,\mathrm{N\,kg^{-1}}$ (3 s.f.), $g = 9.8\,\mathrm{N\,kg^{-1}}$ (2 s.f.)

You will need to use one of these values for g in many of the questions in this chapter. Choose the one that has the same number of significant figures as the data in the question.

2.1 Forces in equilibrium

In this section you will need to

- understand that all forces are pushes or pulls of one body on another
- use the phrase 'the push (or pull) of A on B' when describing any particular force
- understand the meaning of the words weight and tension and of frictional and normal contact forces
- draw free-body force diagrams when analysing problems about bodies which are in equilibrium
- understand that when a body is in equilibrium the sum of the forces acting on it, resolved in any direction, is zero
- understand that forces occur in pairs which act on different bodies and that the push or pull of A on B is always equal in size to the push or pull of B on A
- understand how to resolve forces into two mutually perpendicular components and how to add two forces which are perpendicular to one another.

2.1 A new-born baby is said to be a healthy 7.8 pounds (lb). What is the baby's weight in newtons? Take 2.2 lb = 1.0 kg.

2.2 The bodies shown in each of the following free-body force diagrams are in equilibrium.

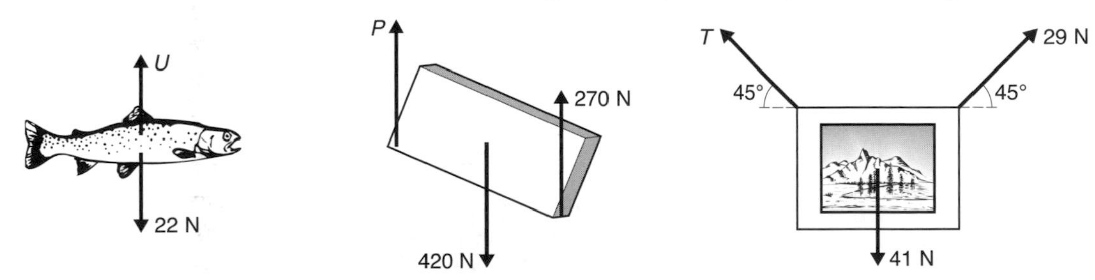

(a) Write down the value of the unknown force in each diagram.
(b) For the hanging picture, draw a closed vector triangle to confirm that the body is in equilibrium.

2.3 A child sits at rest on a swing. The figure shows free-body force diagrams for **(i)** the child **(ii)** the swing seat.

(a) For each of the forces P, W, T, w and P', identify the body which is producing the force. [P' is *not* produced by the Earth.]

(b Write a phrase describing each force as the push or pull of the identified body on **(i)** the child **(ii)** the seat.

(c) In this situation $P = W$, $T = w + P'$ and $P = P'$. Explain each equation in terms of Newton's laws.

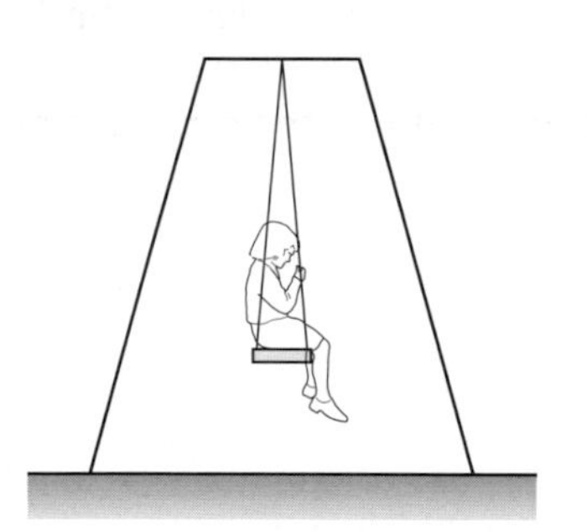

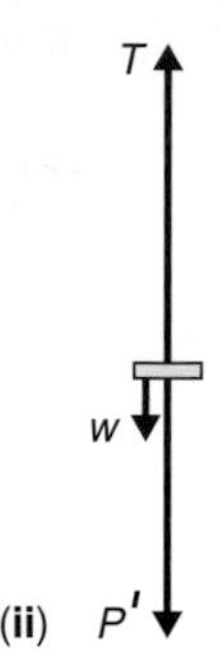

2.4 A child learning to swim is supported in a harness by her instructor who stands on the side of the pool. The forces acting on the child are:

the pull W of the Earth on the child, 300 N
the pull P of the harness on the child
the push U of the water on the child, 250 N.

(a) Draw a free-body force diagram for the child. How big is P?

(b) Newton's third law tells us that there are other forces equal in size to W, P and U. On which bodies do each of these forces act?

(c) Draw a free-body force diagram for the instructor who weighs 800 N. Deduce the normal contact push of the floor on him.

2.5 The diagram shows two skaters. At the moment shown the woman (left) is exerting a horizontal pull on the man who is moving at a constant velocity.

(a) Draw a free-body force diagram for the man.

(b) If the man weighs 700 N and his partner's *horizontal* pull on him is 100 N, what are

(i) the upward push of the ice on the man

(ii) the horizontal frictional force of the ice on the man?

(c) Calculate the total force of the ice on the man.

2.6 A racing car is shown in the diagram together with a free-body force diagram describing the forces acting on it.

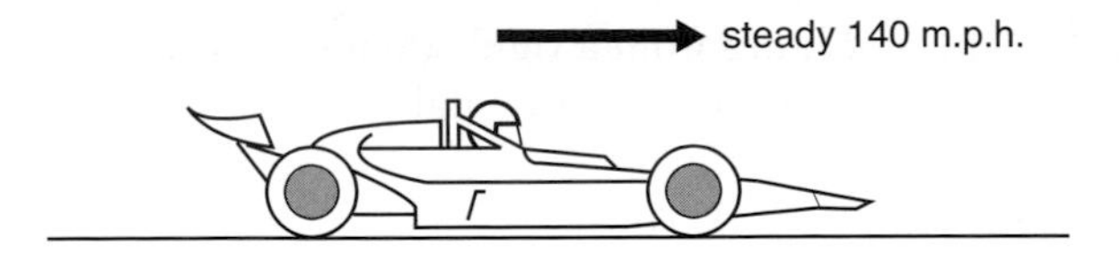

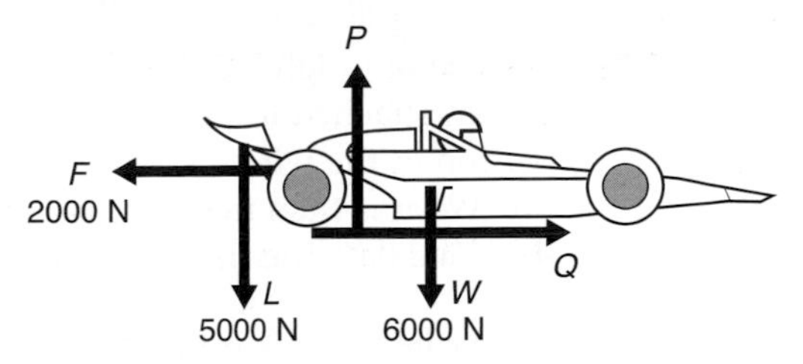

(a) Copy the free-body force diagram and list the forces using phrases which end with 'on the car'. [Q is *not* the push or pull of the engine on the car.]
(b) How big are the forces P and Q?
(c) What would happen to the size of **(i)** L **(ii)** F, if the car was moving more slowly?

2.7 A man is pulled by a dog on a lead. The pull of the lead on the man is 20 N and is inclined at an angle of 15° to the horizontal. What is the size of the resolved part of this force
(a) in the horizontal direction
(b) in the vertical direction?

2.8 A force P is 20 N in a direction N 60° E. What is the resolved part of the force
(a) north
(b) east
(c) N 30° E?

2.9 An empty sledge of mass 5.0 kg slides at a constant speed down a slope that makes an angle of 20° to the horizontal. The diagram shows a free-body force diagram for the sledge.
(a) Sketch a vector triangle showing that these forces are in equilibrium.
(b) Calculate the size of each of the forces.

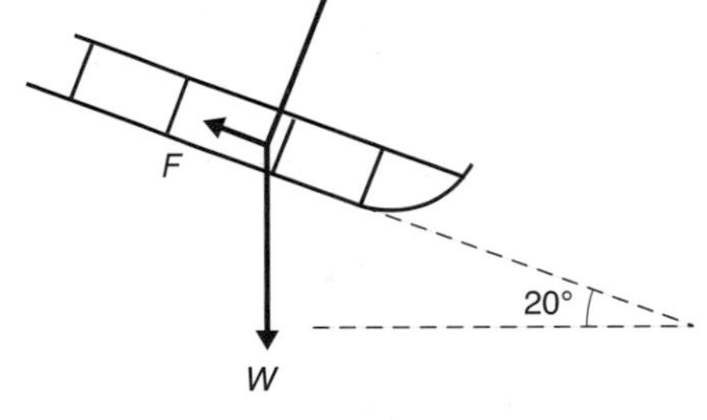

2.10* In designing access ramps for a shopping mall, it is decided that the maximum force required by someone pushing a person in a wheelchair of total mass 90 kg should be 80 N.
(a) Calculate the maximum angle for the ramps.
(b) Discuss whether your answer 'works' for people pushing heavier people in wheelchairs.

2.11 The diagram shows an end-on view of a cable car that has stopped because of high winds. The wind is exerting a steady horizontal sideways force of 5.2 kN on the cable car, which has a total weight of 24 kN. The cable car is in equilibrium.
(a) Representing the cable car as a blob, add the three forces acting on the cable car.
(b) Draw, to scale, a closed vector triangle for the three forces and deduce the angle between the support arm and the vertical.
(c) Is the tension in the support arm very much bigger than its value when there is no wind?

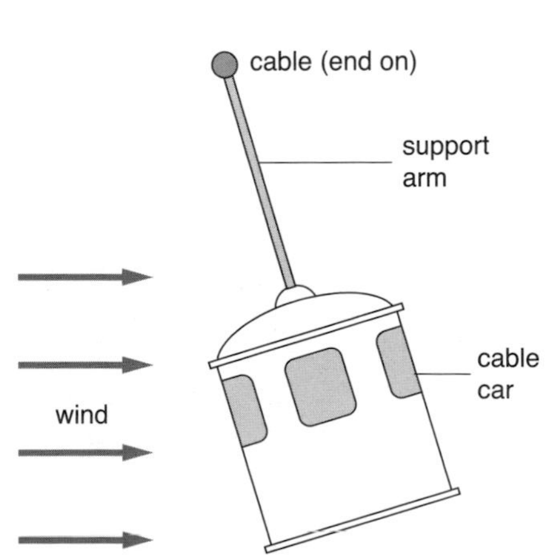

2.12 A stone of weight 32 N is attached to a wire and hung from a rigid support. A string is then attached to the stone and is pulled sideways with a horizontal force P until the tension in the wire is 44 N.
(a) What is then the angle between the wire and the vertical?
(b) Calculate the size of the force P.

2.13 A very heavy sack is hung from a rope and pushed sideways. When the horizontal sideways push is 220 N the rope supporting the sack is inclined at 18° to the vertical.
(a) Calculate the tension in the rope.
(b) Hence find the mass of the sack.

2.14 A gymnast of weight 720 N is holding himself in the cross position on the high rings. He is quite still. A free-body force diagram for the gymnast shows the two upward pulls of the rings on his hands, each of size 380 N.
Calculate the angle θ between the wires supporting the rings and the vertical.

2.2 Forces and moments

In this section you will need to

- use the equation: moment of a force about an axis = the force times the perpendicular distance from the axis to the line of action of the force
- draw free-body force diagrams for extended bodies in equilibrium when solving problems involving the principle of moments
- use the principle of moments: for a body in equilibrium the sum of the moments of the forces acting on it, about any axis, is zero
- understand that couples and torques exert moments on extended bodies.

2.15 The diagram shows a boy B and a girl G on a seesaw plus a free-body force diagram for the seesaw beam.

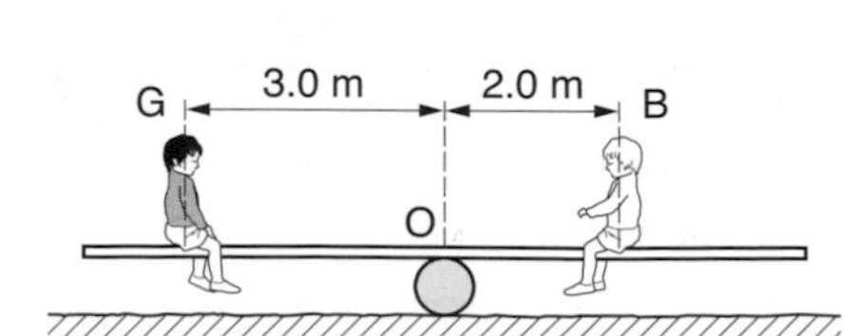

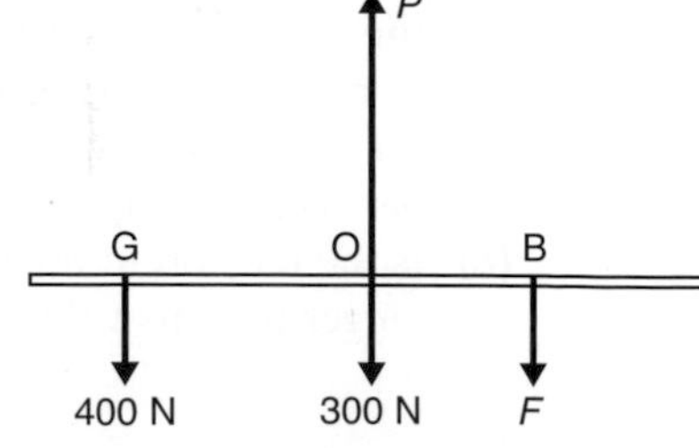

(a) Calculate the moment, about O, of the 400 N push of the girl on the beam.
(b) What is the moment, about O, of the 300 N pull of the Earth on the beam?
(c) Deduce a value for *F*, the push of the boy on the beam.

2.16 Using the diagram from the previous question:
(a) Calculate the moments, *about B*, of the 400 N and the 300 N forces.
(b) Deduce a value for *P*.
(c) Knowing the answer to part **(c)** of the previous question, how else can you deduce a value for *P*?

2.17* A baby-buggy and baby together weigh 140 N. The diagram shows the position of the centre of gravity G of the baby and buggy.
(a) In order to lift the front wheels up a step whilst moving forward the pusher exerts a vertical downward force *F* on the handle. Calculate the value of *F*.
(b) What is the corresponding upward force the pusher needs to exert to lift the back wheels of the pram up a step whilst moving backwards?

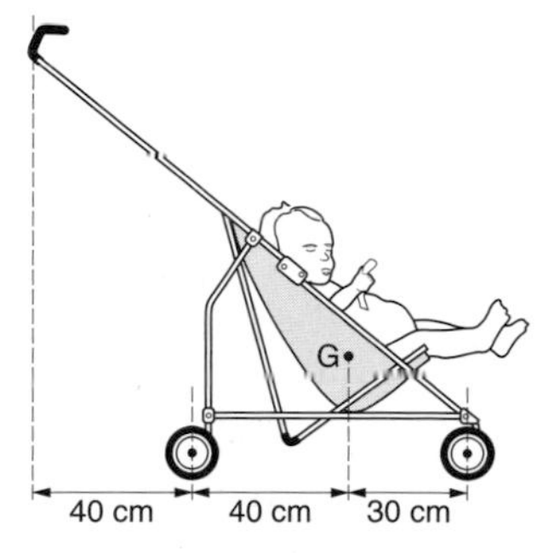

2.18 The diagram shows a free-body force diagram for a travel case of total mass 7.5 kg. The case is being held at rest by a hand at H, and its centre of gravity is at G.
(a) Use Newton's first law to write an equation relating the size of the forces *N*, *W* and *Y* acting on the case.
(b) Use the principle of moments to calculate the size of the upward pull *Y* of the hand on the case.
(c) Explain why repacking the case so as to move the centre of gravity closer to C will reduce the size of the force *Y*.

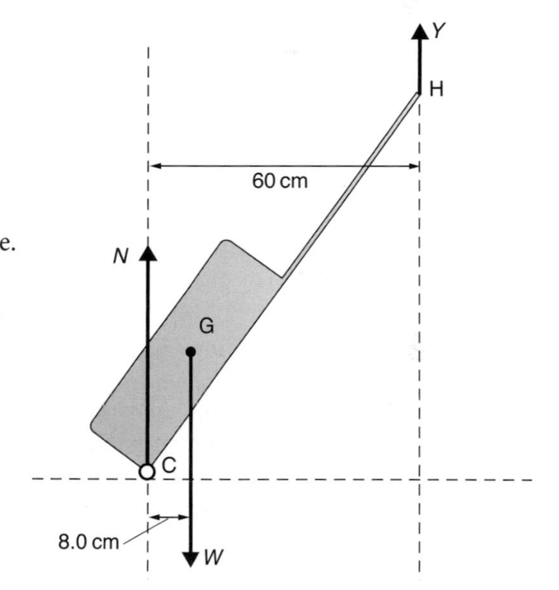

2.19* **(a)** Describe how you would use the apparatus in the diagram to locate the position of the centre of gravity, relative to his feet, of a person with his arms at his sides.
(b) Estimate the likely errors in the measurements you would take and hence the uncertainty which might arise in the experiment.

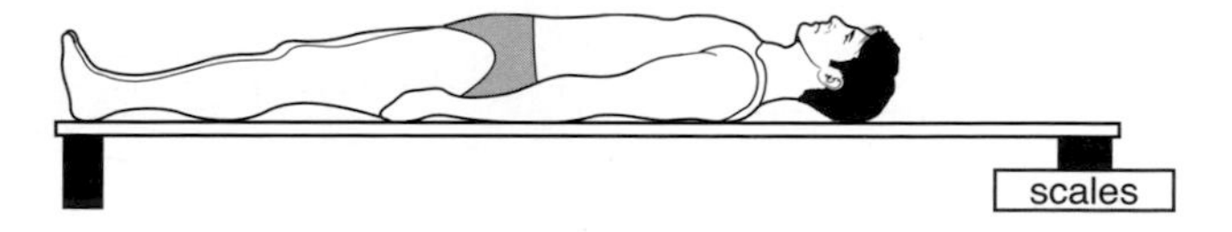

2.20 A uniform 'springboard' diving board 4.0 m long has a mass of 75 kg. It is attached at its inner end A and rests across a fixed support B as shown in the diagram. A diver of mass 40 kg stands at rest on the outer end of the springboard.

(a) Draw a diagram of the springboard showing each of the four forces acting on it, i.e. a free-body force diagram for the board.

(b) Calculate the size of the two vertical forces acting on the board at A and B. Use $g = 10\,\text{N kg}^{-1}$.

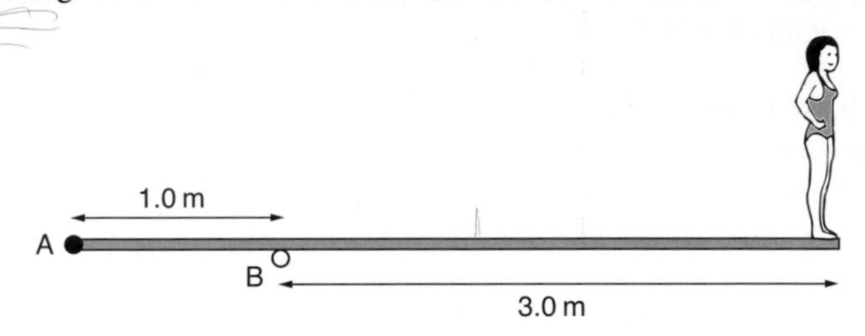

2.21 The diagram shows a lorry of mass 24 000 kg crossing a bridge. The span from A to B is 32 m.

(a) When the lorry is at the centre of the bridge, what is the *increase* in the upward push of each of the bridge supports A and B on the bridge structure?

(b) Where on the bridge will the lorry be when the additional upward push of the bridge support at A is 80 000 N?

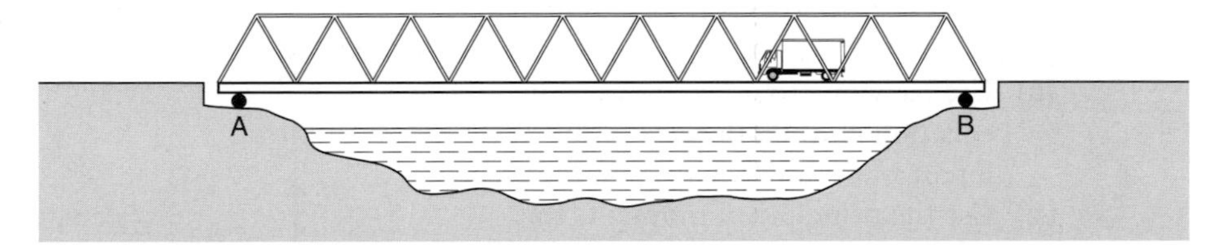

2.22 The diagram shows a simplified model of the forearm alongside a centimetre scale. It is estimated that the centre of gravity of the forearm is 20 cm from the elbow joint.

(a) If the forearm has a mass of 1.7 kg, calculate the moment of the weight of the forearm about the elbow joint.

(b) What is the horizontal distance of the biceps muscle from the elbow joint?

(c) Calculate the tension in the biceps muscle necessary to hold the forearm horizontal.

(d) How big is the vertical contact force between the bones of the elbow joint?

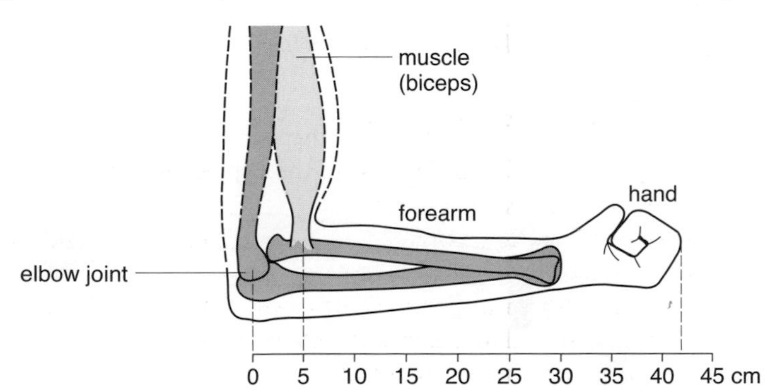

2.23 A ladder of mass 15.0 kg leans with its upper end against a frictionless wall as shown in the diagram.

(a) Describe the forces *N* and *F* shown on the free-body force diagram of the ladder.

(b) By taking moments about the bottom of the ladder, calculate *M*, the normal contact push of the wall on the top of the ladder.

(c) Write down the sizes of the forces *N* and *F*.

(d) Calculate **(i)** the total push of the ground on the bottom of the ladder **(ii)** the angle which this force makes with the vertical.

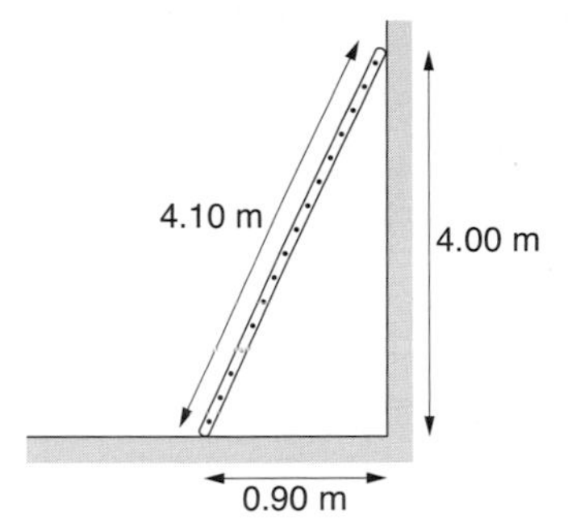

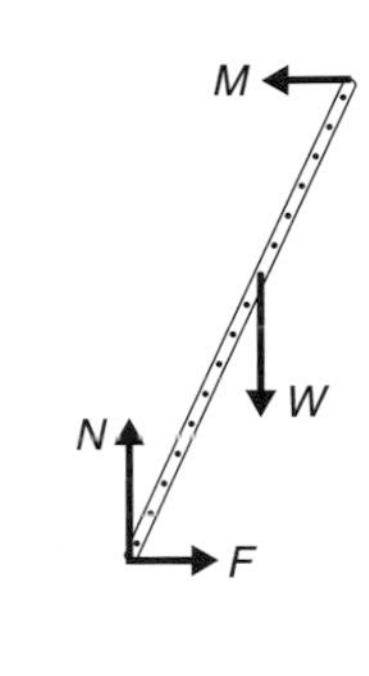

2.24 A uniform horizontal beam of weight 200 N and length 3.00 m is freely hinged to a wall. A rope is attached to its free end. The other end of the rope is attached to the wall at a point vertically above the hinge so that the rope makes an angle of 30° with the vertical.

A free-body force diagram for the beam is shown with the push of the wall on the hinge end of the beam resolved into two parts: horizontal *X* and vertical *Y*.

(a) Show that the perpendicular distance from the hinge to the rope is 2.60 m.

(b) Calculate the tension *T* in the rope.

(c) Hence calculate the sizes of the forces *X* and *Y*.

(d) The resultant push of the hinge on the beam acts at an angle of 60° to the horizontal. Explain why this must equal the angle between the rope and the beam.

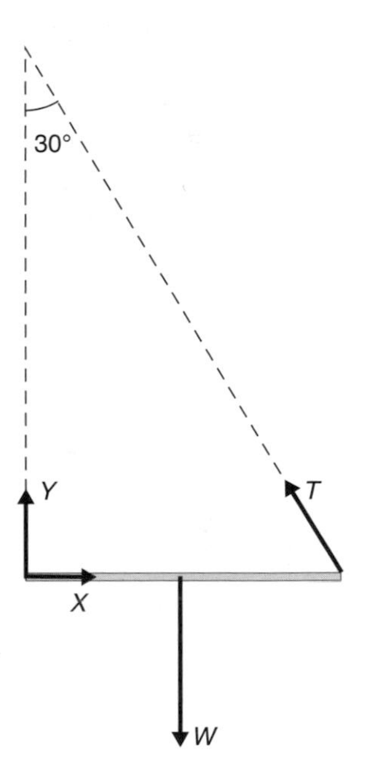

2.3 Forces causing acceleration

In this section you will need to

- draw a free-body force diagram for the accelerated body when solving problems involving Newton's second law
- use the equation, Newton's second law, for accelerated motion: ma = sum of forces resolved parallel to a
- describe experiments to demonstrate the validity of Newton's second law
- understand that all animals and vehicles accelerate forwards by pushing backwards on the ground, some water or the air.

2.25 The diagram shows five bodies **(a)** to **(e)** together with the mass of each. Calculate the acceleration of each body.

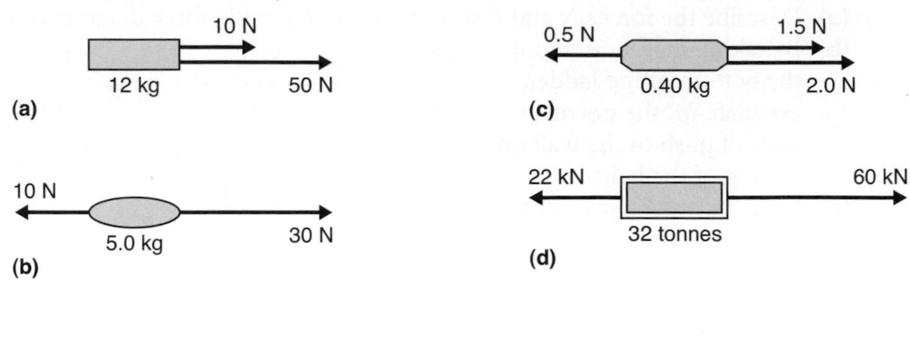

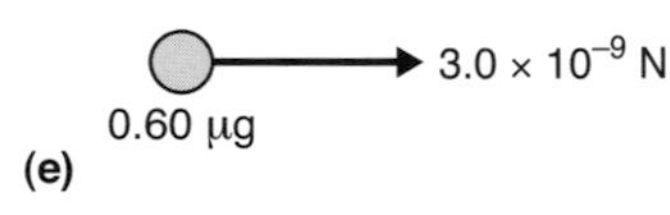

2.26 Calculate the *extra* force, in size and direction, that would need to be added to produce an acceleration of $2.0\,\mathrm{m\,s^{-2}}$ to the right for each object **(a)** to **(e)** shown in the above diagram.

2.27 Two women push a car of mass 800 kg to get it started. Each pushes with a force of 300 N and the resistance forces are equivalent to an opposing force of 160 N. What is the acceleration of the car?

2.28* A person is unlikely to be killed in a car crash if, held by a seatbelt, he or she accelerates at $-250\,\mathrm{m\,s^{-2}}$ or less.

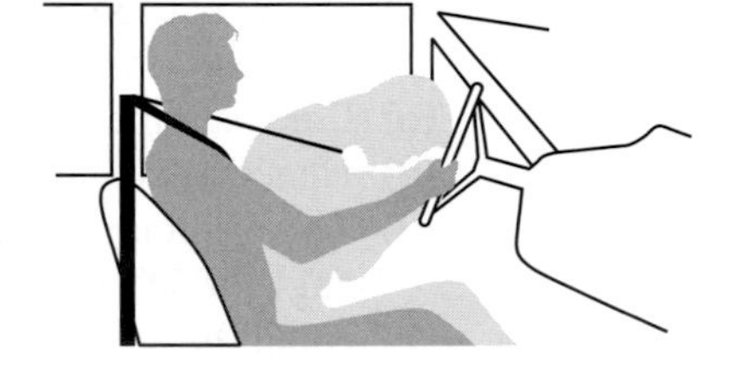

(a) What is the pull exerted by a seatbelt on **(i)** a man of mass 84 kg **(ii)** a child of mass 32 kg, at this maximum safe acceleration?

(b) Express your answers to **(a)** as a multiple of the person's weight.

2.29 A tractor pulls a log of mass 2000 kg. When the tractor is pulling with a horizontal force of 1300 N, the acceleration of the log is $0.050\,\mathrm{m\,s^{-2}}$.
What resistance force does the ground exert on the log?

2.30 Superman slams head-on into a train speeding along at $30\,\mathrm{m\,s^{-1}}$, bringing it to rest in an amazing 0.010 s and saving Lois Lane, who was tied to the tracks ahead of the train.

(a) What is the acceleration of the train?

(b) If the train's mass is 200 tonnes (2.0×10^5 kg), what is the push which Superman exerts on the train?

2.31 In one 10-minute interval during the *Apollo 11* flight to the Moon the spacecraft's speed decreased from $5374\,\mathrm{m\,s^{-1}}$ to $5102\,\mathrm{m\,s^{-1}}$ (with the rocket motors not in use). The mass of the space craft was 4.4×10^4 kg. Calculate the average force exerted on the spacecraft during this time.

2.32 A plane of total mass 42 tonnes (4.2×10^4 kg) lands with a runway speed of $65\,\mathrm{m\,s^{-1}}$. 'Reverse thrust' from its jet engines reduces its speed to $25\,\mathrm{m\,s^{-1}}$ over a distance of 360 m.

(a) What is the average deceleration of the plane on the runway?
(b) Calculate the size of the reverse thrust and explain any assumption you have made in your calculation.

2.33 A free-body force diagram for a rear wheel drive car is shown in the diagram.
(a) Write a phrase describing each of the forces S and T.
(b) How are S and T related when the car is moving **(i)** at a constant velocity v **(ii)** with constant acceleration a?
(c) Describe what happens to the forces S and T when the driver applies the brakes and slows the car down.

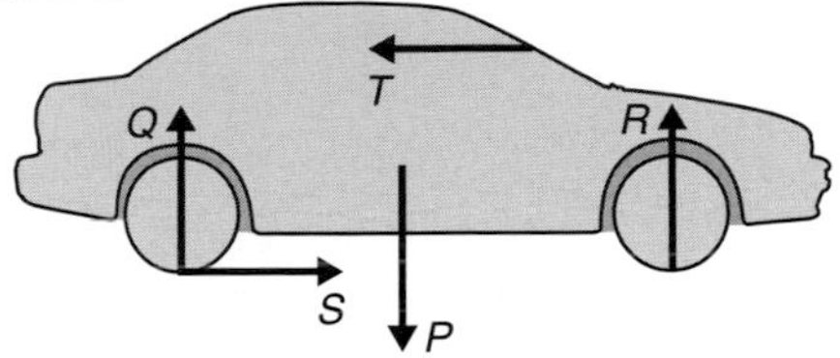

2.34 Just after the gun a sprinter of mass 65 kg is pushing against the starting block with a force of 800 N. This force acts at an angle of 65° to the horizontal.
(a) Calculate **(i)** the resultant horizontal force acting on her, and **(ii)** the resultant vertical force acting on her.
(b) What are **(i)** the forward acceleration of her centre of gravity, and **(ii)** the upward acceleration of her centre of gravity.

2.35 An articulated lorry consists of a tractor unit of mass 4.0 tonnes and a trailer of mass 26 tonnes. The lorry accelerates at 0.20 m s^{-2}.
(a) Ignoring all resistive forces calculate **(i)** the forward push of the road on the driving wheels of the tractor unit **(ii)** the forward pull of the tractor unit on the trailer.
(b) Draw separate free-body force diagrams for the tractor and trailer. Which pair of forces are equal because of Newton's third law?

2.36* Explain how, in the laboratory, you would demonstrate that the acceleration of a body is inversely proportional to its mass for a fixed resultant force.
State how you would process and present any measurements so as to achieve the aim of the demonstration.

2.37 The table gives data for a skydiver during the first phase of the jump:

time from start of jump/s	0	3.0	6.0	9.0
vertical velocity/m s^{-1}	0	28	46	53
vertical acceleration/m s^{-2}	9.8	7.9	4.0	1.0

The skydiver has a mass of 85 kg.
(a) Make a table, for each of the four times in the table, showing **(i)** the resultant force on the skydiver, and **(ii)** the upward push of the air on the skydiver.
(b) In this situation a parachutist is often said to be 'in free fall'. Explain why this is *not* a good description.

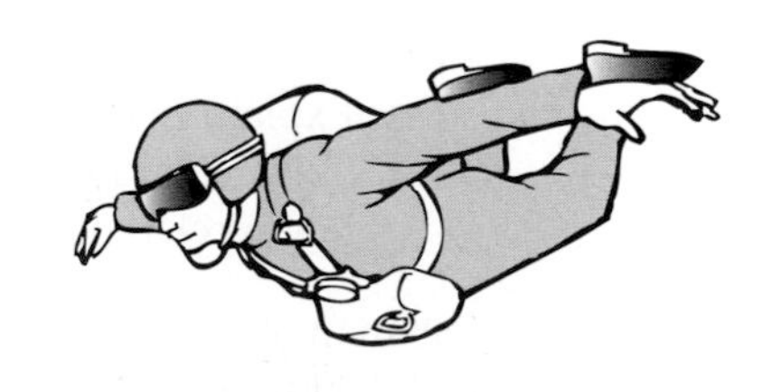

2.38 A lift has a mass of 1200 kg. Calculate the tension in the cable supporting the lift when the lift is

(a) ascending at a constant velocity
(b) ascending with an upward acceleration of $2.0\,m\,s^{-2}$
(c) descending with a downward acceleration of $3.0\,m\,s^{-2}$
(d) descending with an upward acceleration of $3.0\,m\,s^{-2}$.

2.39 The table gives the results of a standing-start acceleration test for a car of mass 1100 kg.

t/s	0	2	4	6	8	10	12
v/m s^{-1}	0	9	15	19.5	23	26	29

(a) Draw a graph of its speed v against time t.
(b) Estimate the resultant force acting on the car when its speed was **(i)** $15\,m\,s^{-1}$ **(ii)** $25\,m\,s^{-1}$. [Hint: You will have to draw tangents to your graph.]

2.40 The diagram shows a trampolinist at the bottom of her jump where she is instantaneously stationary. Her mass is 65 kg and her upward acceleration at this instant is $35\,m\,s^{-2}$.
Calculate the upward push of the trampoline on her at this instant.

2.41 A boy catches a cricket ball of mass 160 g which is moving at $20\,m\,s^{-1}$.
(a) Find the force which he must exert to stop it in **(i)** 0.10 s **(ii)** 0.50 s.
(b) Describe how he can vary the time in this way, and explain the advantage of lengthening the time in which the ball is stopped.
(c) Describe two other situations (as different as possible from this one) in which care is taken to lengthen the time in which a moving object is brought to rest.

2.42* In the UK the *Highway Code* advises that a vehicle travelling on a dry motorway or dual carriageway at 70 m.p.h. ($112\,km\,h^{-1}$) will need 75 m – over 80 yards – to stop once the brakes have been applied.
(a) What braking force is necessary to bring a car and its passengers, with a total mass of 1700 kg, to rest?
(b) Explain carefully what produces this braking force.

2.43 In the diagram the dashed line shows the weight of a parachutist as she falls: her weight is constant. The solid line shows the size of the air resistance (drag) force on her (and her parachute, when open) as she moves downwards. She opens her parachute at time t_1.
(a) Explain the shape of the graph.
(b) Sketch a graph to show the variation of her acceleration with time.

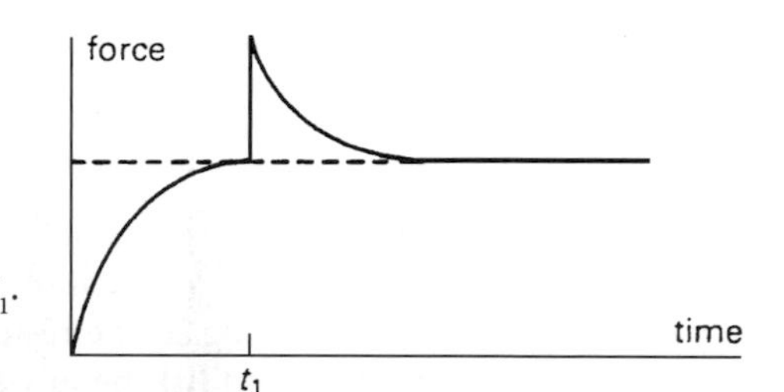

2.44 A man of mass 65 kg stands on a weighing machine in a lift which has a downwards acceleration of $3.0\,m\,s^{-2}$. What is the reading on the weighing machine? Make it clear at what stage you need to use Newton's third law and explain why it does not matter whether the lift is moving up or down.

3 Work and energy

Data: Earth's gravitational field strength near its surface $g = 9.8\,\text{ms}^{-2}$ (2 s.f.), $g = 9.81\,\text{N kg}^{-1}$ (3 s.f.)

1 mile $\equiv$ 1.6 km

3.1 Work

In this section you will need to

- remember that work, energy and power are all scalar quantities
- express units, e.g. the joule, in base units
- use the equation for work done by a constant force: work done by a force = the force times the distance moved in the direction of the force
- understand that the work done by a force can, in certain circumstances, be zero or have a negative value
- understand how to calculate the work done by a variable force by using the average value of the force
- understand that doing work on a body alters its energy.

3.1 To cut a lawn, Mum has to push a lawnmower 80 m.

(a) If her average horizontal push on the mower is 100 N, how much work does she do on the lawnmower?

(b) As the mower has no kinetic energy when she has finished the task, how much work is done *on* the lawnmower by frictional forces?

3.2 During the first 0.60 m of a lift, a weightlifter produced an average upward pull on the barbell of 3800 N.

(a) If the mass of the barbell was 240 kg, calculate

(i) the positive work done by the weightlifter on the barbell

(ii) the negative work done by the pull of the Earth on the barbell.

(b) What can you deduce about the barbell after it has been lifted 0.60 m?

3.3 An airline passenger pulls a wheeled suitcase, exerting a force of 65 N at an angle of 40° to the horizontal. How much work does she do in pulling the case 50 m from a taxi to the check-in counter?

3.4 Two tugs pull a large vessel which has lost power. The tension in each cable is 36 kN.

(a) Resolve each cable pull into forces parallel to and perpendicular to the motion.

(b) How much work is done in pulling the vessel 2.0 km by **(i)** the forces parallel to the direction of motion **(ii)** the forces perpendicular to the direction of motion?

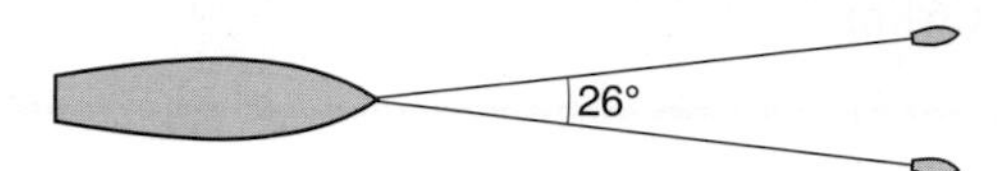

3.5 The diagram shows how the force F needed to stretch a simple (Hooke's law) spring varies with the extension x of the spring. A spring such as this might be used in an exercise machine. Calculate the work done in stretching the spring

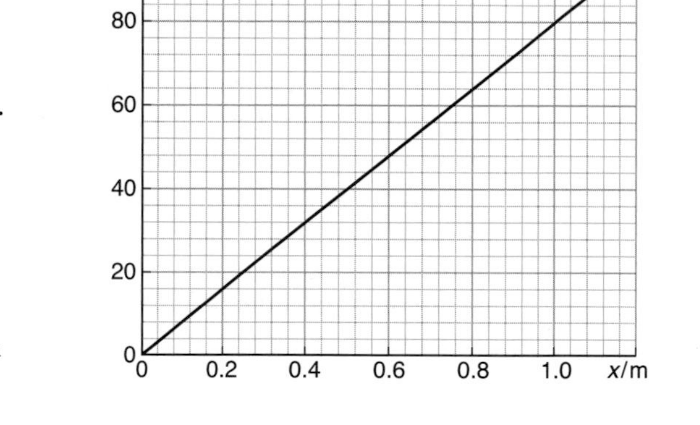

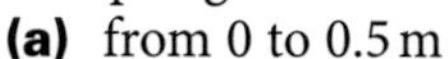

(a) from 0 to 0.5 m

(b) from 0.2 m to 1.0 m

(c) from 0.5 m to 0.9 m.

3.6 A spring of constant stiffness 40 N m^{-1} is fixed at one end. A man pulls the other end horizontally.

(a) How much work does he do in stretching the spring **(i)** 0.20 m from its unstretched position **(ii)** 0.40 m from its unstretched position?

(b) Explain why the answer to **(i)** is not twice the answer to **(ii)**.

3.7 **(a)** For a spring of constant stiffness k, prove that for forces up to F and corresponding extensions up to x, the work done in stretching the spring from its unstretched state is given by $\frac{1}{2}Fx$.

(b) Sketch a graph of F (y-axis) against x (x-axis) and show that the area under the graph represents the work done in stretching the spring.

3.8 A skier is being pulled up a 20° slope by a drag lift. The pulling wire is at an angle of 15° to the slope and the forces acting on the skier are:

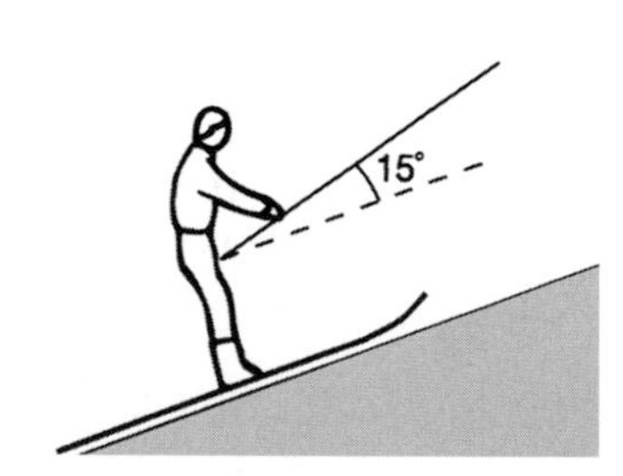

the pull of the drag lift wire, P = 320 N
the perpendicular contact push of the snow, N = 620 N
the frictional push of the snow, F = 53 N
the pull of the Earth, W = 750 N.

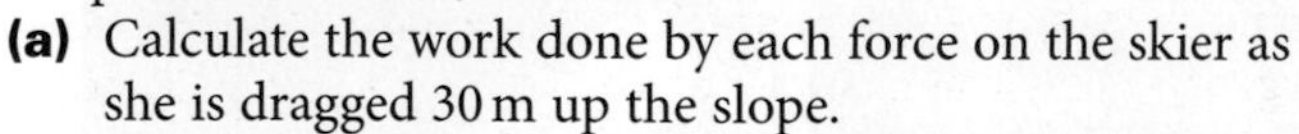

(a) Calculate the work done by each force on the skier as she is dragged 30 m up the slope.

(b) Hence show that the total work done on the skier is zero.

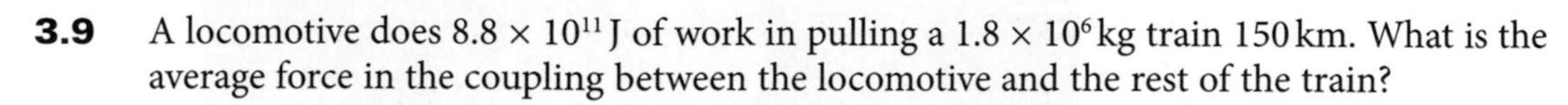

3.9 A locomotive does 8.8 × 10^{11} J of work in pulling a 1.8 × 10^{6} kg train 150 km. What is the average force in the coupling between the locomotive and the rest of the train?

3.10 How much work is done by the pull of the Earth on the Moon as the Moon completes one orbit round the Earth? [Pull of Earth on Moon = 2.0×10^{20} N, radius of Moon's orbit = 3.8×10^{8} m.]

3.11 The graph shows how the Moon's gravitational pull F on a lunar lander varies with the distance h from the Moon's surface. Estimate the work done by the pull of the Moon on the lunar lander as it approaches the lunar surface from a height of 200 km. Explain how you made your estimate.

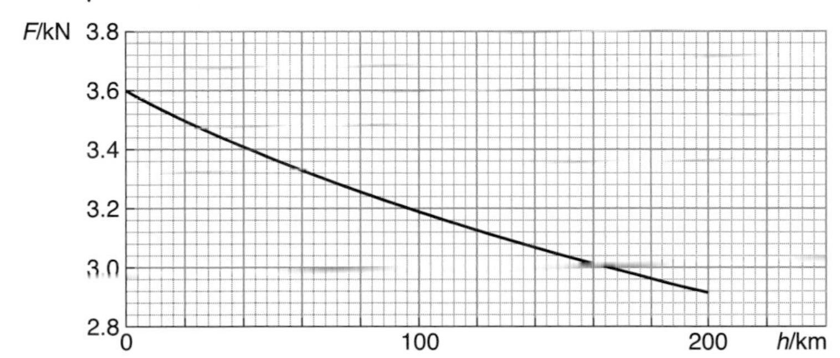

3.2 Power

In this section you will need to

- use the equations for power P: $P = W/t$ and $P = Fv$ where W is the work done or energy transferred
- understand that although energy is always conserved, the effect of transferring energy is nearly always to produce some which is effectively wasted
- remember that the efficiency of energy (or power) transfer is defined as the useful energy (or power) output divided by the total energy (or power) input
- draw energy flow diagrams (Sankey diagrams) to illustrate energy transfer processes.

3.12 The data give the rate at which energy is used by a typical advanced-level student in some common activities.

sleeping	40 W	sitting	80 W
standing	120 W	walking	250 W
running	600 W	eating	170 W

Estimate **(a)** the energy you use in a day **(b)** the heating power of a class of 15 students.

3.13 **(a)** A man digging a trench converts energy at a rate of about 1200 W. He could not keep this up for long, but if he could, how much work would he do in a working day of 8 hours?

(b) Electrical energy is available at a cost of about 20p per kW h (kilowatt-hour).

(i) Show that a kilowatt-hour is equivalent to 3.6 MJ.

(ii) Calculate the cost of the electrical energy equal to the work done by the man and discuss the result of your calculation.

3.14 The power of the electric motor of a locomotive unit pulling a train at a constant speed of 50 m s^{-1} is 2.5 MW. What is the total resistance force on the train?

3.15 Write the units of these expressions in terms of base units:
(a) work ÷ time
(b) force × speed.

3.16 Calculate the average power of
(a) Mum in question 3.1, if she cuts the lawn in 5 minutes
(b) the weightlifter in question 3.2, if she achieves the first 0.60 m of her lift in 0.32 s.

3.17* Two types of bulb give approximately the same illumination.
A: 'old' 60 W filament light bulbs at a unit cost of 50p. Each has a working life of 3000 hours.
B: modern 11 W 'energy-saving' light bulbs at a unit cost of £4.85. Each has a working life of about 10 000 hours.

You buy 18 As and 6 Bs, and use these bulbs to light six areas of your house for 9000 hours (i.e. ≈ 3 hours per day for 3 years). During this period, the average price of electrical energy is 20p per kilowatt-hour.
(a) Discuss the relative costs of **(i)** buying the different types of bulbs **(ii)** running them for a total of 9000 hours.
(b) How do the total costs compare?
(c) What else might you consider when choosing whether to buy the old or the new type of bulb?

3.18 A motor drives a pulley which lifts a box of mass 5.0 kg at a steady speed of $2.0\,m\,s^{-1}$. What is the power output of the motor? [Use data.]

3.19 A piano of mass 300 kg is being lifted to a window 12 m above the ground using a system of pulleys and a diesel motor.
(a) If the motor has a power output of 800 W, how long will it take to raise the piano to the window?
(b) How much chemical energy is converted by the motor during the lift if its efficiency is 20%?
(c) The diagram is an unlabelled Sankey diagram for this process. Copy the diagram and label it to describe the lifting of the piano.

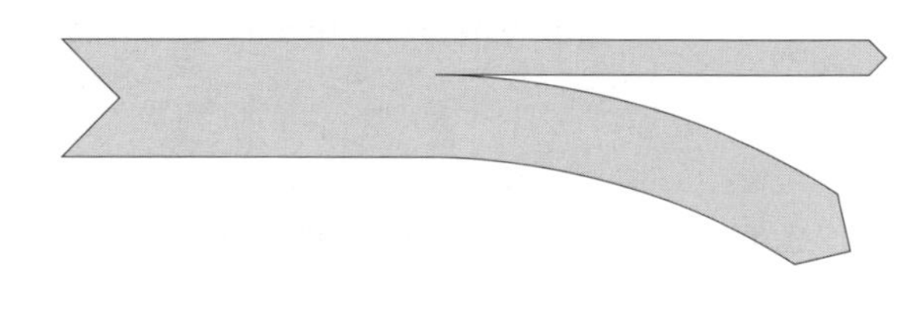

3.20 The diagram, a Sankey diagram, shows the power transfers in a car moving at a steady speed of $18\,m\,s^{-1}$ along a level road.
(a) What percentage of the energy available from the petrol is transferred **(i)** to internal energy in the engine **(ii)** to internal energy overall?
(b) Calculate the effective frictional force opposing the motion of the car produced by **(i)** the air **(ii)** the wheels.

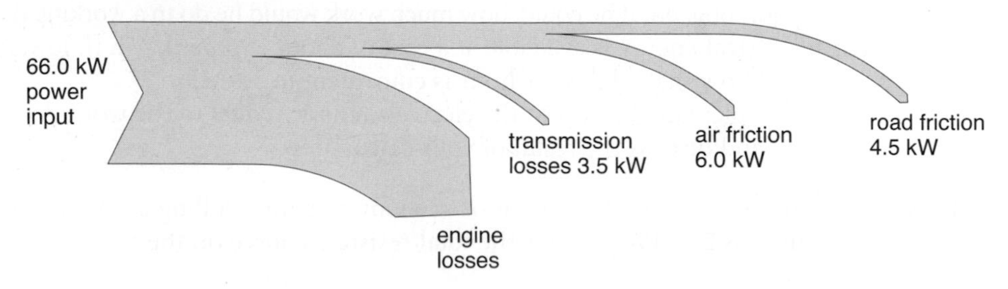

3.21* The data shows the percentages, in 2004, of different renewable energy sources as percentages of the world's total energy supply. The figure for solar energy is for the photovoltaic conversion of sunlight. (Omitted are those that derive from the use of renewable combustible products, e.g. wood.)

hydro	geothermal	solar	wind	tidal
2.2%	0.41%	0.039%	0.064%	0.0004%

(a) In 2004, what total percentage of the world's energy supply was made up from these renewable resources?

(b) Use a search engine to discover how future developments may alter this balance over the next 20 years.

3.22* The radiation received from the Sun per day at the Earth's surface in Great Britain is about $600\,\mathrm{W\,m^{-2}}$ averaged over 8 hours in the absence of cloud.

(a) What area of photovoltaic solar panels would be needed to produce as much energy as a 2000 MW power station in a day? Assume that the solar panels can convert solar radiation to electrical energy with an efficiency of 20%.

(b) What percentage is this area of the total area of Great Britain (which is about $3 \times 10^{11}\,\mathrm{m^2}$)?

(c) If the total power station capacity is about 160 GW, what percentage of the surface of Great Britain would be covered by solar panels if all power stations were replaced?

(d) Comment on the statement: 'Without the ability to store electricity, solar power can never be a major energy source in the UK'.

3.23 The power P of small wind turbines (including those designed for domestic use) is given by the equation

$$P = kA\rho v^3$$

where A is the area swept out by the blades, ρ is the density of the air, v is the wind speed and k is a number less than 1 that represents the efficiency of the system.

(a) Show that the units of the right-hand side are the same as the base units of power.

(b) Sketch, on graph paper, two curves to show how the relative power increases with

(i) the length ℓ of a turbine blade as ℓ increases from zero to 60 m

(ii) the wind speed as it increases from zero to $15\,\mathrm{m\,s^{-1}}$.

3.24* The world reserves of geothermal energy are estimated to be about 5.0×10^{25} J. Of these reserves, only 0.5% is hot enough and accessible enough for use in the generation of electricity. The conversion efficiency might be as low as 3%.
The present consumption of energy in the world is about 3×10^{20} J per year.
Discuss whether geothermal energy could make a significant contribution to this consumption.

3.25 In 1960 the physicist Freeman Dyson described how advanced extraterrestrial civilisations might solve their energy problems by carrying out huge engineering projects. For example, an Earth-like civilisation might dismantle the material of the rocky planets and remould them into a Dyson sphere, a 'shell' of matter entirely surrounding the Sun, and hence collect all the Sun's radiated energy.

(a) Take the total mass of the rocky planets to be 1.2×10^{25} kg with an average density of 5000 kg m^{-3}. Calculate the thickness of the Dyson sphere shown in the diagram at a distance from the Sun equal to the present Earth–Sun distance. (The volume of material in a Dyson sphere is $4\pi r^2$ times the thickness of the shell.)

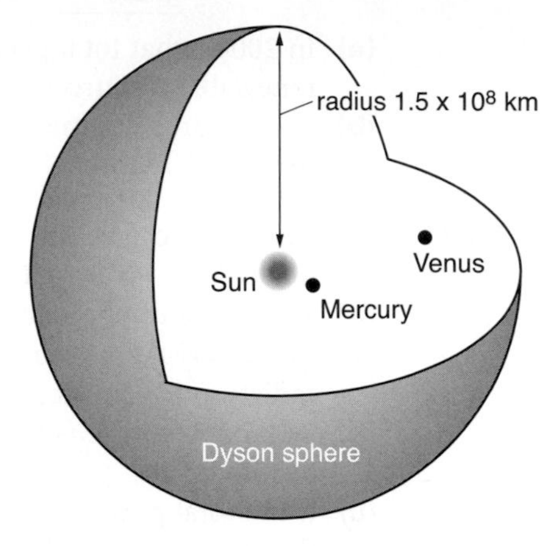

(b) The solar flux at this distance from the Sun is 1400 W m^{-2}. How much energy per second will be available to this future civilisation?

(c) Comment on the practicality of projects such as this based on stars with enough rocky planets to make the thickness of the Dyson sphere of thickness 3 or 4 metres.

3.3 Kinetic energy and gravitational potential energy

In this section you will need to

- use the kinetic energy (k.e.) of a body calculated as $\frac{1}{2}mv^2$
- use the work–energy equation $Fs = \Delta(\frac{1}{2}mv^2)$
- understand that changes in gravitational potential energy (g.p.e.) close to the Earth's surface are calculated as mgh where h is the vertical displacement
- use $\Delta(\frac{1}{2}mv^2) = mgh$ in the simplified form $v^2 = 2gh$ for bodies released from rest in the Earth's gravitational field.

3.26 **(a)** The lift in one of the world's tallest buildings rises 400 m from ground level to the observatory on top of the building. What is the gain in g.p.e. of a passenger in the lift who has a mass of 80 kg?

(b) The speed of the lift is about 6.5 m s^{-1}. Calculate his k.e. while rising up the building.

3.27 Estimate the kinetic energy of

(a) a tennis ball served at Wimbledon at about 130 m.p.h. (a tennis ball has a mass of 58 g)

(b) a world class sprinter (the men's world record for the 100 m is below 10 s)

(c) a family car on a motorway

(d) a two-year-old child running across a room.

3.28 The energy for a 'grandfather' clock is stored in a heavy cylinder. The cylinder, of mass 4.8 kg, gradually transfers g.p.e. to keep the pendulum of the clock swinging as the cylinder descends 1.2 m in seven days. Calculate the power transfer during the descent.

3.29 During a mountain stage of the Tour de France a cyclist travels 1200 m along a road which has an average gradient of 1 in 6, i.e. it rises vertically 1.0 m for every 6.0 m along the road. He has an average speed of 6.5 m s^{-1}. The mass of the cyclist and bicycle is 78 kg.

(a) Calculate the gain of g.p.e. of the cyclist and bicycle.
(b) What is the average output power of the cyclist?

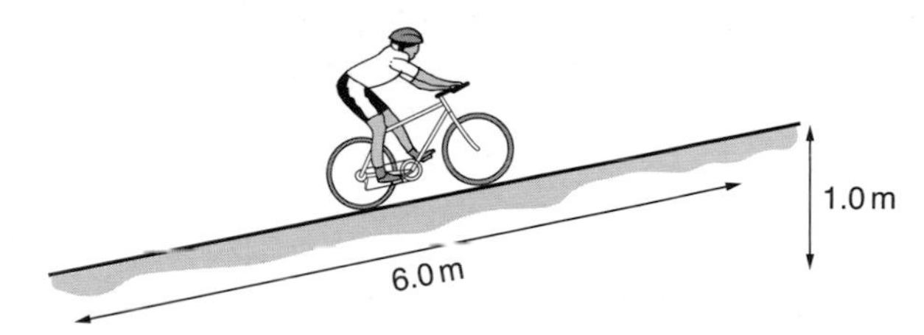

3.30 To push-start a car on a winter's morning (its battery is flat), two people each push with a force of 300 N. After pushing for 15 m the car's engine starts. At that moment the car's kinetic energy is 7500 J.

(a) Calculate **(i)** the work done by the push of the people on the car
(ii) the work done by frictional forces on the car.
(b) Draw an energy flow or Sankey diagram for this process.

3.31 A car which has a mass of 1200 kg is moving at a speed of 18 m s^{-1} (i.e. about 40 m.p.h.). On a dry day the maximum braking force is 8500 N.

(a) **(i)** In what distance can it stop?
(ii) On a wet day the braking force is halved. In what distance can it then stop?
(b) If the car is travelling at 27 m s^{-1} (about 60 m.p.h.) on a wet day, in what distance can it stop?

3.32 A hockey ball is held 2.0 m above the ground and has 3.2 J of g.p.e. It is then dropped.

(a) How much k.e. does it have just before it hits the ground?
(b) It bounces off the ground and is found to have 0.5 J of k.e. as it leaves the ground. How much internal energy has been transferred in the bounce?
(c) When it was momentarily at rest on the ground it had no k.e. and no g.p.e., yet a moment later it had 0.5 J of k.e. Where did this energy come from?
(d) How much g.p.e. will it have when it is at the top of its first bounce?

3.33 Draw an energy flow or Sankey diagram from the data in the previous question.

3.34 The figure shows experiments to compare the power outputs of different sets of muscles in the body. Suppose that the bodyweight of both the boy and the girl is 600 N. Her step-up distance is 25 cm and his pull-up lifts his centre of mass 35 cm. She does 24 step-ups in 50 s and could go on easily; he does 8 pull-ups in 30 s and is exhausted. What are the power outputs of **(a)** the girl and **(b)** the boy in these exercises?

3.35 A climber of mass 58 kg falls vertically off a cliff face. She is attached to a rope which allows her to fall freely for 20 m. The rope then becomes taut, but stretches, bringing her to rest in a further 4.0 m. The Sankey diagram describes her fall.

(a) Calculate her gravitational potential energy and her kinetic energy **(i)** initially **(ii)** when she has fallen 20 m **(iii)** finally.

(b) What is **(i)** the added final internal energy in the rope **(ii)** the average force exerted on the climber by the rope?

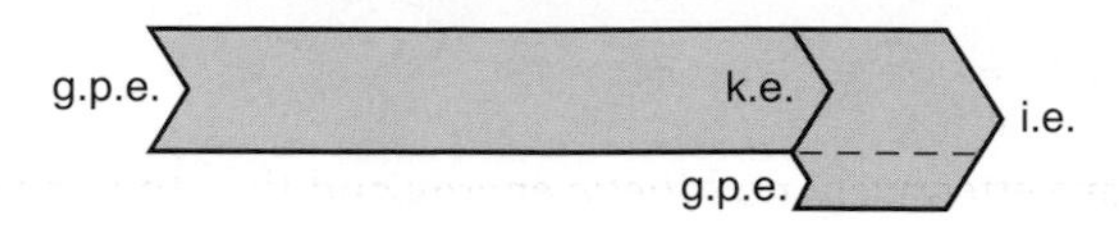

3.36 During a human heart beat, 20 g of blood are pushed into the main arteries. This blood is accelerated from a speed of $0.20\,\text{m}\,\text{s}^{-1}$ to $0.34\,\text{m}\,\text{s}^{-1}$. For a heart pulsing at 70 beats per minute, calculate the average power of the heart pump.

3.37 A skydiver of mass 70 kg reaches a speed of $45\,\text{m}\,\text{s}^{-1}$ after falling 150 m. By finding the loss of g.p.e. and the gain of k.e. determine the work done on the skydiver by the push of the air and hence find the average vertical push of the air on him during the first 150 m of his fall.

3.38 A catapult fires a stone of mass 85 g horizontally from rest. The graph shows how the push of the catapult P on the stone varies with the distance x it has been accelerated by the catapult.

(a) Deduce the work done on the stone by the catapult. Explain your method.

(b) Hence calculate the speed with which the stone is fired.

(c) How high would the stone rise if the catapult fired the stone vertically upwards?

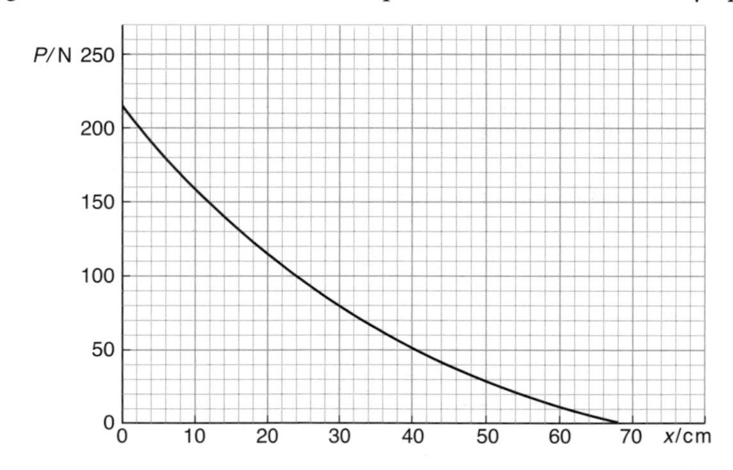

3.39 A baby drops a glass from a high chair onto a vinyl-covered floor. The glass will probably break if it hits the floor at a speed of more than $3\,\text{m}\,\text{s}^{-1}$. Suppose the glass has a mass of 160 g and the chair's tray is 1.0 m above the ground. Calculate

(a) **(i)** the loss of g.p.e. of the glass in falling to the ground

(ii) the k.e. of the glass as it hits the ground.

(b) Hence decide whether the glass will break and discuss whether it matters what the mass of the glass is.

3.40* At the Dinorwig pumped storage facility in North Wales, a 300-MW turbine pumps water at night from a reservoir to an upper lake 600 m above the reservoir.

(a) Assuming the turbine is 100% efficient, calculate the mass of water that is pumped in 6 hours.

The water can later be allowed to fall down through six such turbines to produce electrical energy.

(b) Discuss the purpose of such a facility, which is at best only 75% efficient overall.

3.41 The diagram shows part of a roller-coaster ride at an adventure park. The carriages are pulled from A to B at a steady speed by an electric motor of power output 52 kW. At B they have effectively no kinetic energy and they then run freely down to C. (Assume no resistive forces.) The carriage and passengers have a mass of 3400 kg.

(a) How long do the carriages take to rise from A to B?

(b) Calculate the speed of the carriages at C.

(c) The actual speed at C is found to be $33\,\mathrm{m\,s^{-1}}$. If the track from B to C is 95 m long, calculate the average resistive force acting on this part of the ride.

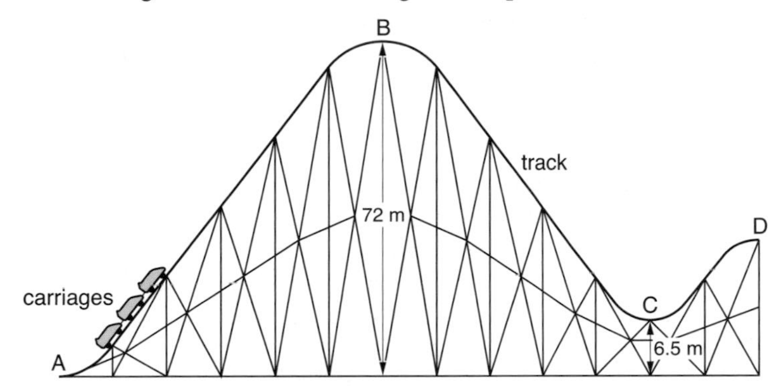

3.42 In order to demonstrate the principle of conservation of energy, a teacher attaches a trolley of mass 1.0 kg to a 50 g mass with a piece of string. Using a pulley attached to the edge of the laboratory bench, as shown in the diagram, he allows the 50 g mass to accelerate the trolley along the bench and measures both the gain of k.e. of the trolley and the loss of g.p.e. of the 50 g mass.

(a) Explain how you would measure the two energies.

(b) It is found that the loss of g.p.e. is 10% greater than the gain of k.e. measured. Suggest why this is so.

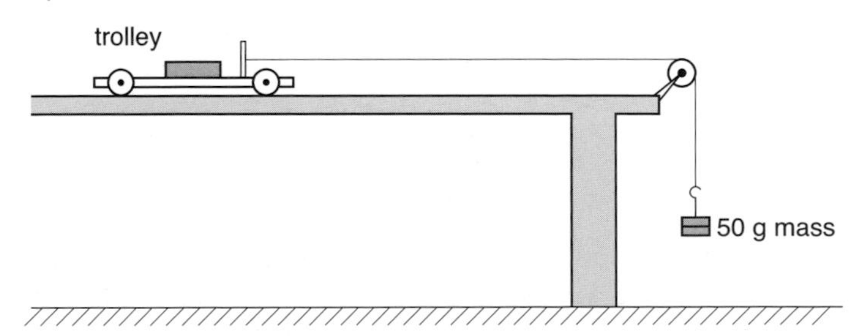

3.43* A student thinks he can improve the experiment in the previous question by placing a further 1 kg on the trolley and attaching a string to the back of the trolley passing over a pulley to the left, i.e. 'behind' it. He suggests now running the experiment with nine 10 g masses over the pulley on the right and one 10 g mass over the pulley to the left. Then move one 10 g mass from right to left and repeat; now move another 10 g mass and repeat, and so on until there are nine 10 g masses over the pulley to the left and one 10 g mass over the pulley to the right. By measuring the change in g.p.e. of the masses over the pulleys and equating this to the *total* gain of kinetic energy in the system (of mass 3 kg), the student believes the conservation of energy principle can be tested. Explain whether or not you agree.

3.44 The diagram shows the sequence of actions as a pole vaulter of mass 70 kg runs up (A), plants his pole (B), swings and pulls on his pole (C) and athletically clears the bar (D).

(a) What energy transfers occur **(i)** from A to B **(ii)** from B to C **(iii)** from C to D?

(b) The speed of the pole vaulter at A is 9.0 m s^{-1} and his centre of gravity at B is 1.2 m from the ground. Estimate the height of the bar that he just clears at D. State any assumptions you make in your estimate. How will they affect the final height the pole vaulter can achieve?

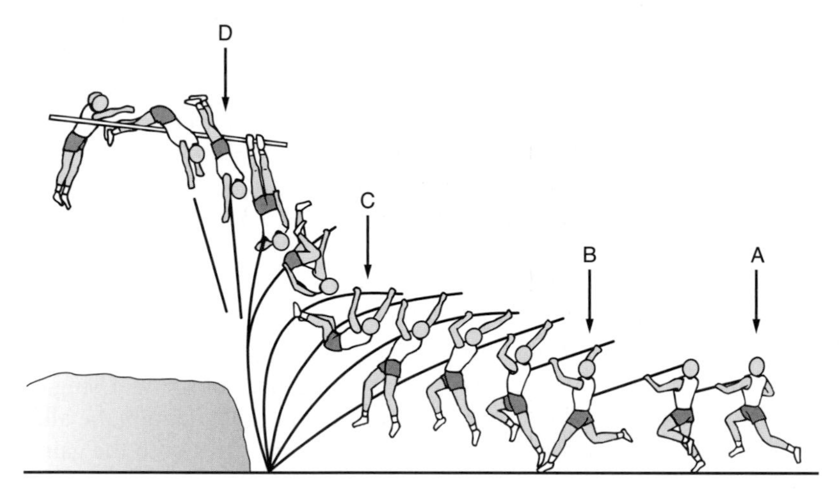

Electricity: charge and energy

Data: electronic charge $e = 1.60 \times 10^{-19}$ C

4.1 Electric current

In this section you will need to

- remember that the current is the rate of flow of electric charge and use the equation $q = It$
- remember that the total current leaving a junction is equal to the total current entering it
- remember that an ammeter can be used to measure current.

4.1 **(a)** Is the current in the circuit shown in the diagram clockwise or anti-clockwise?
(b) Do the electrons in this circuit flow clockwise or anti-clockwise?
(c) Your answers to **(a)** and **(b)** should be different. Explain how this difference arises, and say whether you think it matters.

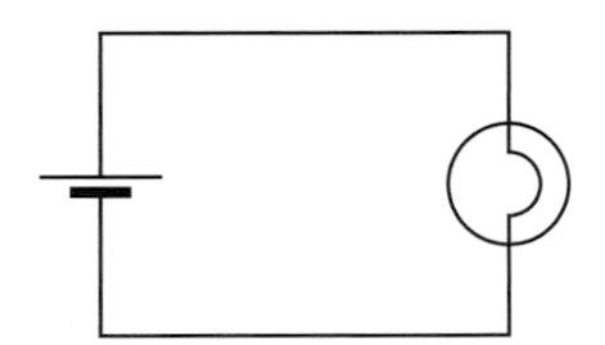

4.2 In the circuit shown the bulbs are identical. The current at P is 0.20 A. What are the currents at **(a)** Q **(b)** R **(c)** S **(d)** T?

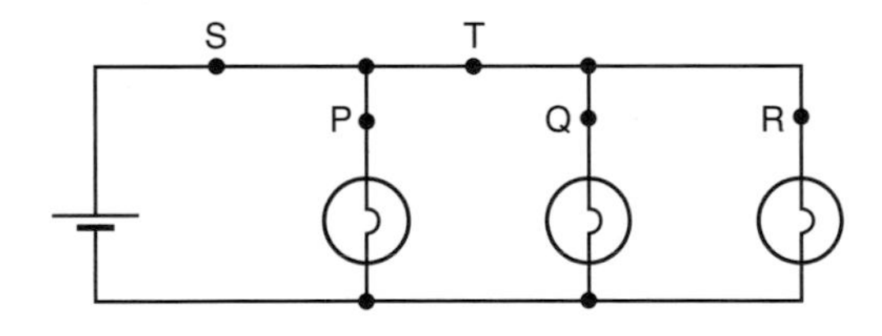

4.3 What is the average electric current in a wire when a charge of 150 C passes in 30 s?

4.4 In a camera flash lamp a charge of 5.0 C passes through the lamp in 10 ms. What is the average current?

4.5 In a lightning flash a typical amount of charge which reaches the Earth is 10 C. If the flash lasts for 0.50 ms, what is the average current?

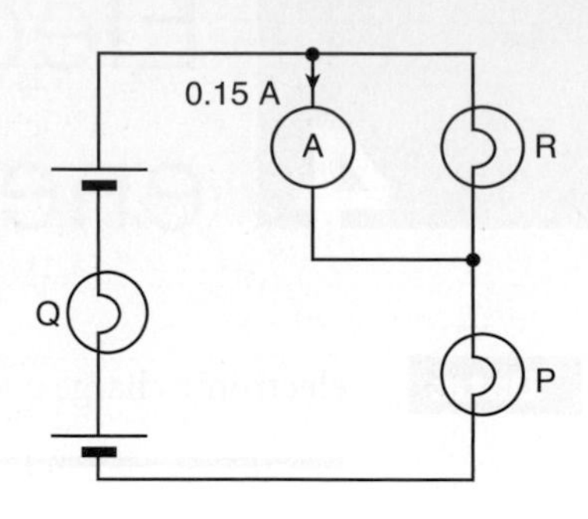

4.6 In the circuit shown the three bulbs are identical and reach full brightness when the current is 0.20 A. The analogue ammeter A has a noticeable resistance. Bulb R is observed to be at full brightness. Which (if any) of the other bulbs will be at full brightness, and what is the current in Q and P?

4.7 The current in a small torch bulb is 0.20 A.

(a) What is the total electric charge which passes a point in the circuit in 12 minutes?

(b) How many electrons pass this point in this time? [Use data.]

4.8 How long will it take 1.0×10^{20} electrons to enter the filament of a torch bulb which carries a current of 0.16 A?

4.9 In a laboratory oscilloscope a stream of electrons hits the screen to form the display. When the beam current is 2.4 mA, how much charge hits the screen in 30 s?

4.10 The graphs show how the current through an ammeter varies with time in three different situations. Calculate the net electric charge passing through the ammeter in **(a)** 15 s **(b)** 10 s **(c)** 6.0 s.

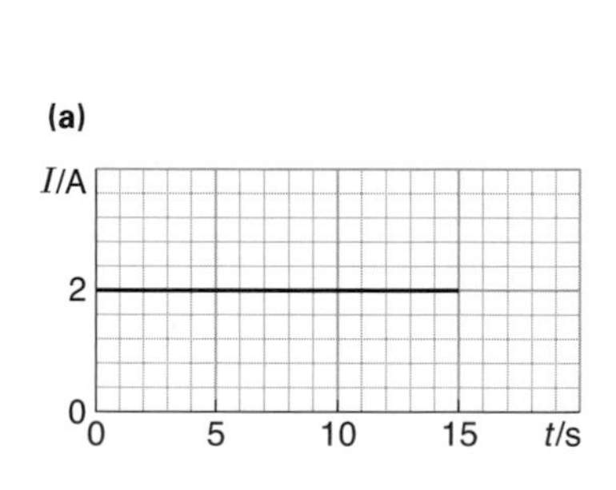

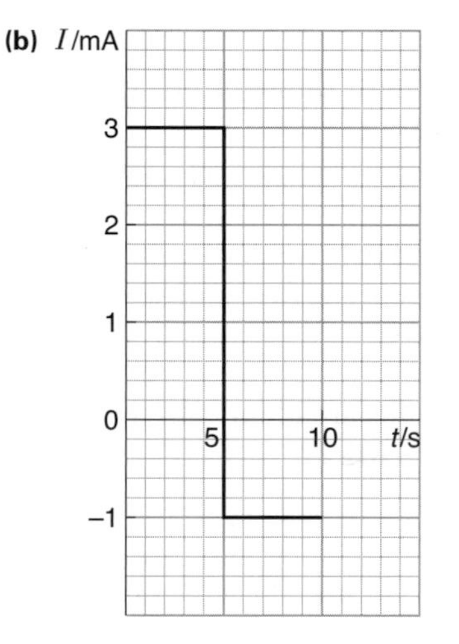

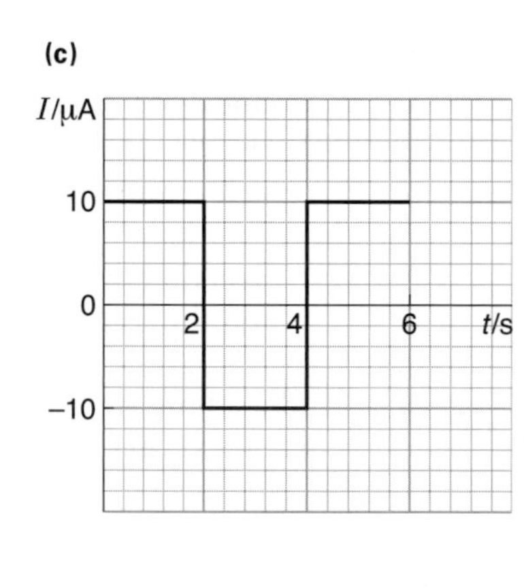

4.11 A new electric cell was joined in series with a bulb and an ammeter. The initial current was 0.30 A. At subsequent intervals of 1 hour the readings on the ammeter were: 0.27 A, 0.27 A, 0.26 A, 0.25 A, 0.23 A, 0.19 A, 0.09 A, 0.03 A and at 9 hours the ammeter reading had become negligibly small.

(a) Plot a graph of current against time.

(b) What does the area under the graph represent?

(c) How much electric charge would pass through the circuit if a current of 0.10 A passed for an hour?

(d) How much electric charge would pass through the circuit if the current passed for 9 hours?

4.12 A car battery sends a current of 5.0 A through each of two headlamps and a current of 0.50 A through each of two side-lamps.
(a) Draw a circuit diagram for the battery and the lamps.
(b) In 20 minutes how much charge passes through **(i)** each headlamp **(ii)** each side-lamp **(iii)** the battery?

4.13* A car battery is rated at 80 ampere-hours, that is, it can supply a current of 80 A for 1 hour, 40 A for 2 hours, etc., before it becomes discharged. If you accidentally leave the headlights on until the battery discharges, how much charge moves through the battery?

4.14 A doorbell is powered by a battery and is to be operated by a push-switch at the front door and also by a push-switch at the back door. Draw the circuit.

4.15* The graph shows a current surge I against time t through a conductor. Estimate the total charge that flows during the 4.0 s of the surge in two ways:
(a) **(i)** by counting the small squares under the graph and using the fact that each small square is equivalent to 0.10 A × 0.20 s = 0.020 C
(ii) by approximating the curve to a single triangle of base b and height h and then using area $= \frac{1}{2}bh$.
(b) Comment on the difficulty and precision of the two methods.

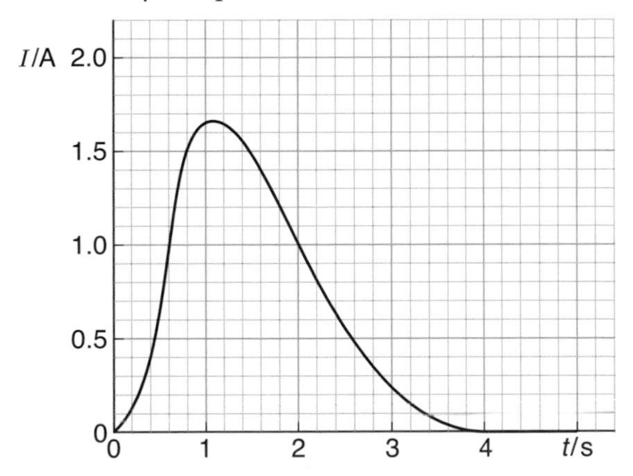

4.16* The following data describe how the charge on a capacitor varies as it discharges. (Capacitors – devices for storing charge – are covered later in Chapter 19.)

q/mC	28	21	15.5	11.5	8.5	5.0	2.5
t/ms	0	2.5	5.0	7.5	10	15	20

(a) Plot a graph of q against t.
(b) Explaining your method, deduce the current at **(i)** 5.0 s **(ii)** 15.0 s.

4.17 The Stanford Linear Accelerator produces 180 pulses of electrons per second, each pulse containing about 6×10^{11} electrons.
What is the average current in this linear accelerator? [Use data.]

4.2 Currents in solids and fluids

In this section you will need to

- use the equation $I = nAqv$ for the current in a conductor
- remember that there is a difference between the drift speed of the charge carriers and the speed of transmission of the electric field
- understand that the current in liquids and gases is caused by the movement of electrically charged particles such as ions and electrons.

4.18 Some copper fuse wire has a diameter of 0.22 mm and is designed to carry currents of up to 5.0 A. What is the mean drift speed of the electrons in the fuse wire when it carries a current of 5.0 A? [Use data.]

4.19 Write down the units for each of the physical quantities on the right-hand side of the equation $I = nAqv$.
Hence show that the units on the right reduce to amperes (A).

4.20 A copper wire joins a car battery to one of the tail lamps and carries a current of 1.8 A. The wire has a cross-sectional area of $1.0\,mm^2$ and is 6.0 m long. Calculate how long it takes an electron to travel along this length of wire. [Use data.]

4.21* 'I'm sure the answer to the last question must be wrong. It can't possibly take 15 hours for the tail lamps to come on!' What would you say to this?

4.22 Fuse wire which is labelled '15 A' will melt when it carries a current of 1.5 times that current. Its diameter is 0.51 mm. What is the maximum drift speed of the electrons in this wire? [Use data.]

4.23 In an ionic solution, 3.8×10^{15} ions, each carrying a charge $+2e$, pass to the right each second; 3.1×10^{15} ions, each carrying $-e$, pass to the left in the same time. What is the net current?

4.24 In a gas discharge tube containing hydrogen the current is carried partly by hydrogen ions (carrying a single positive charge) and partly by electrons. An ammeter in series with the tube indicates a current of 3.2 mA. If the rate of passage of electrons past a particular point in the tube is $1.4 \times 10^{16}\,s^{-1}$, find the rate at which hydrogen ions pass the same point.

4.25 Two copper wires of diameter 2.0 mm and 1.0 mm are joined end to end.
(a) What is the ratio of the average drift speeds of the electrons in the two wires when a steady current passes through them?
(b) In which wire are the electrons moving faster?

4.26 A shallow trough has the shape shown in the diagram and contains a liquid with positive and negative ions. Electrodes A and D are fixed to the ends of the trough as shown, and connected to a battery. The width of the trough at A is twice the width of the trough at D.
(a) What is the direction of the current in the liquid?

(b) In which direction do **(i)** the positive ions move **(ii)** the negative ions move?
(c) If the speed of the positive ions is $1.2 \times 10^{-7}\,\mathrm{m\,s^{-1}}$ between A and B, what is their speed when they are moving between C and D?
(d) Draw a graph of current against distance from A, to show how the average speed of the positive ions varies along the line ABCD.

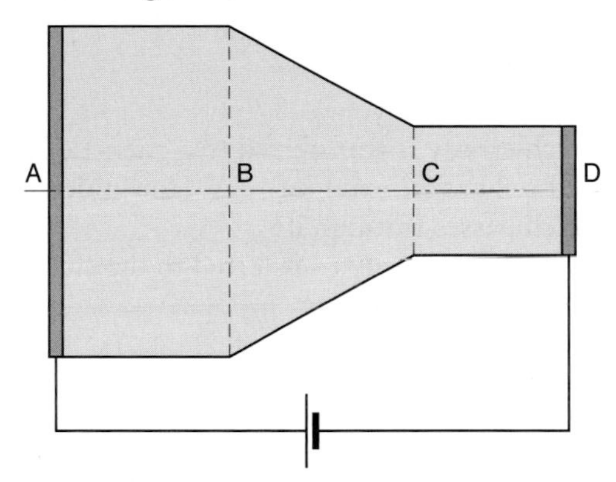

4.27 A piece of n-type germanium is 2.0 mm wide and 0.25 mm thick. At a certain temperature the number of conduction electrons per cubic metre is $6.0 \times 10^{20}\,\mathrm{m^{-3}}$. What is the average drift speed of the electrons when a current of 1.5 mA flows? [Use data.]

4.3 Electric cells and e.m.f.

In this section you will need to

- understand that the e.m.f. $\mathcal{E}$ of a cell is defined by the equation $\mathcal{E} = W/q$ where W is the energy transferred to electrical energy from some other form of energy, and q is the charge passing through the cell
- understand that for cells in series the total e.m.f. is the sum of the separate e.m.f.s
- understand that for cells of e.m.f. $\mathcal{E}$ in parallel the total e.m.f. is $\mathcal{E}$ (but that the cells can now supply a larger current)
- remember that a voltmeter can be used to measure e.m.f.
- understand what is meant by the capacity of a battery
- draw a circuit diagram to show how a secondary cell may be recharged.

4.28 How much chemical energy is transferred to electrical energy when
(a) a charge of 10 C flows through a cell of e.m.f. 1.5 V
(b) a charge of 30 C flows through a cell of e.m.f. 1.5 V
(c) a charge of 10 C flows through two cells, each of e.m.f. 1.5 V, connected in series?

4.29 The diagram shows a circuit with four 1.5 V cells connected in series.
(a) How much energy is gained when a charge of 3.0 C flows through **(i)** A **(ii)** A and B **(iii)** A, B, C and D?
(b) What is the net e.m.f. of these four cells connected in series?

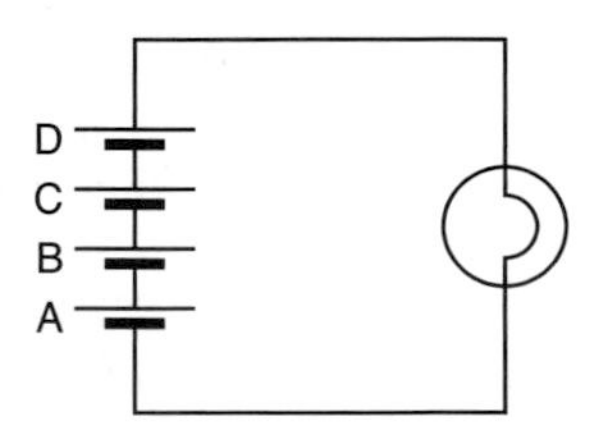

4.30 Five identical cells each provide an e.m.f. of 1.5 V. What does a voltmeter read when connected between A and B when the cells are arranged as shown in the diagram?

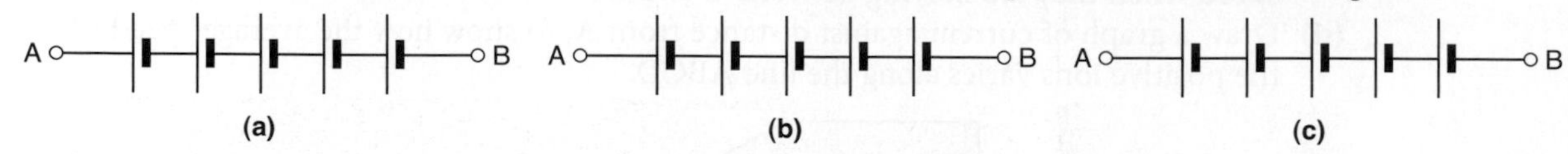

4.31 A cell has an e.m.f. of 1.2 V.

(a) How much energy is transferred to a charge of 10 C which passes through it?

(b) A second cell has an e.m.f. of 1.5 V. How much energy is transferred to a charge of 10 C which passes through it?

(c) What is the total energy transferred to the charge of 10 C when it passes through the two cells in succession?

4.32 Explain why the combined e.m.f. of two cells each of e.m.f. $\mathcal{E}$ is still $\mathcal{E}$ when they are connected in parallel.

4.33 A battery of e.m.f. 6.0 V passes a current of 0.30 A through a torch bulb for 5 minutes. How much energy is transferred from the cell?

4.34 A lead–acid cell of e.m.f. 2.0 V can drive a current of 0.50 A round a circuit for 10 hours.

(a) How much chemical energy is transferred to electrical energy in this time?

(b) How long would you expect the same cell to maintain a current of 0.20 A?

4.35 A cell is said to have an e.m.f. of 1.5 V and a 'capacity' of 10 A h, e.g. it can pass a current of 1.0 A for 10 hours.

(a) How much charge would pass?

(b) How much chemical energy is stored in the cell?

4.36* Button lithium–manganese cells come in different capacities, but all have a nominal e.m.f. of 3 V. Cells of capacity 38 mA h, 160 mA h and 280 mA h cost £2.80, £3.00 and £3.20, respectively.
What are the costs of 100 J of energy stored in each of these three lithium–manganese cells?

4.37 A battery manufacturer makes three different D-size 1.5 V cells. Their capacities are 5.2 A h, 7.9 A h and 16 A h.

(a) Calculate the energy stored in each of these cells, assuming that the cells provide an e.m.f. of 1.5 V throughout their lives.

(b) If the total masses of the cells are 79 g, 100 g and 131 g, respectively, calculate the energy per unit mass for each of the cells.

4.38 A secondary cell, such as a lead–acid battery, can be recharged if a source of higher e.m.f. is connected to it.

(a) Why would it be more correct to say that the battery can be 're-energised'?

(b) A simple circuit for recharging a battery is shown in the diagram. The resistor is there to limit the current that flows to a safe value. Write down the type of energy transfer that takes place in

(i) A, the battery of higher e.m.f.

(ii) the resistor

(iii) B, the battery of lower e.m.f.

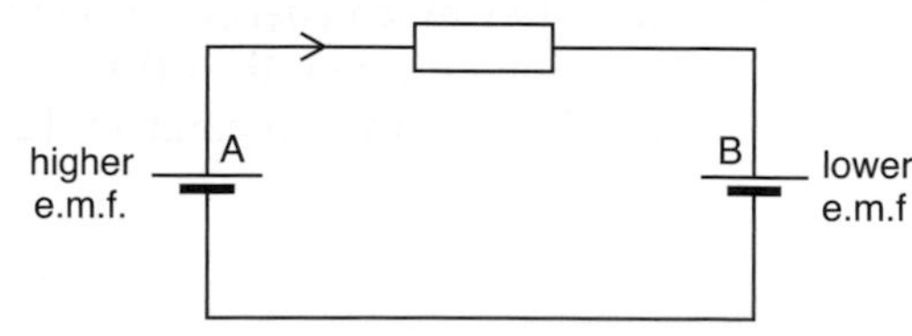

4.39* Photovoltaic cells, mounted on roofs, can be used to replace some of the electricity from the mains and hence lower costs.
Describe some of the difficulties associated with this form of electricity generation in the UK. You may wish to use the internet to provide relevant information.

4.4 Electrical energy and potential difference

In this section you will need to

- understand the meaning of potential difference V and use the equation $W = Vq$
- understand that when two components are connected in series, the total p.d. is equal to the sum of the separate p.d.s
- understand that when two components are connected in parallel, the p.d. between the ends of each of them is the same
- understand that we can assign values of potential to points in a circuit if we decide to fix the value of the potential at some point (usually calling the cell's negative terminal zero).

4.40 When you connect a voltmeter between two points in a circuit it tells you how much energy is transferred per coulomb of charge passing between those points. You can assume that in the circuit shown below, the connecting wire and the ammeters have negligible resistance. Between which of the following pairs of points would a voltmeter read zero (or nearly zero): A and B, B and C, C and D, D and E, E and F, F and G, G and H, H and A? Explain your answers.

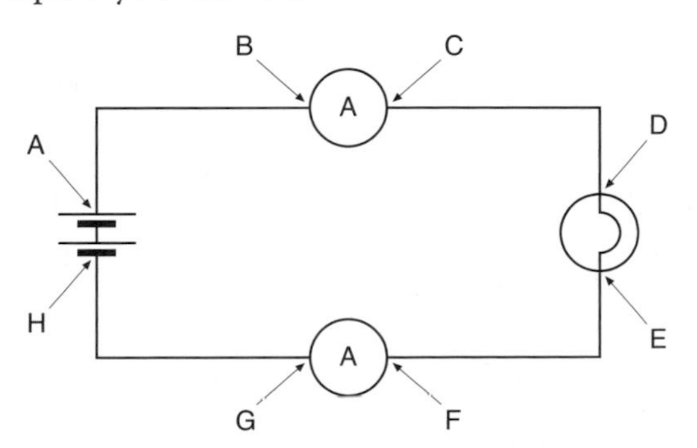

4.41 When 1.0 C passes through each of the lamps in the circuit 4.0 J of electric p.e. is transferred to internal energy in the lamp. What is

(a) the p.d. across each bulb
(b) the charge which passes through the battery
(c) the energy supplied by the battery
(d) the p.d. across the battery?

If it took 20 s for this to happen, calculate

(e) the current in each bulb
(f) the current in the battery
(g) the rate at which energy is supplied by the battery.

4.42 A teacher wants her students to understand what potential difference means, so she connects up the circuit shown in the diagram. Before the switch is closed, the joulemeter (J) reads 66 100 J. She closes the switch and while the lamp is lit the ammeter reads 2.0 A. After 5 minutes she opens the switch and notes that the joulemeter reading is now 73 300 J.

(a) How much charge passed through the lamp in this experiment?
(b How much energy was transferred?
(c) What was the potential difference across the lamp?
(d) How would you use this experiment to explain to someone what potential difference means?

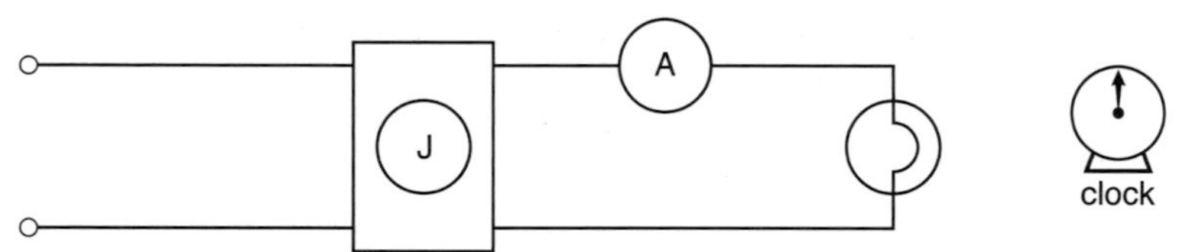

4.43 The circuit shows a lamp, a resistor and a motor connected in parallel across a battery. A voltmeter connected across the battery reads 2.89 V. What does it read when connected across

(a) the lamp
(b) the resistor
(c) the motor?

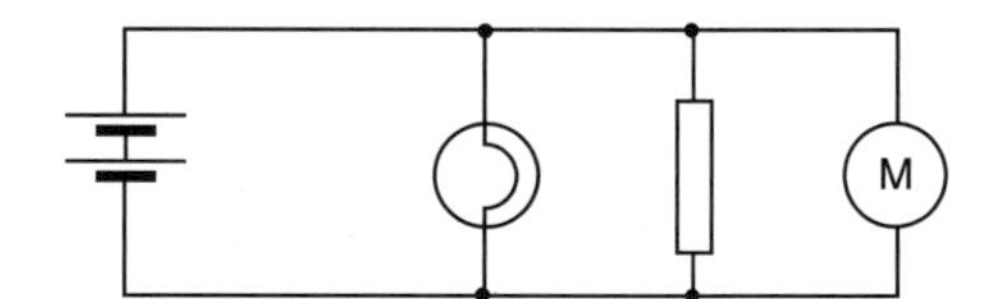

4.44 The diagram shows a horizontal 'rail' at 6.0 V and another at 0 V. A number or resistors of resistance R or $2R$ have been connected between the rails. The potential difference between points a and c, d and g, h and j, k and n is 6.0 V in every case.

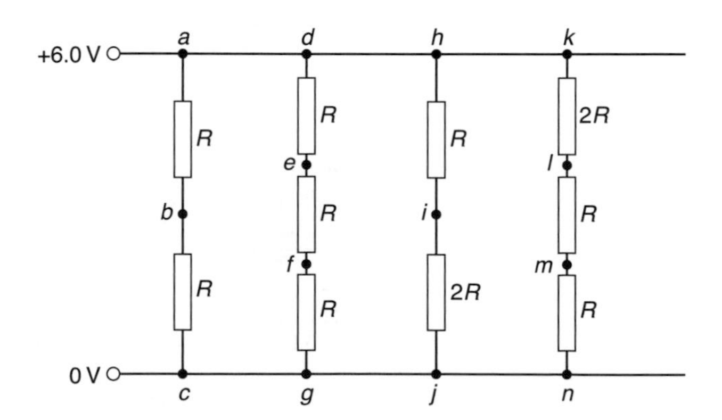

(a) Write down – do *not* think of what the currents might be as you are not given the value of R – the p.d.s between

(i) a and b, b and c

(ii) d and e, e and f, f and g

(iii) h and i, i and j

(iv) k and l, l and m, m and n

(b) Make up further resistor chains, using only low multiples of R, and challenge your friends to write down similar p.d.s. (Do not be too unkind!)

4.45 In the diagram for the previous question, the p.d between b and e is minus 1.0 V, i.e. −1.0 V.

(a) Write down the p.d.s between

(i) b and f

(ii) e and i

(iii) h and l

(iv) g and l

(v) a few extra stated pairs of points on *different* resistor chains.

(b) Between which pairs of points is the p.d.

(i) +3.0 V **(ii)** +4.0 V **(iii)** +1.5 V **(iv)** −4.0 V?

4.46 The diagram shows a battery connected to a length AB of 'resistance' wire (i.e. wire which has much more resistance than the other wires or the battery). The length of the wire is 0.50 m. The battery maintains a p.d. of 6.0 V between A and B. If B is earthed, how far from B is the point on the wire which has a potential of **(a)** 1.8 V **(b)** 5.4 V?

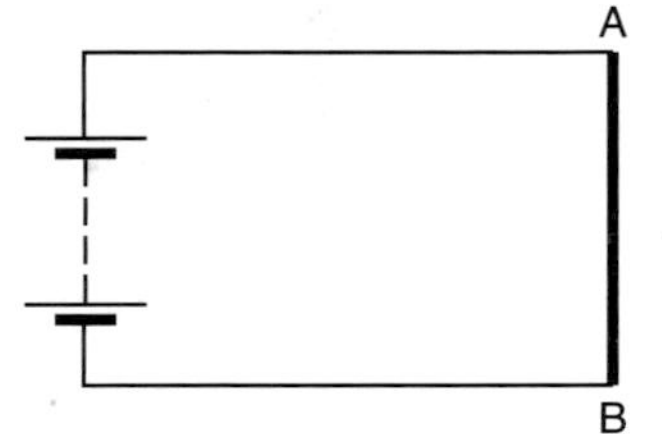

4.47 Draw circuit diagrams to show four lamps connected to a 12 V battery so that

(a) each has a p.d. of 12 V across it

(b) each has a p.d. of 6.0 V across it

(c) each has a p.d. of 3.0 V across it.

4.48 The diagram shows a vertical 'rail' at +6.0 V and another at −6.0 V. A number of resistors of resistance R, $2R$ or $3R$ have been connected between the rails.

(a) Write down the p.d.s between

(i) p and q, q and s, s and t

(ii) u and v, v and w, x and y.

(b) Make up further resistor chains, using only low multiples of R, and challenge your friends to write down similar p.d.s.

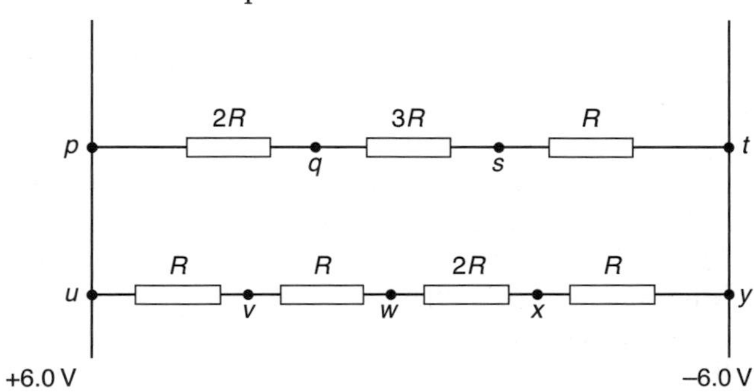

4.49 In the diagram for the previous question:

(a) write down the p.d.s between **(i)** q and w **(ii)** s and x **(iii)** a few extra stated pairs of points across the resistor chains.

(b) Between which pair of points is the p.d. **(i)** +4.8 V **(ii)** −1.6 V?

4.50 In this circuit the battery p.d. is 6.0 V and the p.d.s across A and B are measured to be 2.0 V and 4.0 V. What are the potentials of X, Y and Z if

(a) Z is earthed

(b) X is earthed?

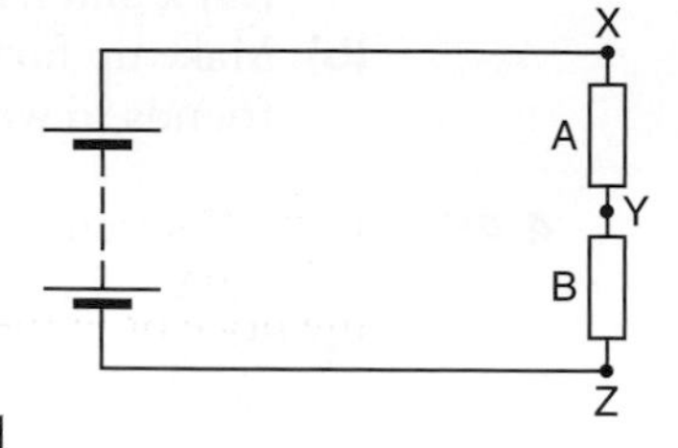

4.51 **(a)** It is estimated that the average electric charge carried in a lightning flash is 5 C. If the p.d. between the cloud and the ground is about 800 MV, approximately how much energy is transferred in the flash?

(b) In a typical thunderstorm lightning flashes strike the ground at intervals of about 3 minutes. Over the whole surface of the Earth the total current carried in this way between the atmosphere and the ground averages 1800 A. Estimate the average number of thunderstorms taking place at any instant over the whole Earth.

4.5 Electrical power

In this section you will need to

- use the equations $W = Pt$ and $P = W/t$
- use the equations $P = \mathcal{E}I$ and $P = VI$ to calculate the power of a cell or the power of another circuit component.

4.52 Electrical engineers often use the unit volt-amp when referring to the power of an electrical installation e.g. 1200 volt-amp for a toaster. Explain why the volt-amp is equivalent to a watt.

4.53 A lead–acid battery of e.m.f. 12 V is supplying a current of 10 A to some car headlamps. What is the rate of transfer of chemical energy to electrical energy?

4.54 The power rating of a hand-held tape recorder is 2.4 W and the current it needs is 0.4 A. How many 1.5 V cells would be needed, and would they be connected in series or in parallel?

4.55 An electric toaster is labelled 800 W. How much energy is transferred in it in 3 minutes?

4.56 Electrical energy is priced in kilowatt-hours, abbreviation kW h.

(a) A 3.6 kW electric grill is switched on for 15 minutes.

(i) Show that the grill uses less than one kilowatt-hour of electrical energy.

(ii) Calculate how much it costs to use the grill for this time when the price of a kW h is 19.2p.

(b) **(i)** How many joules of energy did the grill use?

(ii) What is the price of 1 MJ of electrical energy?

4.57* ‘Energy-saving’ light bulbs rated at 11 W are said to produce the same light as their equivalent filament bulbs of 60 W.

(a) Compare the currents in these two bulbs when each is connected to a 230 V supply.

(b) Compare the energy transformed by each if both are switched on for an average of 4 hours per day for 1000 days (about 3 years).

(c) Advertisements for these ‘energy-saving’ light bulbs state that they are ‘80% cost saving’. Comment on this statement.

4.58* The advertised ‘lives’ of the two light bulbs in the previous question are: energy-saving (E) 5000 hours; filament (F) 1000 hours. However, the price of E is much greater than the price of F: the bulbs cost £4.95 and 50p respectively.

(a) Calculate the total cost (buying and using) for the two bulbs over a period of 5000 hours. Assume that 1 kW h of electricity costs 20p.

(b) Hence show that the ‘old’ filament bulbs are about four times more expensive to run over a 5000-hour period than the ‘new’ energy-saving bulbs.

4.59* Use the internet to research information about a variety of energy-saving electric light bulbs for use at 12 V. Follow through similar calculations to the previous two questions in order to compare their energy consumption and relative costs over a period of 5000 hours.

4.60 The current in a small immersion heater is 3.8 A and the p.d. across its terminals is 11.9 V. How much electrical energy is transferred to internal energy in 20 minutes?

4.61 The energy of a single flash of light from a stroboscopic lamp is 0.60 J. The p.d. across the bulb is 240 V.

(a) How much charge passes through the lamp during the flash?

(b) If the flash lasts for 10 μs, what is the average current?

(c) What is the average power?

4.62 A car has two headlamps, two side-lamps and two tail-lamps. The electrical circuit for these lamps is shown in the diagram. The earthing signs to the bottom show that the metal frame of the car acts as the return wire to complete the circuit.

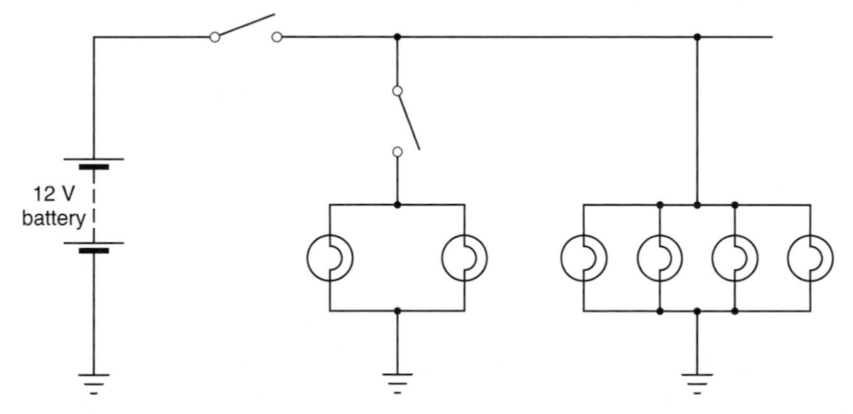

(a) Copy the diagram and label the switches and lamps.

(b) Explain why the headlamps cannot be on unless the side-lamps and tail-lamps are on.

(c) What will happen if, with the headlamps on, one of the side-lamps fails?

(d) If each side- and tail-lamp carries a current of 0.50 A, what is the power transfer from the battery with only the side- and tail-lamps on?
(e) The headlamps are known to have a power of 48 W each. What total current passes through the battery when all the lamps are on?
(f) A car also has a heater for the rear window. Add this to your circuit diagram.

4.63* In the UK each 'standby' feature on TV screens and computer monitors drains an average of 50 mA from the 230 V mains. What total power is needed to keep all such devices on standby, assuming that there are a hundred million of them in the UK?

4.64 A power station generates electrical energy at a p.d. of 25 kV and an average rate of 500 MW.
(a) What is the current in the cables leaving the power station?
(b) How much energy is generated in one day?
(c) If the efficiency of the power station is 40%, how much internal energy is delivered to the surroundings in one day?

4.65 The Hubble space telescope (see question 13.46) has solar panels to supply it with the 5.0 kW of electrical power that it needs. The solar panels convert energy with a efficiency of 25% and are able to face the Sun 80% of the time.
Given that sunlight has an intensity of 1.4 kW m^{-2}, what is the minimum area for the solar panels?

5 Electrical resistance

Data: electronic charge $e = 1.60 \times 10^{-19}$ C

gravitational field strength $g = 9.81$ N kg^{-1}

material	copper	aluminium	steel	nichrome
resistivity/10^{-8} Ω m	1.7	3.2	14	130

5.1 Resistance

In this section you will need to

- use the equation $R = V/I$
- understand that Ohm's law states that the resistance of a metallic conductor is constant if physical conditions are constant
- describe how to check on whether Ohm's law is obeyed for a particular circuit component
- describe how to plot characteristics (i.e. graphs of current I against p.d. V) for circuit components
- draw the characteristics for a lamp filament, a thermistor and a silicon diode
- use the equations $R_{ser} = R_1 + R_2$ and $\frac{1}{R_{par}} = \frac{1}{R_1} + \frac{1}{R_2}$
- use the equations $P = I^2R$ and $P = V^2/R$ to calculate the power transfer in a resistor
- explain why the resistance of two resistors in parallel is always less than the resistance of the smaller of the two.

5.1 There is a current of 0.20 A in a wire when the potential difference between its ends is 5.0 V. What is its resistance?

5.2 The opposite faces of a sheet of polythene are covered with metal foil. When the potential difference between the two layers of foil is 12 V, the current through the polythene is 1.4×10^{-10} A. What is the resistance of the polythene?

5.3 What potential difference must be applied to a resistor of resistance of 10 MΩ to drive a current of 5.0 μA through it?

5.4 A lamp bulb is connected to the mains supply by a cable consisting of two wires. Each wire has a resistance of 0.025 Ω per metre. If the length of the cable between the supply and the lamp is 8.0 m, what is

(a) the resistance of each wire

(b) the p.d. between the ends of each wire when the current in it is 0.60 A

(c) the power in the cable then?

5.5 The graph shows the current I in a conductor for various values of the applied p.d. V.

(a) What is the resistance of the conductor when $V = 200\,\text{V}$?

(b) What is the resistance of the conductor when $V = 400\,\text{V}$?

(c) Estimate the value of the resistance when V is very small.

(d) Was the conductor made from metal or from carbon?

5.6 The diagram shows the kind of rheostat, or variable resistor, often found in laboratories. The resistance wire wrapped round the long cylinder has insulation between its coils, but the slider contact makes electrical contact with the top of the coils as it moves along.

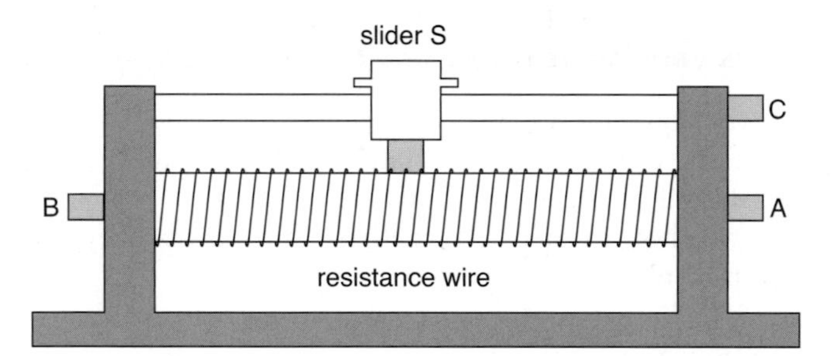

The coil of wire is connected to A and B, while the slider is connected to C. Suppose the coil has a resistance of 100 Ω.

(a) Copy and complete the table to show the resistance between the different terminals (e.g. R_{AC} means the resistance between A and C) as the slider is moved away from C. Three have been done for you.

		R_{AC}/Ω	R_{BC}/Ω	R_{AB}/Ω
(i)	slider at C	0		
(ii)	slider $\frac{1}{4}$ way to B		75	
(iii)	slider $\frac{1}{2}$ way to B			100
(iv)	slider $\frac{3}{4}$ way to B			
(v)	slider at B			

(b) Write an equation that links R_{AC}, R_{BC} and R_{AB}.

5.7 A house uses an average of 120 kW h of electrical energy each week. This energy is supplied at 230 V. Calculate

(a) the average power delivered to the house

(b) the average resistance that the house presents to the power line.

5.8 A rheostat consists of resistance wire uniformly wound on a former of length 300 mm. The resistance of the wire is 100 Ω. Initially the slider is at the centre of the rheostat (so that its resistance is 50 Ω), and the rheostat is connected in series with a battery which provides a constant p.d. of 6.0 V across the rheostat.

(a) What is the current in the circuit initially?
(b) The slider is now moved 30 mm so as to reduce the resistance of the rheostat. What is now the resistance of the rheostat, and the current in the circuit?
(c) The slider is now moved 30 mm on three more occasions, each time in the direction which reduces the resistance. What are the new resistances, and the currents?
(d) Comment on the suitability of the rheostat for adjusting the current in the circuit.

5.9* A kettle and a reading lamp are each connected to a 230 V supply. The kettle has a much greater power than the lamp.
(a) Explain which has the greater resistance.
(b) Suggest powers for the kettle and the lamp and calculate values for their resistances.

5.10 Two resistors, of resistance 3.3 Ω and 4.7 Ω, are connected first in series and then in parallel to a power supply which provides a constant p.d. of 6.0 V. Which resistor has the greater power, and what is that power, when the resistors are **(a)** in series **(b)** in parallel?

5.11 A car headlamp bulb has a power of 60 W when it is connected to a potential difference of 12 V.
(a) What is the resistance of the filament?
(b) If its resistance remained constant, what would be the power if a p.d. of 6.0 V were connected across it?
(c) Its resistance will not remain constant: how will it change, and will the power of the bulb be larger or smaller than the power calculated in **(b)**? Explain.

5.12 The resistance of some wire is 14 ohms per metre. What length is needed to provide a power of 20 W when a p.d. of 12 V is available?

5.13* A set of Christmas tree lights consists of 20 lamps in series, and is designed to be connected to a 240 V supply. Each bulb is rated at 1.2 W.
(a) Draw a circuit diagram showing the lamps connected to the supply. What is the main disadvantage of this way of connecting the lamps? Can you think of any advantage?
(b) Calculate **(i)** the p.d. across each bulb **(ii)** the current in each lamp **(iii)** the current drawn from the supply.
(c) When one lamp fails the others do not go out. How might this result be achieved? Assuming that your solution is correct, what would happen if several lamps failed?

5.14 Resistances of 10 Ω and 15 Ω are joined **(a)** in series **(b)** in parallel. What is the total resistance in each case?

5.15 Calculate the combined resistance of each of the arrangements of resistors shown in the figure.

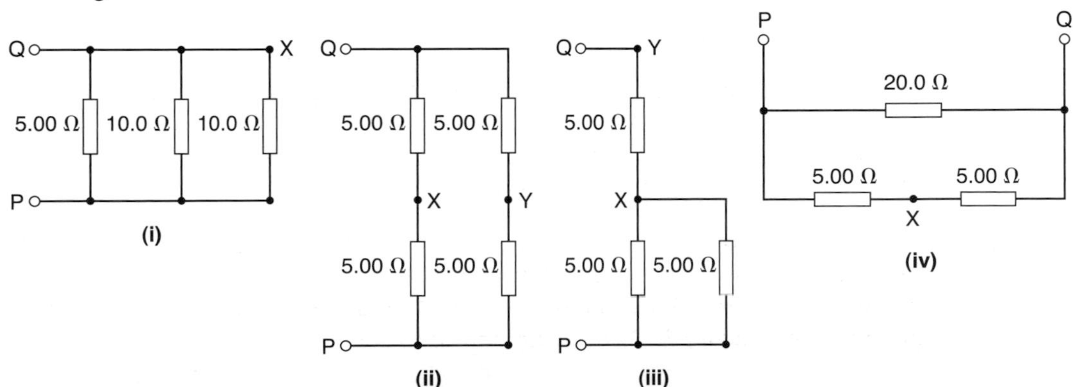

5.16 Each of the arrangements of resistors in question 5.15 is joined to a power supply, so that P is earthed and Q is at a potential of 6.0 V.

(a) What is the potential at X in each case?

(b) In **(ii)** and **(iii)** in question 5.15, what are the potential differences between Y and X?

5.17 A connecting lead of length 1.0 m used in a laboratory consists of 55 strands of wire. Each strand has a resistance of 2.3 Ω. What is the resistance of the complete wire?

5.18* The diagram shows fifteen 100 Ω resistors mounted in a small package. The pins are labelled 1 to 16. Where there is a single straight line there is no resistance. What is the resistance between the following pins (the '&' sign means that the points are connected together)?

(a) 1 & 2 and 16 **(b)** 1 & 2 and 15

(c) 1 & 2 and 15 & 16 **(d)** 1 & 2 & 3 & 4 and 16

(e) 1 & 2 & 3 & 4 and 15 **(f)** 1 & 2 & 3 & 4 and 15 & 16

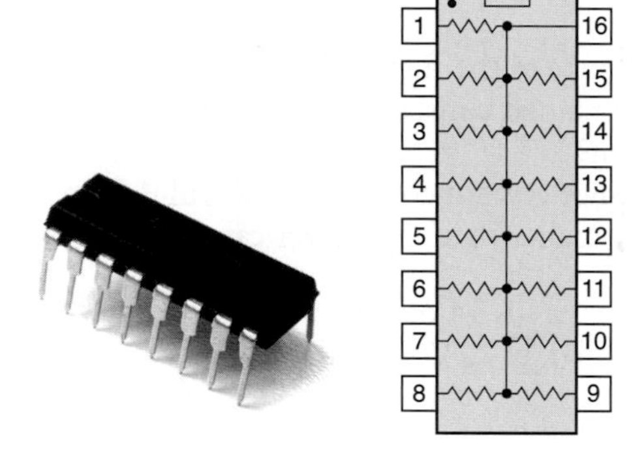

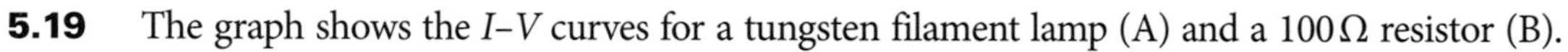

5.19 The graph shows the I–V curves for a tungsten filament lamp (A) and a 100 Ω resistor (B).

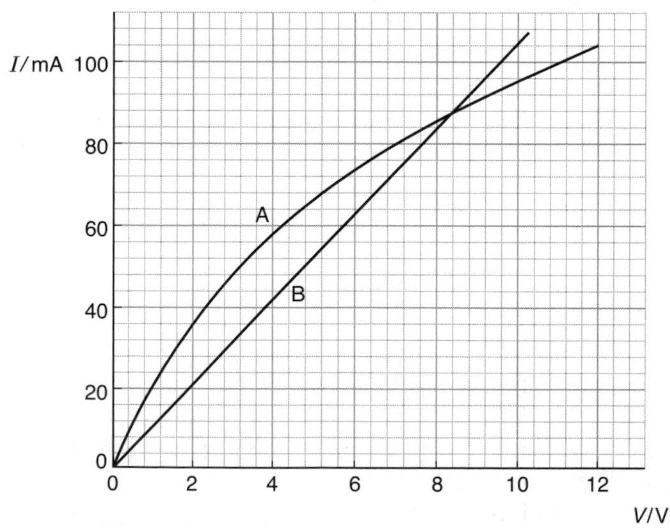

(a) What will be the total current I when A and B are connected in parallel to a p.d. of 6.0 V?

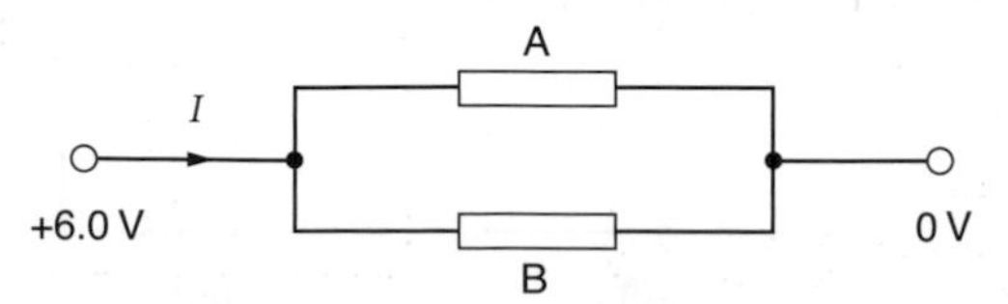

(b) A and B are now connected in series, and a p.d. of 12 V is applied to them. Remembering that the p.d.s across A and B must add up to 12 V, explain why the current in them is 67 mA.

5.20 In the two arrangements **(a)** and **(b)** of resistors shown in the figure, is the combined resistance about 100 Ω, between 1 Ω and 100 Ω, or less than 1 Ω? What general rule could you state for calculating a rough value of the resistance of resistors connected in parallel?

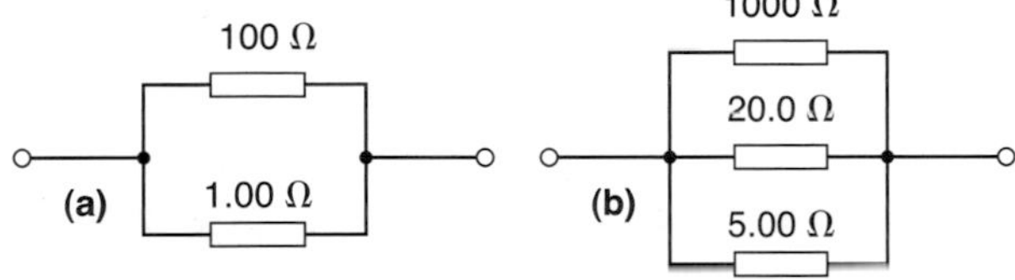

5.21 What is the current in each of the resistors shown in the diagram?

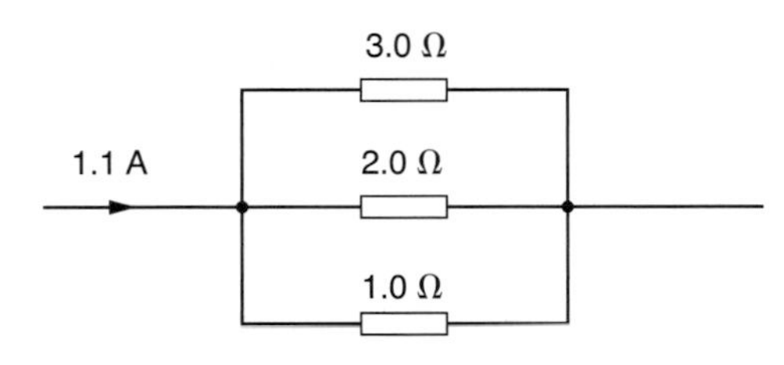

5.22 Refer to the graph in question 5.19.

(a) With A and B connected in series, what p.d. V is needed to pass a current of 40 mA in them?

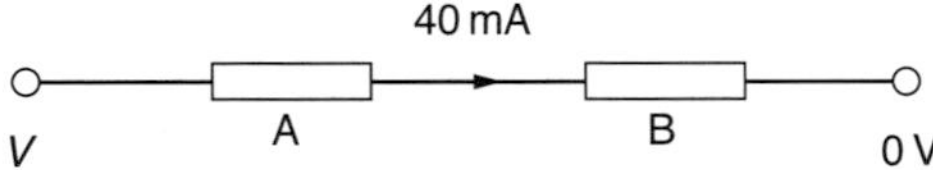

(b) A and B are now connected in parallel and the total current is found to be 192 mA. Remembering that this current is the sum of the currents in A and B, find the p.d. across them.

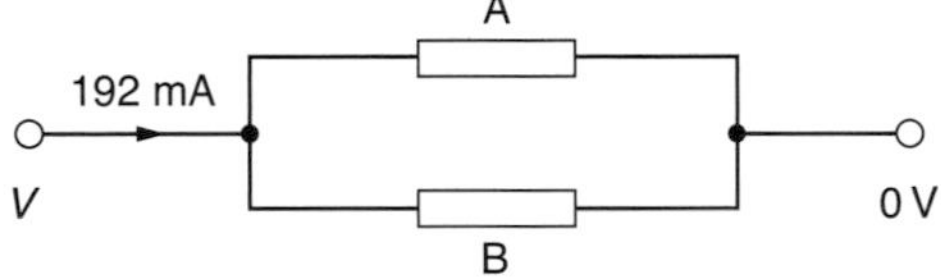

5.23 In the first circuit in the diagram a power supply maintains a p.d. V across two resistors R_1 and R_2 in parallel. The currents which flow in the circuit are I, I_1 and I_2. In the second circuit the two resistors have been replaced by a single resistor which draws the same current I from the power supply.

(a) What is the relationship between I, I_1 and I_2?

(b) Write down equations connecting

(i) V, I_1 and R_1

(ii) V, I_2 and R_2

(iii) V, I and R.

(c) Hence show that $\frac{1}{R} = \frac{1}{R_1} + \frac{1}{R_2}$.

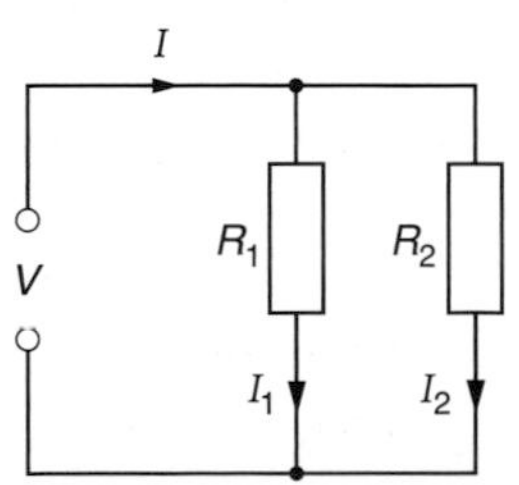

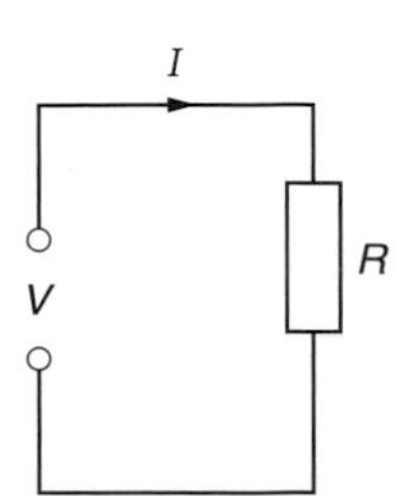

5.24 You are given a box in which there are three 10 Ω resistors connected in series, as shown in the diagram. Connections can be made to the ends of any resistor, but they cannot be removed from the box. You are asked to obtain a resistance of **(a)** 30 Ω **(b)** 5 Ω **(c)** 15 Ω. Make three copies of the diagram and add any connections you would need to do this. Label X and Y, the points between which the network has the required resistance.

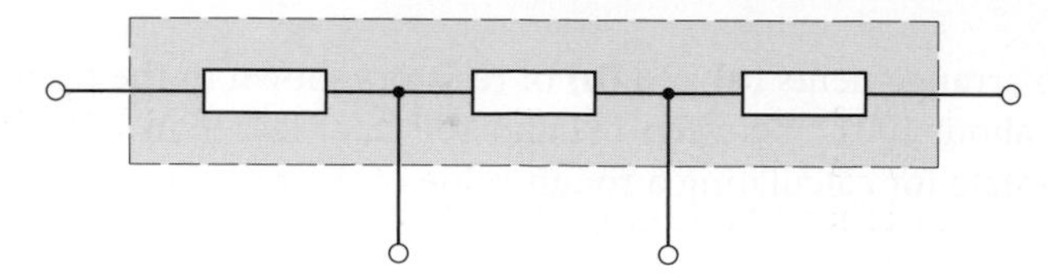

5.25 A small clothes iron (for use when travelling) is bought in Britain and has a power of 600 W when connected to a p.d. of 230 V. If it is taken to the USA, and used on a mains power supply whose p.d. is 110 V, what is its power? Assume that its resistance is constant.

5.26 A manufacturer makes resistors. Each resistor, of whatever resistance, is designed to transfer safely a maximum power of 25 W.
Calculate the maximum p.d. which may be applied to one of these resistors of resistance **(a)** 22 Ω **(b)** 220 Ω **(c)** 2.2 kΩ.

5.27 The maximum current in a particular light-emitting diode (LED) is 30 mA. The maximum power which can be transferred in this type of l.e.d. is 100 mW.
(a) Show that the LED cannot be connected by itself to a 5.0 V power supply.
(b) Which of the following resistors would you place in series with the LED to limit the current to a safe value: 33 Ω, 47 Ω, 68 Ω, 82 Ω?

5.28 The diagram shows a graph of p.d. across a silicon diode against the current in it.
(a) What is the resistance of the diode when the p.d. across it is
(i) 0.25 V
(ii) 0.64 V
(iii) 0.74 V?
(b) Estimate the resistance of the diode when the p.d. across it is
(i) 0 V
(ii) 10 V.
(c) Draw a graph of resistance against p.d. for values of V between 0.55 V and 0.75 V.

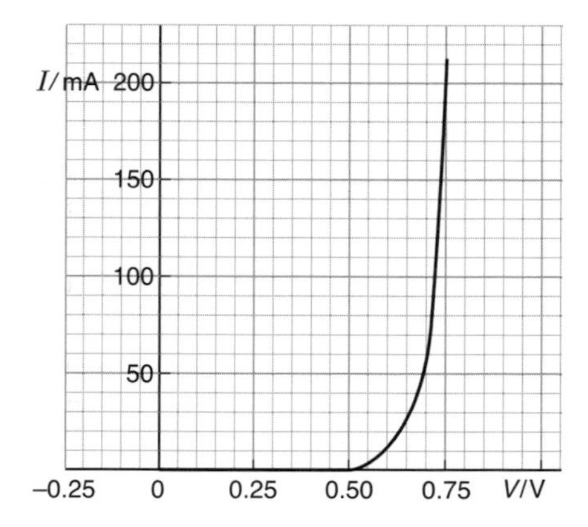

5.29 A silicon diode is connected to two resistors as shown in the diagram.
(a) First assume that the silicon diode has zero resistance in the forward direction, and infinite resistance in the reverse direction. The p.d. between A and B is varied from –3.0 V to +3.0 V. Draw a graph to show how the current between A and B varies with p.d.

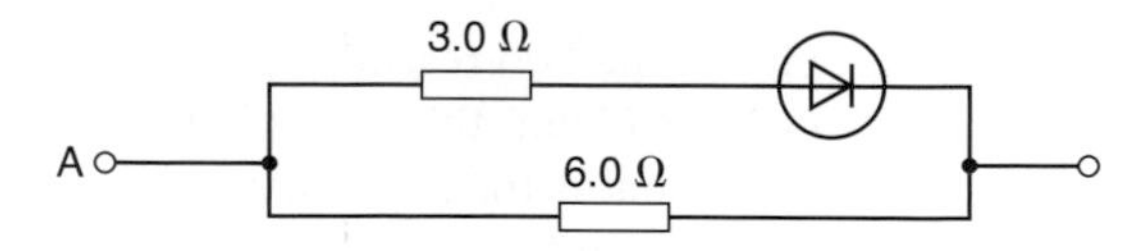

(b) Now assume that the diode behaves as a 'real' diode, like the diode in question 5.28. Sketch (no detailed calculations required) on the same axes how the current between A and B varies with p.d.

5.30* A power station delivers 250 kW of power to a factory through 4.0-Ω lines. How much less power is wasted if the electricity is delivered to the 4.0-Ω lines at 40 kV rather than at 10 kV?

5.31 The unit of resistance, the ohm, is named after Georg Simon Ohm, a German physicist who formulated his law in 1827 as $I \propto V$.
Copy this proportional relationship and add a constant k to the left-hand side. What unit does k have? Express the unit in base SI units, i.e. in terms of kg, m, s and A. [Remember that 1 joule ≡ N m.] What do we now call the constant k?

5.2 Resistivity

In this section you will need to

- use the equation $R = \rho\ell/A$ which defines resistivity ρ
- remember that the resistivity of metals increases with temperature but the resistivity of semiconductors decreases with temperature.

5.32 What are the units of resistivity ρ?

5.33 Each of the copper wires in a three-core power cable has a cross-sectional area of 0.50 mm^2. What is the resistance of a 10 m length of one of these wires? [Use data.]

5.34 Consider two pieces of copper wire. The first one has a length of 200 mm and a diameter of 0.50 mm. The second one has a length of 100 mm and a diameter of 0.25 mm.
(a) Explain whether these wires have the same resistance.
(b) If they do not, explain which has the greater resistance and calculate how many times greater it is.

5.35 Conducting putty is a material which is similar to Plasticine but it is an electrical conductor. A student rolls some of the putty into a cylindrical shape which is 60 mm long and has a diameter of 20 mm. He then rolls it into a new cylindrical shape which has a diameter of 10 mm.
(a) What is the new length?
(b) What is the ratio (new resistance)/(old resistance)?

5.36 A power cable (for the grid system) consists of six aluminium wires enclosing a central steel wire. The purpose of the steel wire is to give the cable strength: the current in it may be assumed to be negligible. If each of the aluminium wires has a diameter of 4.0 mm, calculate
(a) the cross-sectional area of each wire, in mm^2
(b) the resistance of 1.0 km of one of the aluminium wires
(c) the resistance per km of the whole cable. [Use data.]

5.37* Manganin is a metal alloy used for making resistors. The table gives the values of resistance per metre of manganin wire of different cross-sectional areas.

A/mm^2	0.66	0.25	0.11	0.078	0.043
$r/\Omega\,\text{m}^{-1}$	0.65	1.73	3.82	5.45	9.90

(a) By calculating the five values of Ar, show that the resistance per unit length is inversely proportional to the cross-sectional area.

(b) A graph of A against $1/r$ should be linear and pass through the origin, so, as a second test, make a table for values of $1/r$ and draw such a graph.

5.38 The heating element of an electric toaster consists of a ribbon of nichrome which is 1.0 mm wide and 0.050 mm thick. What length of ribbon is needed to provide a power of 800 W when the element is connected to a p.d. of 230 V? [Use data.]

5.39 A slice of silicon which measures 30 mm by 30 mm and is 0.50 mm thick, as shown in the diagram below, has conducting strips fitted to two opposite edges AB and CD.

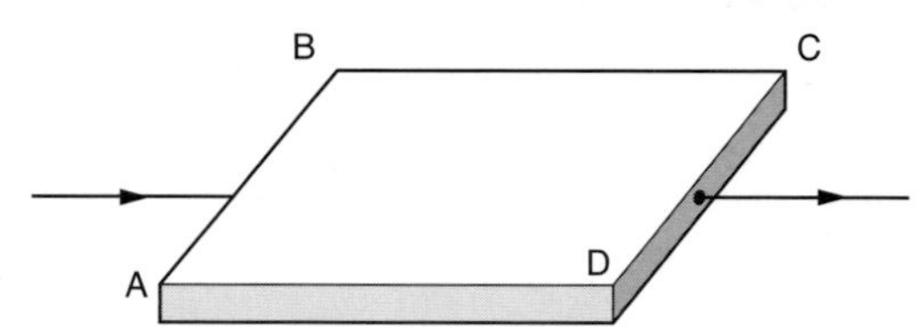

(a) If the resistivity of silicon is $4.0 \times 10^3\,\Omega\,\text{m}$, calculate the resistance of the sheet measured between AB and CD.

(b) What would be the resistance of a similar sheet, measuring 15 mm by 15 mm, of the same thickness?

5.40 A heater used on the rear window of many cars might consist of five strips of resistive material connected in parallel between two vertical conductors (shaded) of negligible resistance.

The power of the heater was 31.0 W when the p.d. between the vertical conductors was 11.5 V.

(a) Calculate **(i)** the total resistance of the heater **(ii)** the resistance of each strip.

(b) Each strip has a cross-sectional area of $5.8 \times 10^{-8}\,\text{m}^2$ and is made from nichrome. Calculate the length of each strip. [Use data.]

5.41 The resistance of the tungsten filament of an electric lamp is found to be 27 Ω at 0 °C (= 273 K). If the resistivity of tungsten increases by 0.0056 of its resistivity at 0 °C, for each degree rise in temperature, calculate

(a) the resistance of the filament at its working temperature of 2400 K

(b) the ratio $\dfrac{\text{power of lamp when first switched on}}{\text{power of lamp at working temperature}}$

5.3 Internal resistance

In this section you will need to

- understand what is meant by the internal resistance r of a cell
- use the equations $\mathcal{E} = I(R + r)$ and $\mathcal{E} = V + Ir$ to analyse circuits in which the internal resistance r of the cell has to be taken into account
- describe how to measure the internal resistance of a cell
- explain that a cell delivers maximum power when the resistance of the external circuit is equal to the cell's internal resistance.

5.42 A bulb of resistance 14.0 Ω is connected to a dry cell of e.m.f. 1.50 V and internal resistance 0.80 Ω. Calculate
(a) the current in the circuit
(b) the p.d. across the bulb
(c) the p.d. across the cell.

5.43 Two cells, each of e.m.f. 1.50 V and internal resistance 0.50 Ω, are connected **(a)** in series **(b)** in parallel. In each case what is the combined e.m.f. and internal resistance?

5.44 A battery has an e.m.f. of 3.0 V and is connected to a bulb. The current in the bulb is 0.30 A and the potential difference between its ends is 2.8 V.
(a) In 5 minutes how much chemical energy is transferred to electrical energy in the battery?
(b) In the same time how much electrical energy is transferred to internal energy in the bulb?
(c) How do you account for the fact that these two answers are not the same?

5.45 A student sets up the circuit shown in the diagram below. He expects the ammeter to read 0.25 A but finds that it reads 0.20 A.
(a) Explain **(i)** his expected reading **(ii)** why the actual reading is lower.
(b) Calculate the internal resistance of the cell, assuming that the ammeter and the leads have negligible resistance.
(c) Calculate the rate at which
(i) the cell is transferring chemical energy to electrical potential energy
(ii) the resistor is transferring electrical potential energy to internal energy
(iii) the cell is transferring electrical potential energy to internal energy.

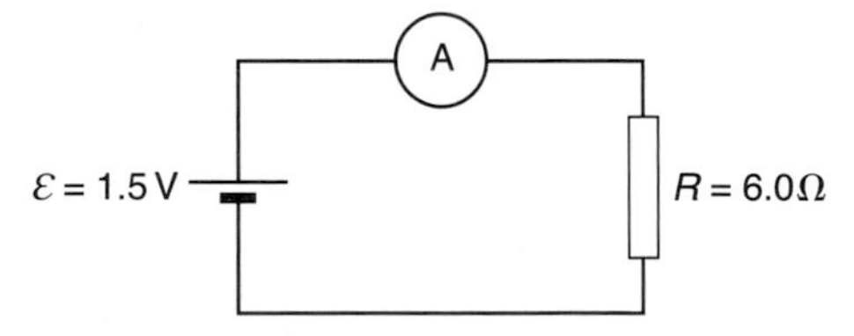

5.46 A student uses the circuit shown in the diagram to find the internal resistance of a cell.

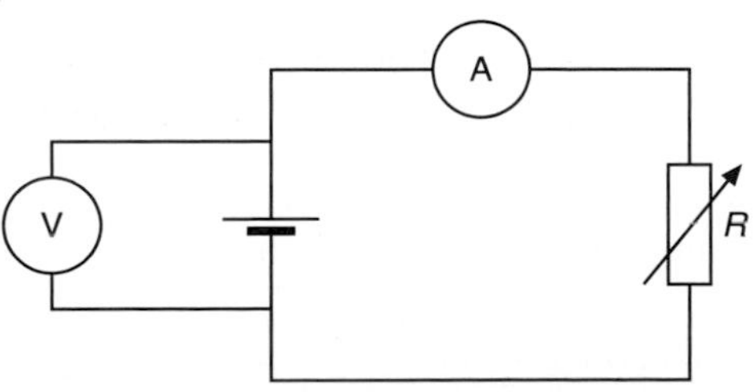

The student plots her values for I and V producing the graph shown.

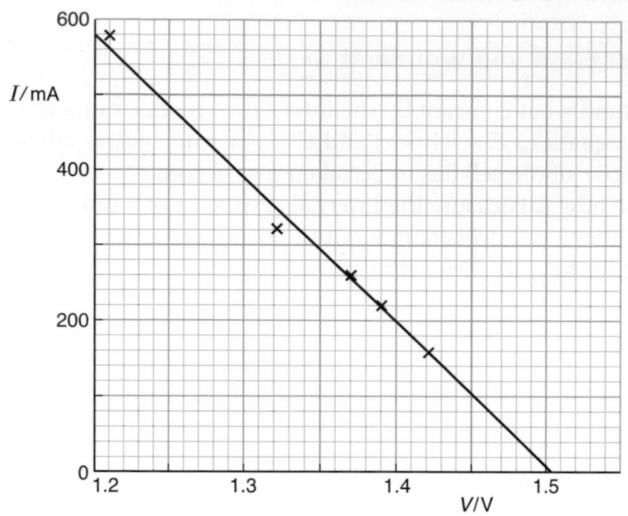

(a) Use $\mathcal{E} = V + Ir$ to find r. (Beware! V is plotted along the x-axis.)
(b) What is the e.m.f. $\mathcal{E}$ of the cell?

5.47 A small torch bulb is marked '2.5 V 0.3 A'.
(a) What is its resistance when it is working normally?
(b) Show that, if only a 6 V battery is available, a resistance of about 12 Ω connected in series would enable the torch to run normally. Draw a circuit diagram.
(c) When the torch is run directly using a 3.0 V battery, it is found to run normally, i.e. the current is 0.30 A. What is the internal resistance of the 3.0 V battery?

5.48 The internal resistance of a single photovoltaic cell, such as those used to power a communications satellite, is 40 Ω. The satellite's electrical system is powered by an array of 18 000 such cells arranged in 300 rows of 60 by connecting 60 cells in series and joining 300 such rows in parallel. The total active area of each cell is $0.0012\,m^2$.
(a) Show that the combined resistance of the 18 000 cells in this arrangement is less than 10 Ω.
(b) What e.m.f. will this arrangement produce if the e.m.f. of one cell is $\mathcal{E}$?
(c) When fully illuminated by sunlight of intensity $1.4\,kW\,m^{-2}$, the array produces 4.5 kW of electrical power. Calculate the efficiency of transfer of solar to electrical energy.

5.49 A car battery has an e.m.f. of 12.0 V and an internal resistance of 5.0 mΩ. What is the p.d. across its terminals when it supplies
(a) a current of 0.50 A to each of two side-lamps and two tail-lamps
(b) a current of 0.50 A to each of these four lamps and a current of 5.0 A to each of two headlamps
(c) a current of 800 A to a starter motor?

5.50 A car driver switches on the car's side-lamps and headlamps before starting the engine. Explain why the headlamps dim when he operates the starter switch.

5.51 Corrosion at a car's 12-V battery terminals results in increased resistance, and is a frequent cause of difficulty in starting cars. In an effort to diagnose the problem, a

mechanic measures the voltage between the battery terminal and the wire carrying current to the starter motor.
While the motor is turning, the current in the motor is 180 A and the measured voltage is 4.2 V. Explain why the corrosion resistance at the battery terminal is 23 mΩ.

5.52 A 12-V car battery is recharged by passing a current I through it in the reverse direction using a 15-V battery charger. The internal resistances of the battery and charger are 0.72 Ω and 0.04 Ω, respectively.

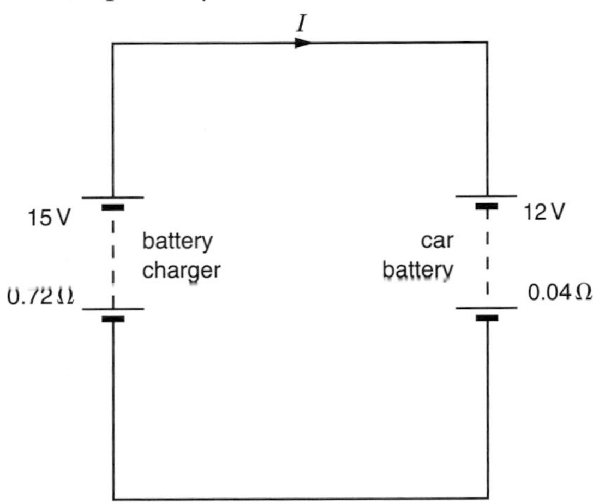

(a) Show that the current I is about 4 A.
(b) At what rate is energy being stored in the car battery?

5.53 A laboratory power supply is designed to provide p.d.s of up to 6.0 kV; it is provided with an internal resistance of 50 MΩ. Suppose the power supply is set at 6.0 kV.
(a) What will be the current if the power supply is connected to a resistor of resistance 100 MΩ?
(b) What will be the current if the terminals are short-circuited?

5.54* High-voltage power supplies in schools and colleges can provide up to 6.0 kV. This is much higher than the mains voltage, yet they are considered safe to use.
(a) Describe what is done to make them 'safe'.
(b) Under what circumstances might they not be safe?
(c) 'I don't see the point of providing such a big voltage if you're going to stop it providing a big current.' Does it matter that the current they can provide is very small?

5.55* The diagram shows a power supply of 250 mV with an internal resistance of 4.0 Ω connected to a resistor, the resistance R of which can be set at any integral value between 2 Ω and 8 Ω.
A student sets out to measure the current I in the circuit for various values of R, and to calculate the power transfer P to the resistor in each case. His readings and calculations are shown in the table on the next page.

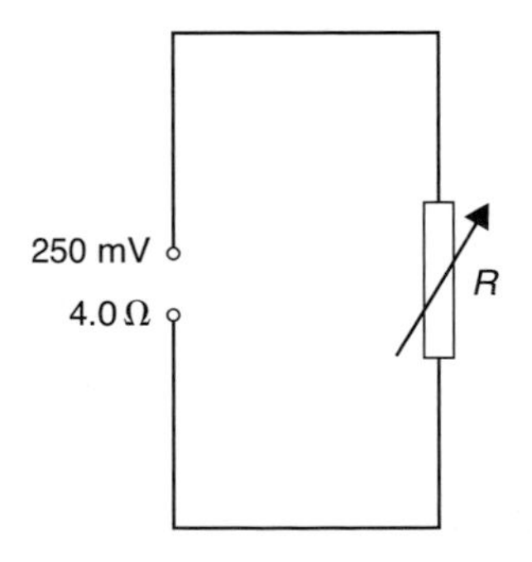

R/Ω	2.00	3.00	4.00	5.00	6.00	7.00	8.00
I/mA	41.7	35.7	31.3	27.8	25.0	22.7	20.8
P/mW	3.48	3.82	3.92	3.86	3.75	3.61	3.46

(a) Check the correctness of the values given in any one column. (Or two, if you fail first time!)
(b) Predict the value of P when **(i)** R is zero, and **(ii)** R is infinitely large.
(c) Plot a graph of P (y-axis) against R (x-axis) and determine at what value of R the power P is a maximum. How does your value relate to the above circuit?

5.56 A battery is connected to an electric motor which is used to raise a load. An ammeter shows that the current in the motor is 2.6 A when the motor raises a load of 3.2 kg at a steady speed of 0.43 m s^{-1}. Assuming that there are no transfers of energy into internal energy, what is the e.m.f. of the battery? [Use data.]
Explain whether the actual e.m.f. of the battery would be greater or less than this calculated value if transfers of energy into internal energy cannot be neglected.

5.57 A dry cell, of e.m.f. $\mathcal{E}$ and internal resistance r, is connected in series with a switch and a resistor of resistance 2.0 Ω. A voltmeter is connected across the cell.
(a) Draw a circuit diagram of this arrangement.
With the switch open the initial reading on the voltmeter is 1.50 V. The switch is closed for 6 minutes and then opened. The readings on the voltmeter during a 12-minute period are shown on the graph.

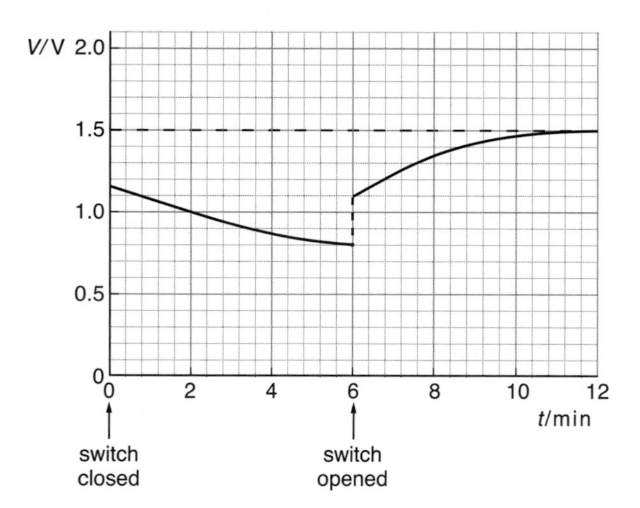

(b) What is the e.m.f. of the cell **(i)** before the switch is closed **(ii)** after 12 minutes when the switch is again open?
(c) Calculate the current in the circuit immediately after the switch is first closed, i.e. at $t = 0$, and hence deduce the initial internal resistance of the cell.
(d) What is the e.m.f. of the cell immediately after the switch is opened, i.e. at $t = 6$ min? Hence determine the internal resistance of the cell after it has been driving current round the circuit for 6 minutes.

6 Electrical circuits

Data: In this chapter you will need to use the definitions and equations in the lists for Chapters 4 and 5.

Note also the following:

ammeters – digital ammeters have effectively zero resistance

voltmeters – digital voltmeters have effectively an infinite resistance

circuit diagrams – *always draw or copy a circuit diagram before making any circuit calculations.*

6.1 Circuit calculations

In this section you will need to use what you have learnt in Chapters 4 and 5.

6.1 A catalogue states that when a particular light-emitting diode (LED) is used with a 5.0 V supply a 270 Ω resistor must be connected in series with it to limit the current to 10 mA. Calculate
(a) the p.d. across the resistor
(b) the resistance of the LED in these conditions.

6.2 A power supply which provides a constant p.d. of 6.0 V is connected in series with a resistor of constant resistance 100 Ω and a thermistor. The resistance of the thermistor is 380 Ω at 25 °C but falls to 28 Ω at 100 °C. Calculate the p.d. between the ends of the resistor at **(a)** 25 °C **(b)** 100 °C.

6.3 Three resistors are connected as shown in a circuit.

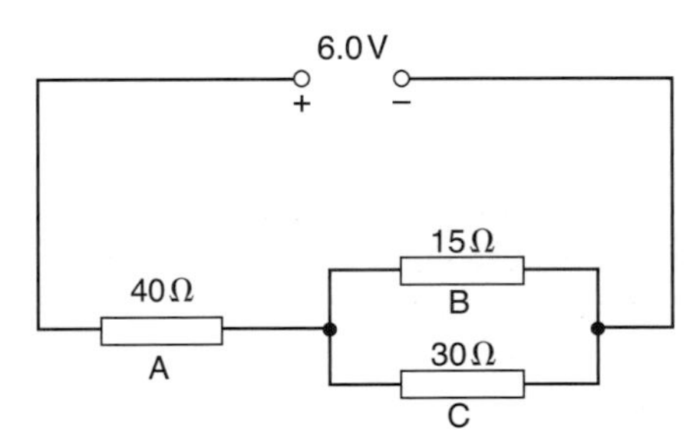

(a) Calculate the p.d. across each resistor in the circuit.
(b) Calculate the current in each resistor in the circuit.

6.4 Three resistors, of resistance 4.7 Ω, 10 Ω and 15 Ω, respectively, are connected to a battery as shown in the figure. A voltmeter connected across the battery reads 5.7 V.

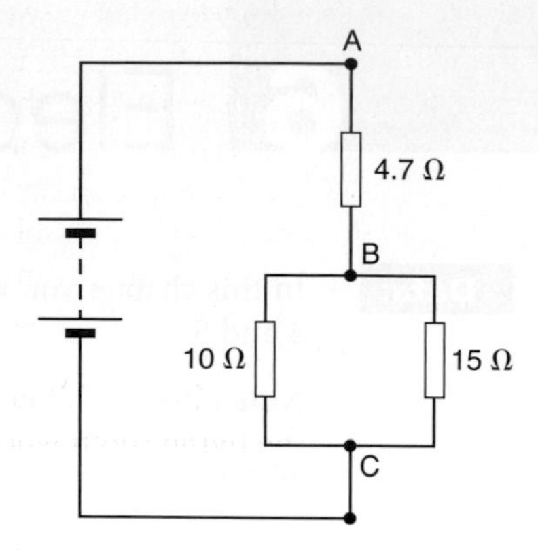

(a) Calculate the resistance of the circuit between B and C.

(b) Calculate the p.d. between
- **(i)** A and B
- **(ii)** B and C.

(c) Calculate the current in
- **(i)** the 4.7 Ω resistor
- **(ii)** the 10 Ω resistor
- **(iii)** the 15 Ω resistor.

6.5 The diagram shows two identical resistors connected to the same cell in two different ways. In which case, **(a)** or **(b)**, is the total power greater, and how many times greater is it? Assume that the cell has no internal resistance, and that the resistors have constant resistance.

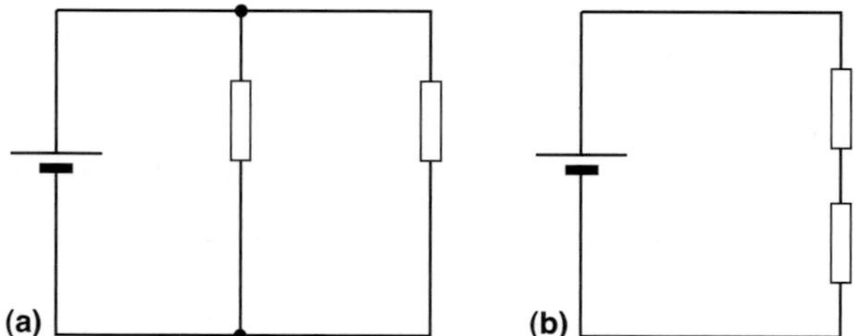

6.6 An electric blanket working from a 230 V mains supply contains two heating elements, each of resistance 460 Ω. The blanket has low, medium and high settings.

(a) How are the heating elements connected in each case? (You are not asked to draw a circuit diagram showing the switching mechanism.)

(b) Calculate **(i)** the total resistance of the blanket in each setting **(ii)** the current drawn from the mains on each setting **(iii)** the electrical power of the electric blanket in each case.

6.7 The figure shows a circuit in which two cells provide a p.d. of 2.9 V. The p.d. across the 22 Ω resistor is found to be 1.2 V.

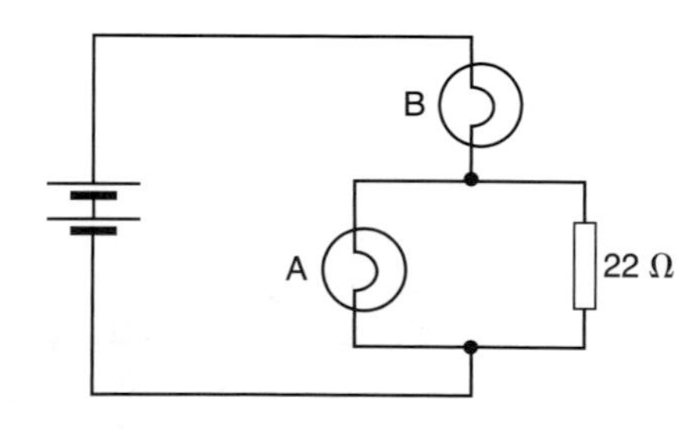

(a) What is the p.d. across
- **(i)** lamp A
- **(ii)** lamp B?

(b) What is the current in the resistor?

(c) If the lamps have the same resistance, how many times greater is the current in B than the current in A (leave your answer as a fraction)?

(d) What is the current in the cells?

6.8 In the circuit in the previous figure the fixed resistor is replaced by a rheostat of maximum resistance 22 Ω. Explain what happens to the brightness of each lamp as the resistance of the rheostat is reduced from 22 Ω to zero.

6.9 Your electric hair dryer is not working and you need to draw a circuit diagram to help you find out what is wrong. All you know is that there are three simple on–off switches. One switches on the hair dryer without heating, and the other two provide different powers. You assume that the circuit is arranged so that the heating cannot be switched on if the hair dryer is not on, and that there are two separate heating elements.

(a) Draw a possible circuit diagram.

(b) The maximum power of the hair dryer is stated as 460 W when connected to a 230 V supply. Calculate the resistance of each resistor, assuming they are the same.

6.10* A family with a hearing-impaired child installed a doorbell that also lights up an LED in the living room when the bell is rung.

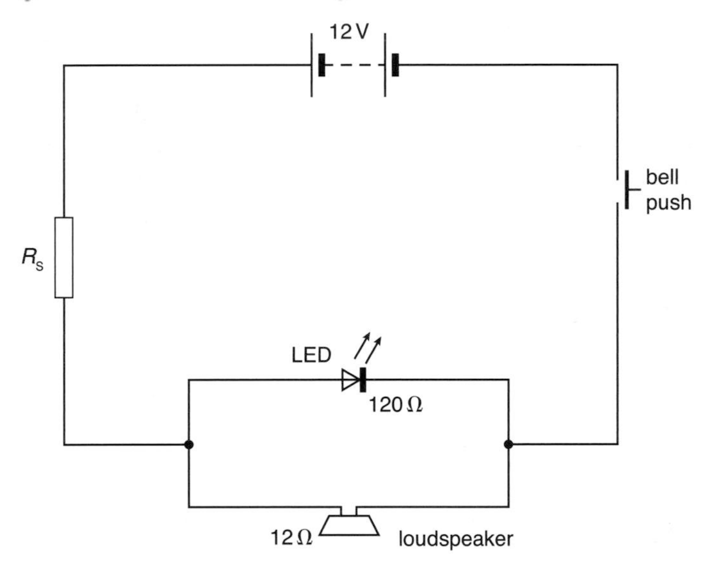

The maximum permitted current in the loudspeaker is 200 mA.

(a) Calculate

(i) the maximum allowed p.d. across the loudspeaker

(ii) the maximum current in the LED

(iii) the maximum current in the 'safety' resistor R_S.

(b) Hence explain why the resistance of the safety resistor must be more than 44 Ω.

6.11 An energy-saving lamp labelled '230 V 11 W' and a car headlamp bulb labelled '12 V 60 W' are connected in series to a 230 V supply.
What will happen? [Hint: Calculate the resistances of each of the lamps when they are in normal use.]

6.12 An indicator lamp which is labelled '2.5 V 0.2 A' is connected in series with a rheostat of maximum resistance 50 Ω and a battery. Assume that the battery has a constant p.d. of 3.0 V between its terminals.

(a) Calculate the resistance of the indicator lamp when it is working normally.

(b) Assume that the resistance of the indicator lamp is constant, and calculate the minimum current which may be obtained by adjusting the rheostat.

(c) The resistance of the indicator lamp is not constant. Explain how its resistance varies if the minimum current is actually larger than your answer to **(b)**.

6.13* In measuring resistances either circuit P or Q could be used.

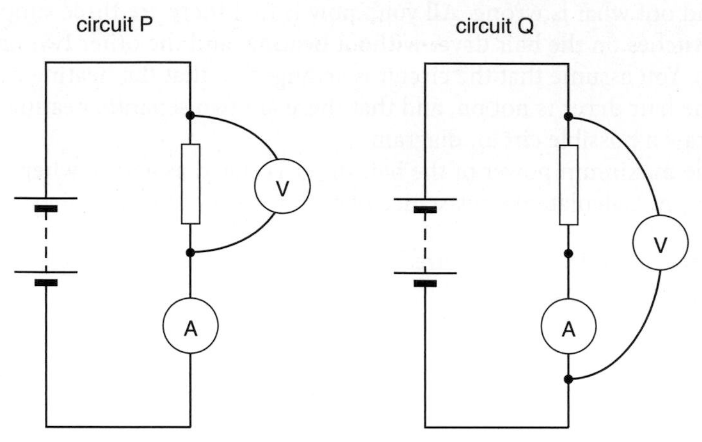

(a) Explain why, with 'perfect' instruments, either arrangement would give a valid answer.

(b) When measuring resistances of above 500 kΩ, discuss which arrangement would be preferable

(i) if your ammeter had a noticeable resistance, such as 1.5 kΩ

(ii) if your voltmeter did not have infinite resistance, but a resistance such as 1.0 MΩ.

6.14 The graph shows how the current and voltage output from a single solar cell changes as the variable resistor to which it is connected is varied from 0 Ω to 200 Ω. The intensity of light shining on the solar cell is kept constant throughout.

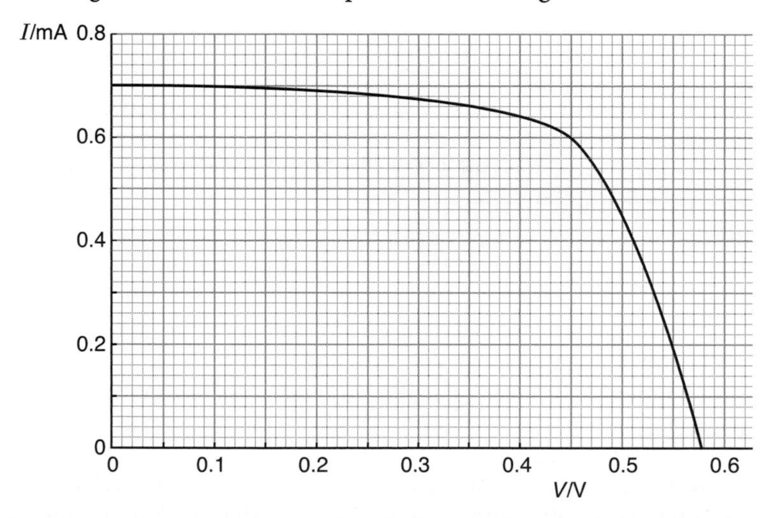

(a) Make a table showing the variation of current I against the voltage output V for values of

V/V 0 0.20 0.40 0.45 0.50 0.55 0.58

(b) Add to your table values for the power output P/W of the solar cell, and show that the maximum P is about 0.3 mW.

6.2 Circuits for measurement and sensing

In this section you will need to

- know how to connect an ammeter and a voltmeter in a circuit to measure current and p.d.
- describe how to measure the resistance and power of a resistor
- understand how a rheostat can be used as a current limiter, using two terminals
- understand how a rheostat can be used as a potential divider, using three terminals, and what the advantage of this is
- understand that thermistors and light-dependent resistors (LDRs) are sensors which can be used to control other devices
- understand how potential dividers can be used with sensors.

6.15 The circuit shows a potential divider arrangement using an 'input' voltage of 6.0 V.

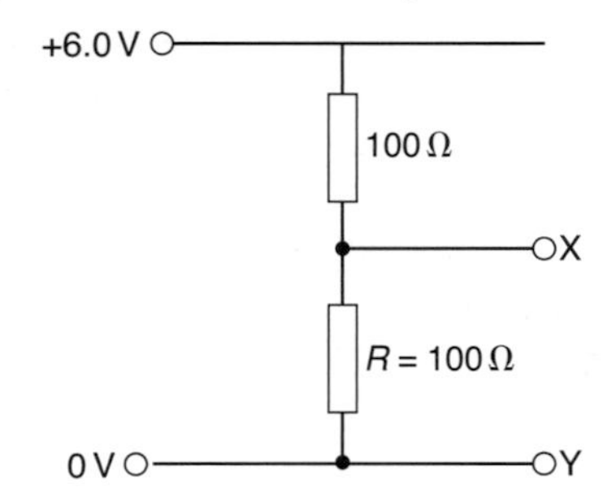

(a) Calculate the 'output' voltage, i.e. the p.d. V_{out} between X and Y.
(b) What will be the output voltage when separately **(i)** the input voltage is doubled to 12 V **(ii)** R is increased to 500 Ω **(iii)** R is increased to 2000 Ω?

6.16 The diagram shows a potential divider in which a rheostat is connected to a p.d. of 6.0 V. A voltmeter is connected to measure the output voltage V_{out} as the slider bar is moved from end A to end B of the rheostat.

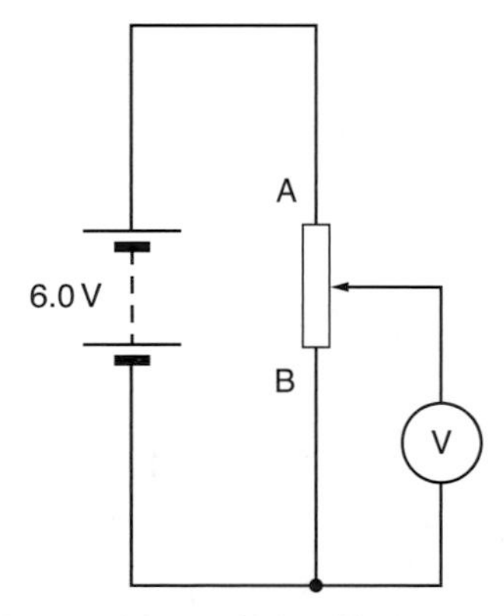

(a) State and explain the readings of V_{out} when the rheostat slider is connected to **(i)** A **(ii)** B.
(b) Sketch a graph of V_{out} against the position of the slider it moves from A to B.

6.17 The diagram shows a potential divider ABC where the output can be connected across a signal lamp labelled '3.0 V 0.25 A'.

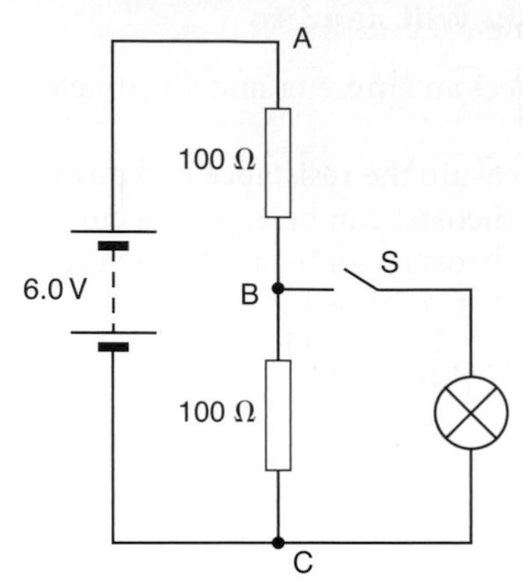

(a) Show that the working resistance of the lamp is 12 Ω.
(b) Assuming the resistance of the lamp remains 12 Ω, calculate, when the switch S is closed **(i)** the resistance between B and C **(ii)** the p.d. across the lamp.
(c) Hence explain why the lamp will not then light.

6.18 A light-dependent resistor and a fixed resistor (of resistance 10 kΩ) are connected as shown in series between the terminals of a power supply which provides a constant p.d. of 5.0 V. The negative terminal of the power supply is earthed, i.e. may be taken to be 0 V. In the dark the potential of the point X is 0.21 V; when more light falls on the LDR the potential of X rises to 4.2 V. Calculate the resistance of the LDR in these two situations.

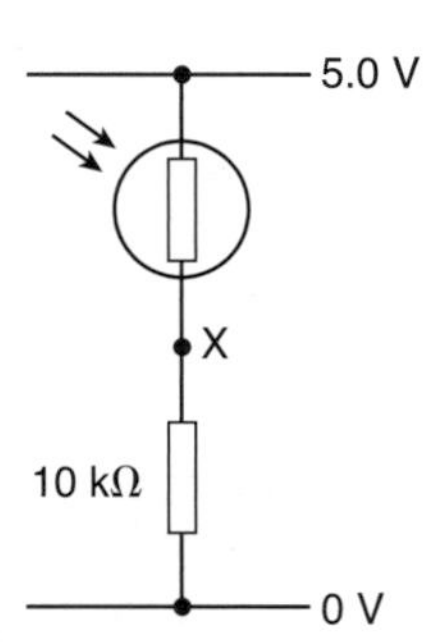

6.19 In order to measure the internal resistance r of a dry cell, a student connects it in series with a 4.70 Ω resistor and a switch S. She has already measured the e.m.f. of the cell to be 1.49 V. She now draws the circuit diagram as shown in order to stress that the internal and external resistors form a potential divider. On closing the switch she notes the voltmeter reading to be 1.41 V.
What value does she deduce for r?

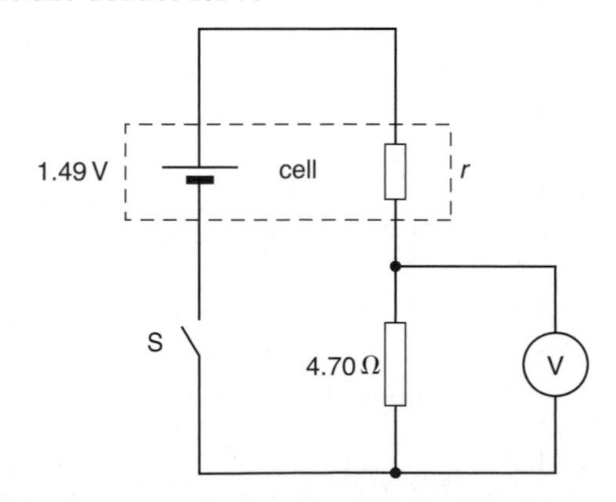

6.20 A rheostat of maximum resistance 100 Ω is connected as a potential divider, as shown in the diagram, to a power supply which provides a constant p.d. of 12 V. The slider is moved until the voltmeter reads 3.0 V.

(a) Copy the diagram and show the position of the slider.

(b) A 47 Ω resistor is now connected across the output of the potential divider, i.e. in parallel with the voltmeter. The voltmeter reading changes. Explain whether it rises or falls.

(c) What is the new voltmeter reading?

(d) Calculate the current in the power supply.

(e) What is then the current in **(i)** the 47 Ω resistor **(ii)** the part of the rheostat in parallel with the 47 Ω resistor?

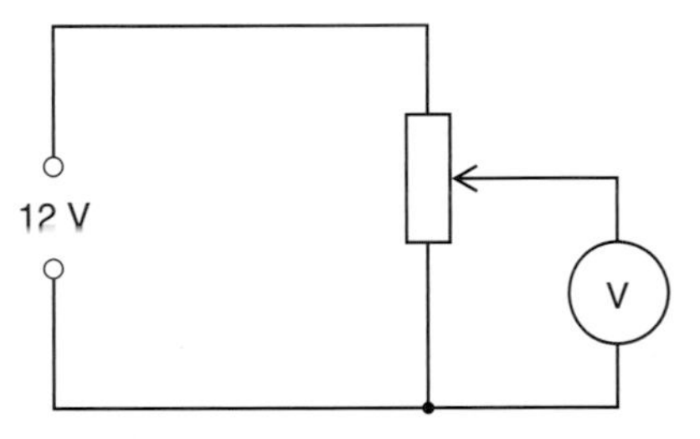

6.21* The diagram shows a possible control circuit for a cooling system such as a refrigerator. The cooling system, which can be assumed to have infinite resistance, is switched on when the voltage across it rises above 3 V.
At temperatures of $\theta/°C = 0, 2, 4, 6$, the resistance of the thermistor is
$R_t/k\Omega = 0.65, 0.32, 0.16, 0.10$.

(a) Calculate the p.d. V_c across the cooling system at each of these temperatures.

(b) Discuss whether this switching circuit can be used to switch on the cooling system in a domestic refrigerator when the temperature inside the fridge rises above 3 °C.

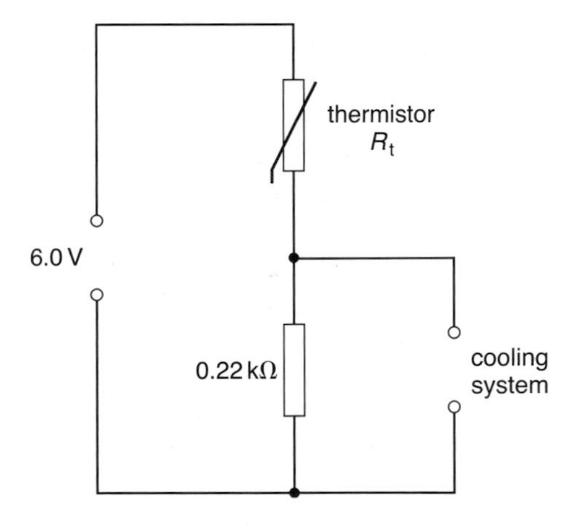

6.22 The diagram shows a circuit in which an LDR is being used as a sensor. A lead at X is connected to a device called a transistor (not shown). When the potential at this point rises above about 0.7 V, the transistor 'turns on' and operates a relay which switches on a lamp.

(a) On a dull day, the resistance of the LDR is 500 kΩ, and the rheostat is set to provide maximum resistance in series with the fixed resistor of 300 Ω.

(i) What is the potential at X?

(ii) Is the transistor 'on'?

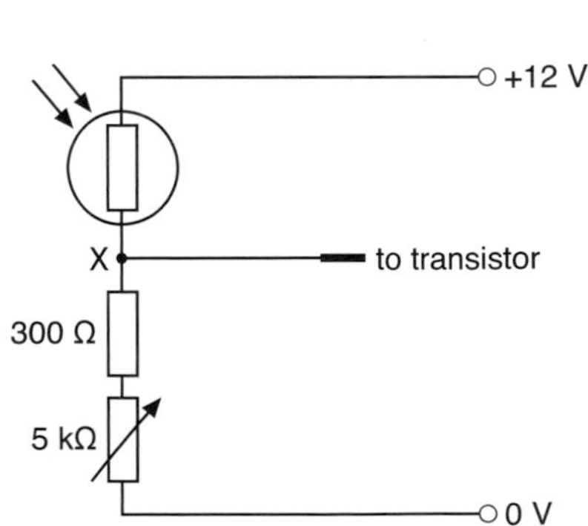

The resistance of an LDR decreases as the illumination increases.

(b) Calculate the resistance of the LDR when the transistor first turns on.

(c) Why is the rheostat included in the circuit?

6.23 The diagram shows a thermistor connected in a potential divider circuit. The graph shows how the resistance of the thermistor varies with temperature.

(a) Calculate the values of the p.d. measured by the voltmeter at temperatures of θ/°C = 20, 30, 40, 50.

(b) Draw a graph to show how the p.d. measured by the voltmeter varies with temperature.

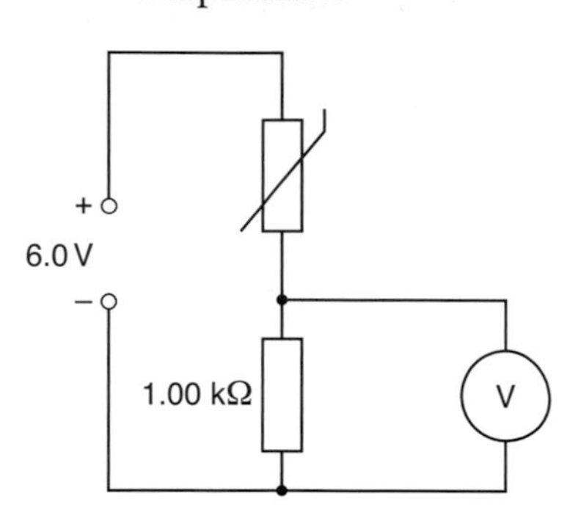

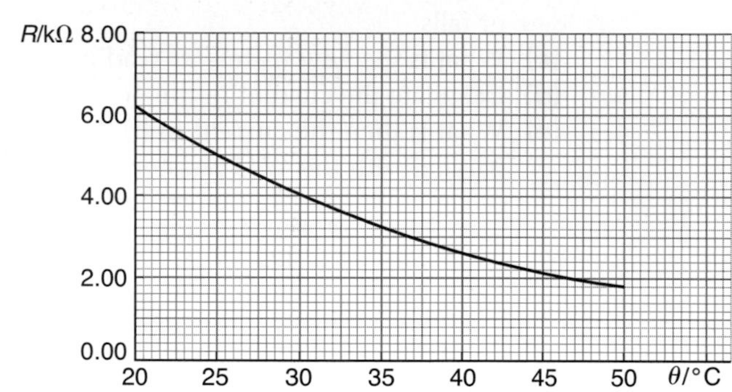

6.24 The diagram shows a circuit containing a battery of e.m.f. 6.0 V and negligible internal resistance, and four resistors *P*, *Q*, *R* and *S*. The negative terminal of the battery is earthed, so the potential at that part of the circuit is zero.

(a) What are the potentials at X and Y if

(i) $P = Q = R = S = 10\,\Omega$

(ii) $P = Q = 10\,\Omega$, $R = S = 20\,\Omega$

(iii) $P = 5\,\Omega$, $Q = R = 10\,\Omega$, $S = 20\,\Omega$?

(b) What can you say about the values of *P*, *Q*, *R* and *S* if the potentials at X and Y are to be the same?

(c) What circuit component would you need to add in order to test whether the potentials at X and Y were the same?

(d) Suppose $P = 20\,\Omega$, $Q = 10\,\Omega$, *R* is a resistor of unknown size, and *S* is a variable resistor. If the potentials at X and Y are the same, what is *R* if $S = 23\,\Omega$?

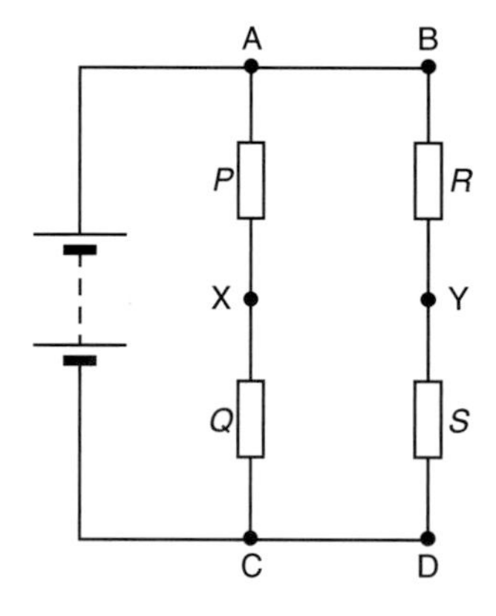

6.25 The circuit in the previous diagram was once used as a method of measuring resistance. It is called a bridge network. The bridge is said to be 'balanced' when the potentials at X and Y are the same.

(a) The resistances of *P*, *R* and *S* are 15 Ω, 33 Ω and 22 Ω. Show that the resistance of *Q* must be 10 Ω if the bridge is to be balanced.

(b) Explain why, if *P* and *R* are unchanged, and *Q* and *S* are increased to 100 Ω and 220 Ω, the bridge is still balanced.

(c) If the resistance of *Q* is now increased slightly from 100 Ω, explain in which direction the current will flow in an ammeter connected between X and Y.

6.3 Meters and oscilloscopes

In this section you will need to

- understand why an ammeter must have a much lower resistance than the circuit in which it is placed
- understand why a voltmeter must have a much higher resistance than the component across which it is measuring the p.d.
- understand the advantages and disadvantages of digital meters, data loggers and electronic meters
- describe how to use an oscilloscope to measure p.d.s and how the sensitivity (or gain) control is used
- describe how to use an oscilloscope to measure the frequency of an alternating p.d. and how the time-base control is used.

6.26* A student wants to find the resistance of a filament lamp that he knows to be about 20 Ω. He has a 6.0-V battery of negligible internal resistance and two 'old' analogue meters: an ammeter with a resistance of 10 Ω and a voltmeter with a resistance of 10 kΩ. He tries connecting up the two circuits shown.

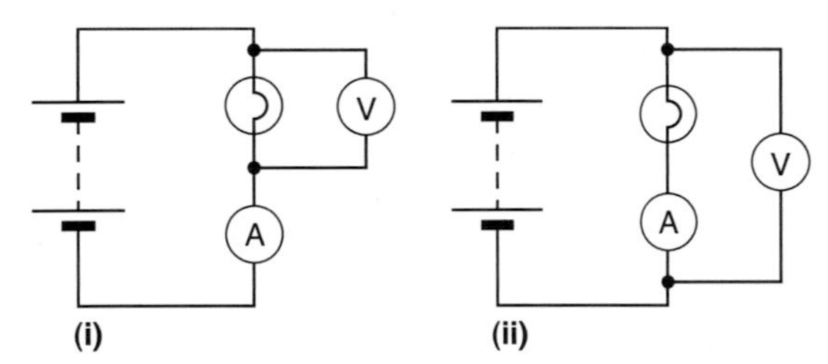

(a) Assuming his lamp does have a resistance of 20 Ω, what readings does he get in circuit **(i)** and in circuit **(ii)**?

(b) Comment on the problems associated with using analogue meters.

6.27 A student wishes to investigate how the current varies with time when a filament lamp is switched on. Its resistance, when operating normally, is about 100 Ω. He decides to use a data logger with a circuit which includes a 0.47 Ω resistor, as shown in the diagram.

(a) Why is it sensible to choose a resistor with such a small resistance?

(b) The graph shows the trace he obtains from the data logger. When the lamp is operating normally, what is the p.d. across the resistor?

(c) What is the current in the lamp 1.0 ms after switching on?

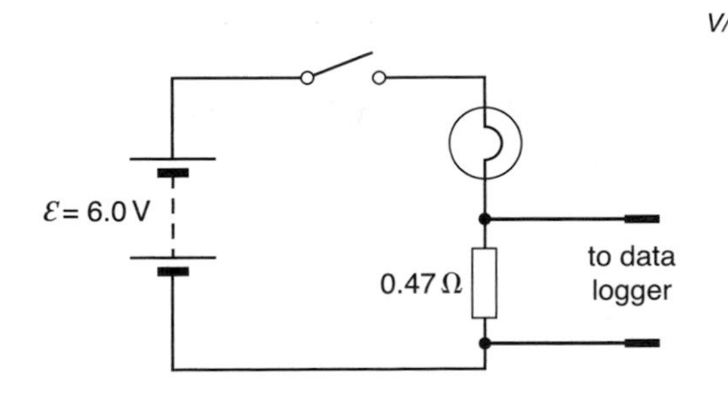

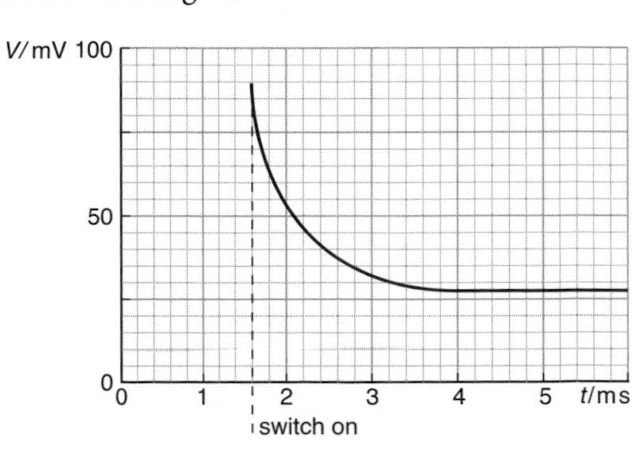

6.28 The diagram shows some traces on an oscilloscope screen. What is the frequency of the alternating p.d. if in each case the time base speed is $5.0\,\text{ms div}^{-1}$?

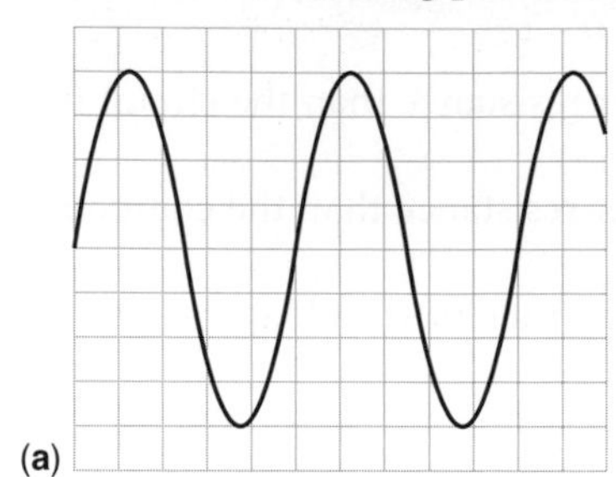
(a)

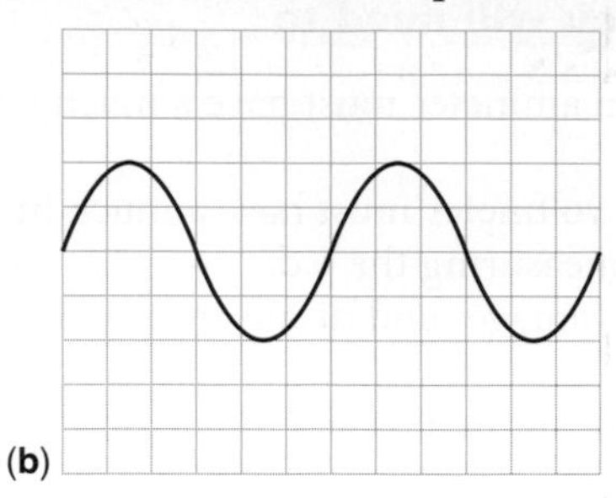
(b)

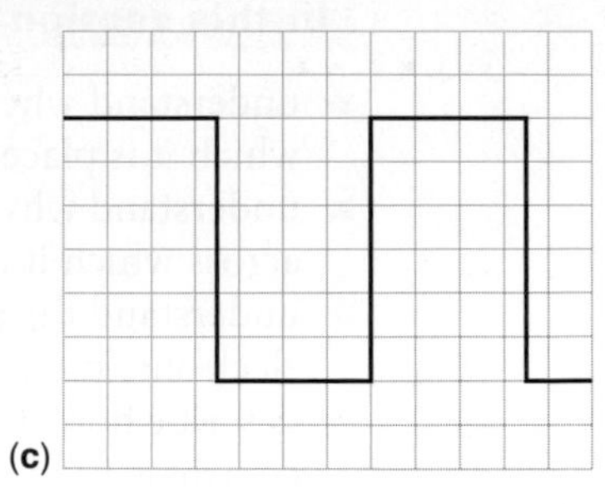
(c)

6.29 A small a.c. generator and battery, of negligible internal resistance, are joined in series, and the circuit is completed by a $100\,\Omega$ resistor. An oscilloscope connected across the resistor gives the trace shown. The horizontal line is the trace with no signal from the a.c. generator.

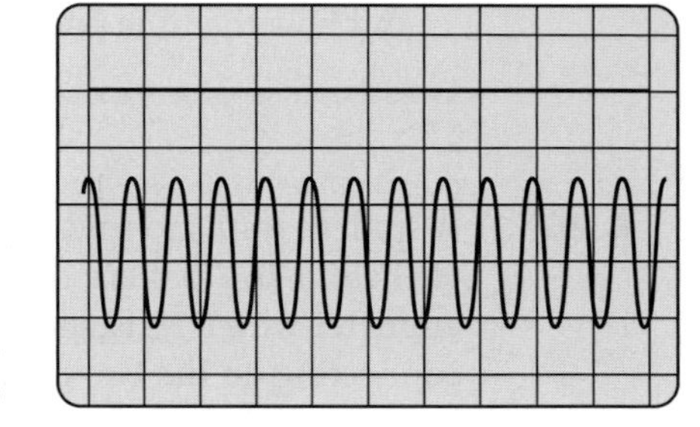

The settings of the oscilloscope are: *y*-amplifier control $1.0\,\text{V div}^{-1}$; time base $10\,\text{ms div}^{-1}$. Calculate

(a) the peak output voltage of the a.c. generator

(b) the frequency of the a.c. generator

(c) the e.m.f. of the battery

(d) the maximum and minimum currents in the resistor.

6.30 With the time base switched off, the vertical movement of the spot on an oscilloscope screen corresponds to the change in p.d. across its terminals. Knowing the sensitivity enables you to measure the p.d. The oscilloscope can also be used to measure current. The circuit diagram shows suitable connections in an experiment to measure the current in a circuit when it is varied using a $2.2\,\text{k}\Omega$ variable resistor. The e.m.f. of the cell is $1.50\,\text{V}$ and it has negligible internal resistance.

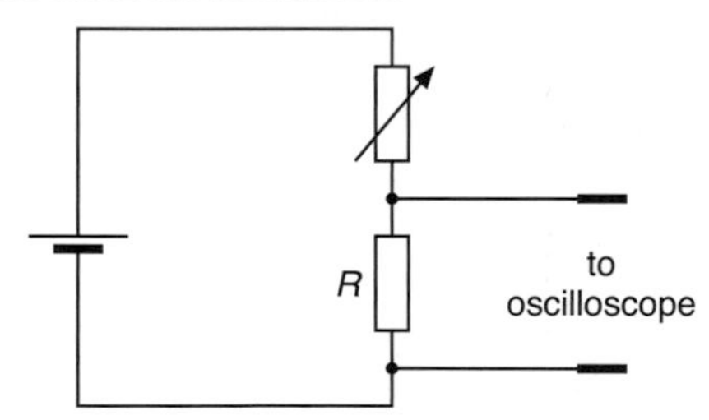

(a) The fixed resistor R has a known resistance of $10.0\,\Omega$. Why is it necessary to include this fixed known resistance in the circuit?

(b) What is the p.d. across it when the variable resistor is set to its maximum resistance?

(c) Which sensitivity would you use on the oscilloscope: $100\,\text{mV div}^{-1}$, $20\,\text{mV div}^{-1}$, $5\,\text{mV div}^{-1}$, $1\,\text{mV div}^{-1}$?

(d) What would be your answers for **(b)** and **(c)** if the variable resistor's resistance is reduced to $50\,\Omega$?

(e) Later, the oscilloscope spot is found to move 6.3 divisions with the sensitivity set to $10\,\text{mV div}^{-1}$. What was the p.d. across the $10\,\Omega$ resistor?

(f) What was the current in the circuit?

7 Physics of materials

Data:

$g = 9.81\,\mathrm{N\,kg^{-1}}$

normal atmospheric pressure = 101 kPa

densities: $\rho_{air} = 1.29\,\mathrm{kg\,m^{-3}}$

$\rho_{water} = 1.00 \times 10^3\,\mathrm{kg\,m^{-3}}$

$\rho_{mercury} = 13.6 \times 10^3\,\mathrm{kg\,m^{-3}}$

$\rho_{steel} = 7.70 \times 10^3\,\mathrm{kg\,m^{-3}}$

viscosities: $\eta_{water} = 1.1\,\mathrm{N\,s\,m^{-2}}$

$\eta_{air} = 1.8 \times 10^{-5}\,\mathrm{N\,s\,m^{-2}}$

7.1 Density, pressure and flow

In this section you will need to

- use the equation $\rho = m/V$, which defines density ρ
- use the equation $p = F/A$, which defines pressure p
- understand that atmospheric pressure is caused by the layer of air which is attracted to the Earth by gravitational forces
- understand that the pressure in a fluid exerts a force at right angles to any surface with which the fluid is in contact
- use the equation $\Delta p = \rho g(\Delta h)$
- calculate pressure differences measured using manometers and barometers
- understand how Archimedes' principle can be used to calculate the upthrust on a body immersed in a fluid
- understand what it means to say that a fluid is viscous
- understand the difference between frictional and viscous forces
- understand the difference between streamline and turbulent flow, and the conditions under which each may occur
- use the equation $F = 6\pi r\eta v$ (Stokes' law)
- use the drag force equation $F = \frac{1}{2}AC_D\rho v^2$ where the symbols have their usual meaning
- understand that in a pipe fluid flows faster where the pipe is narrower, and that the pressure there is less
- understand how the Bernoulli effect applies to situations where air flows over a curved surface (aerofoils, hydrofoils, sails, windsurfing, paragliding)
- be able to use the lift force equation $F = \frac{1}{2}SC_L\rho v^2$ where the symbols have their usual meanings.

7.1 Copy and complete the following table:

	material	mass/kg	volume/m^3	density/kg m^{-3}
(a)	aluminium	160	0.060	
(b)	lead		0.032	11×10^3
(c)	steel	60		7.7×10^3

7.2 The average radius of the Earth is 6.4×10^6 m. Its mass is 6.0×10^{24} kg. What is its average density?

7.3 A scaffolding pole has an external diameter of 40 mm and an internal diameter of 32 mm. It is made of steel. What is the mass of a 5-metre length of pole? [Use data.]

7.4 Copy and complete this table. Each object is in equilibrium on a horizontal surface.

	object	weight/N	contact area/m^2	pressure/N m^{-2}
(a)	elephant	5.5×10^4	0.14	
(b)	skier	6.9×10^2		2.5×10^3
(c)	tractor	1.5×10^4		1.2×10^4
(d)	tray	8.0	0.20	
(e)	pavement slab		0.50	8.0×10^2

7.5 A building brick has a mass of 2.8 kg and measures 225 mm by 112.5 mm by 75 mm. What pressure does it exert when stood, in turn, on each of its three faces on a horizontal surface?

7.6* Refer to question 7.3. Suppose that during building one of these scaffolding poles rests vertically on the ground, and the contact force is 12 kN.

(a) What pressure would the pole exert on the ground?

(b) In practice a horizontal steel plate is placed between the pole and the surface it is resting on. If the plate measures 15 cm by 15 cm, what is the new pressure on the ground? (Ignore the weight of the plate.)

(c) When the ground is particularly soft, the plate may rest on a scaffolding board. If the board measures 3.0 m by 0.25 m, what is the pressure between the board and the ground? (Ignore the weight of the board and the plate.)

7.7 Referring to the pressure which the objects produce when they are used correctly, explain the construction of **(a)** skis **(b)** drawing pins **(c)** football boots.

7.8 Show that the units of both sides of the equation $\Delta p = \rho g(\Delta h)$ are the same.

7.9 Pressures, and pressure differences, are sometimes given in heights of a stated liquid. For example, it might be said that 'atmospheric pressure is 760 mm of mercury' or '760 mmHg'.

(a) Explain how it is possible to express a pressure in terms of a height of liquid.

(b) Express 760 mm of mercury as a pressure in kPa. [Use data.]

7.10* The heart pumps blood through our arteries. As it does so, the pressure in the arteries rises and falls. The usual method of measuring the pressure of the blood in the arteries is to wrap a rubber cuff round the patient's upper arm and increase the pressure in it until the blood flow stops (as indicated by the pulse in the wrist), as shown in the photo.

This gives the maximum blood pressure (the systolic pressure), which is recorded in mmHg; the air in the cuff is then released until the blood begins to flow again, and the pressure (the diastolic pressure) when this happens is again recorded on the pressure scale on the right. (All these pressures are relative to atmospheric pressure.)

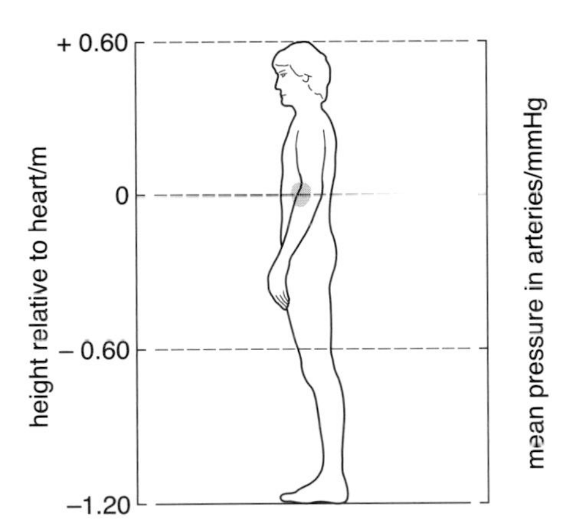

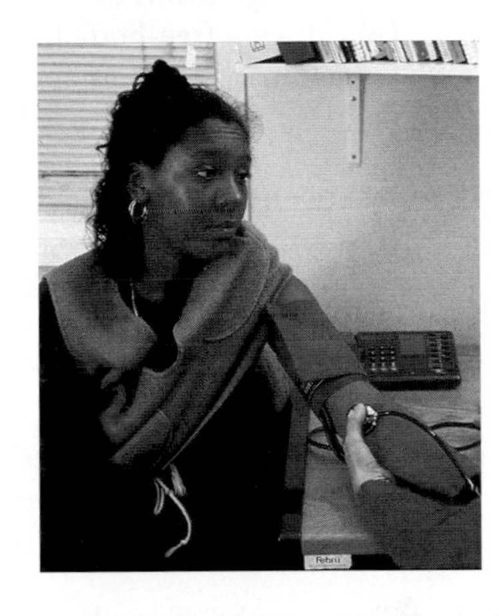

(a) If your blood pressure measurements are '120/80', what are these pressures in kPa? [Use data.]
(b) What is your mean arterial blood pressure in mmHg?
(c) Copy the diagram of a human being. Assume that his mean arterial blood pressure is the same as yours and on the pressure scale on the right mark the mean arterial blood pressures at the four heights shown on the height scale. The density of blood may be assumed to be the same as that of water.
(d) What is the systolic pressure in this person's feet?
(e) Why is blood pressure usually measured on your upper arm?

7.11 Oak has a density of about 700 kg m^{-3}. Draw a free-body diagram for a piece of oak floating in water, and explain fully why 0.70 of its total volume is submerged.

7.12 The airship R101 (which burst into flames in 1930 when the hydrogen in it ignited after a crash) had a volume of 1.38×10^5 m^3. Using data given above, calculate
(a) the upthrust on it
(b) the weight of the gas in it if it was filled with **(i)** hydrogen of density 0.0880 kg m^{-3} **(ii)** helium of density 0.176 kg m^{-3}
(c) the differences between the upthrust and these two weights.

7.13* The diagram shows the Plimsoll lines which are painted on the sides of ships. They show where the water line should be for different situations.

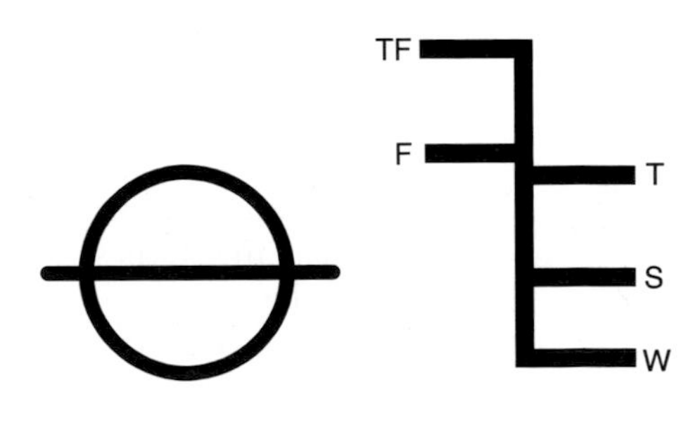

(a) If the water is at level F for a ship in dock, and then rises to TF, has the ship been loaded or unloaded? Explain.
(b) TF stands for 'tropical fresh water', T stands for 'tropical sea water'. Why is TF higher than T?
(c) S stands for 'summer', W stands for 'winter'. Why is S higher than W?

7.14 Use Stokes' law to show that the units of viscosity η are $N\,s\,m^{-2}$.

7.15 A steel ball-bearing of radius 2.0 mm reaches a terminal velocity v when falling through a liquid (glycerol) of viscosity $1.5\,N\,s\,m^{-2}$.

(a) Draw a free-body force diagram for the falling sphere and write an equation showing the equilibrium of these forces.

(b) Use Stokes' law to calculate the terminal velocity of the sphere. [Use data.]

7.16* Just over 100 years ago, in 1910, Robert Millikan measured the charges on a number of tiny sprayed drops of oil of density $7.8 \times 10^2\,kg\,m^{-3}$.
In order to determine the radius r of each drop, he measured the terminal speed v at which each drop fell in air.

(a) Show that

$$r^2 = 9\eta v \div 2g\rho_{oil}$$

(Ignore the Archimedean upthrust of the air on the drop.)

(b) Calculate r for a drop that falls 3.2 mm in 75 s. [Use data.]

(c) What is the mass of this drop?

7.17 Photographs of windsurfers often show the sail with a curved shape, but the wind is blowing across the surface of the sail, not into it, as might have been expected: see the diagram. Explain why the sail is curved in this way.

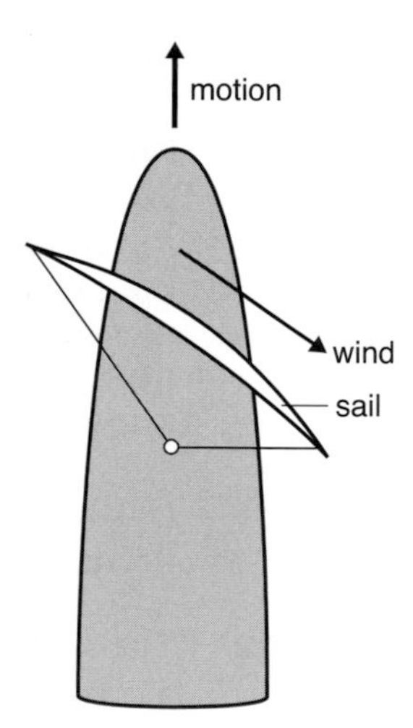

7.18 **(a)** Draw a diagram which shows a cross-sectional view of an aerofoil, and draw streamlines to show the flow of air past it when it is placed in a wind tunnel.

(b) Hence explain why there is an upward push on the aerofoil.

(c) How does the upward push depend on the area of the surface of the aerofoil, the density of the air, and the speed of the air?

7.19 Use the equation $F = \frac{1}{2}SC_L\rho v^2$ to calculate the upward push of the air flowing past the wings of an air liner which has a total wing area S of $500\,m^2$ and a speed of $280\,m\,s^{-1}$ in a region where the density of air is $0.025\,kg\,m^{-3}$. Take C_L to be 0.50.

7.2 Materials in tension and compression

In this section you will need to

- understand what is meant by the tension in a stretched wire, spring or rod and that it is the same throughout
- use the equation $F = kx$, which defines the stiffness k of a spring
- calculate the stiffness of two identical springs of stiffness k when arranged in series or in parallel
- use the equation $\sigma = F/A$, which defines tensile stress σ
- use the equation $\varepsilon = \Delta\ell/\ell$, which defines tensile strain ε
- understand what is meant by saying that a wire or spring obeys Hooke's law
- understand what is meant by the limit of proportionality, and the elastic limit of a material
- use the equation $E = \sigma/\varepsilon$, which defines the Young modulus E

- understand what is meant by stiffness, and how it differs from strength
- understand what is meant by elastic and plastic behaviour
- explain where tension and compression occur in a beam or cantilever
- draw stress–strain graphs for typical ductile and brittle metals, rubber and other polymers
- describe how to measure the Young modulus for a material in the form of a wire.

7.20 **(a)** Explain what is meant by the phrase 'the tension in a rope' as used by a physics teacher.
(b) Two tug-of-war teams each pull on a rope with a force of 5000 N. The rope is horizontal. What is the tension in the rope at its mid point?

7.21 A mass of 6.0 kg is placed at the lower end of a vertical wire; the upper end is fixed to a ceiling. What is the tension in the wire **(a)** at its lower end **(b)** at its upper end? What assumption do you have to make to be able to answer **(b)?**

7.22 A mass of 6.0 kg is supported by a uniform rope fixed to a beam. The rope has a mass of 1.0 kg. What is the tension in the rope **(a)** at its lower end **(b)** at its upper end **(c)** at its mid-point?

7.23 The graphs in the diagram show how the force F needed to produce an extension x varies for three different springs A, B and C.
(a) Calculate the gradient of each line and hence find the stiffness of each spring.
(b) Use the graphs to find the force needed to produce an extension of 0.12 m in **(i)** A **(ii)** B **(iii)** C.
(c) Use the graphs to find the extension produced by a force of 16 N in **(i)** A **(ii)** B.

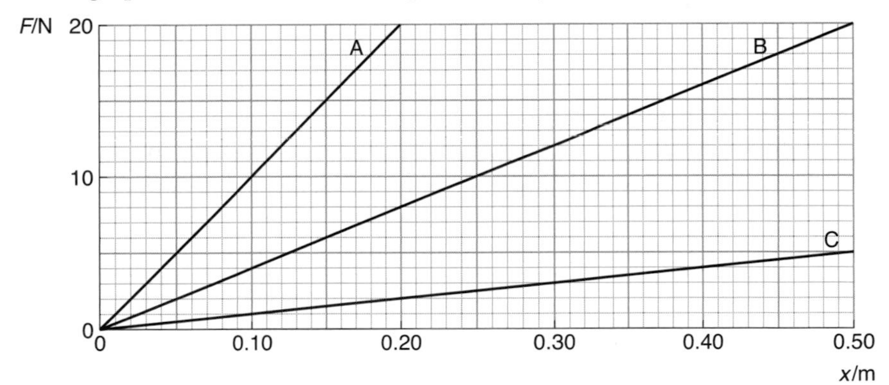

7.24 A spring has an unstretched length of 12 cm and a stiffness of 50 N m^{-1}. What force is needed to **(a)** double its length **(b)** treble its length?

7.25 The common 'expendable' springs often found in laboratories have a stiffness of 30 N m^{-1}. What is the stiffness of
(a) three of these springs connected end-to-end
(b) two of these springs connected side-by-side
(c) a spring consisting of one half of one of these springs?

7.26 For Hooke's law springs of stiffness k, write down the stiffness, in terms of k, of two springs **(a)** in parallel **(b)** in series.

7.27* The diagram shows an experimental set-up for measuring the length ℓ of a rubber band as the force F stretching it increases. A vertical rule is also needed. The graph shows the result of the experiment (the band snaps at X).

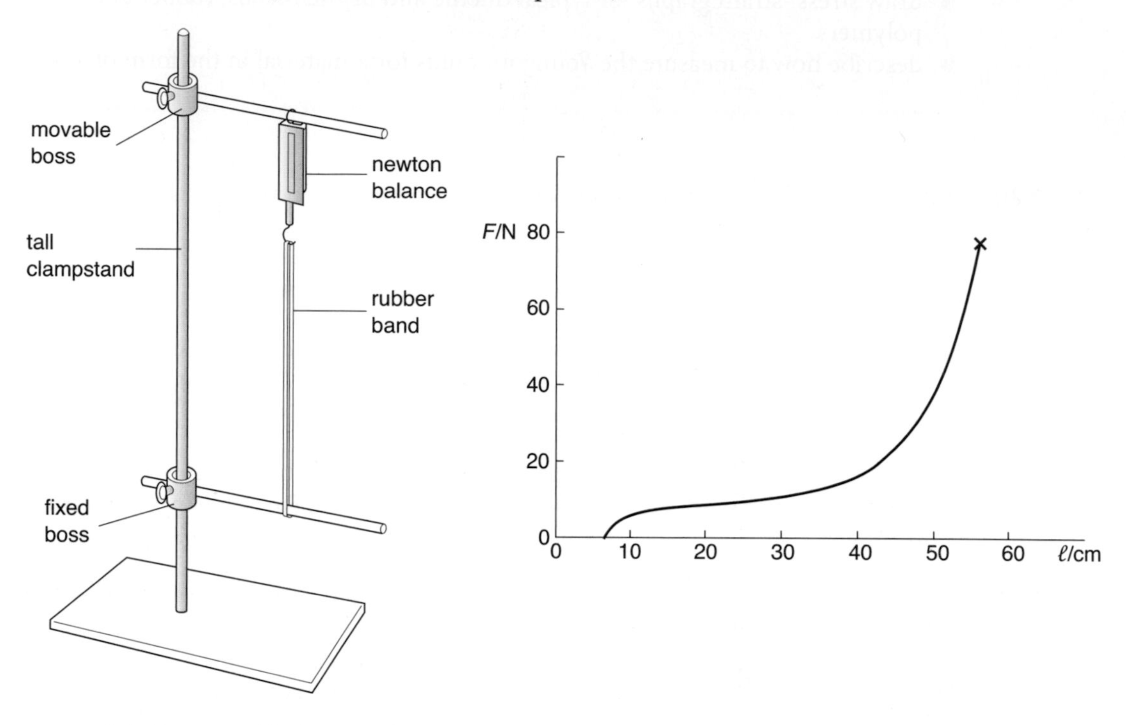

(a) Estimate **(i)** the initial length of the rubber band **(ii)** the force needed to treble its length **(iii)** the strain when it snaps.

(b) In what way is this experimental set-up better than simply hanging a rubber band from a fixed point and hanging weights of known mass from it?

7.28 What is the tensile strain when

(a) a copper wire of length 2.0 m has an extension of 0.10 mm

(b) a rubber band of length 50 mm is stretched to a length of 150 mm?

7.29 Copy and complete the following table:

	length	extension	strain
(a)	2.0 m	4.0 mm	
(b)	20 cm	50 cm	
(c)	10 m		3.0×10^{-3}
(d)	3.4 m		5.2×10^{-3}
(e)		0.57 mm	1.6×10^{-4}

7.30 The diagram shows an experiment in which a copper wire was stretched using controlled loads. The graph shows the result of such an experiment (the wire broke at X).

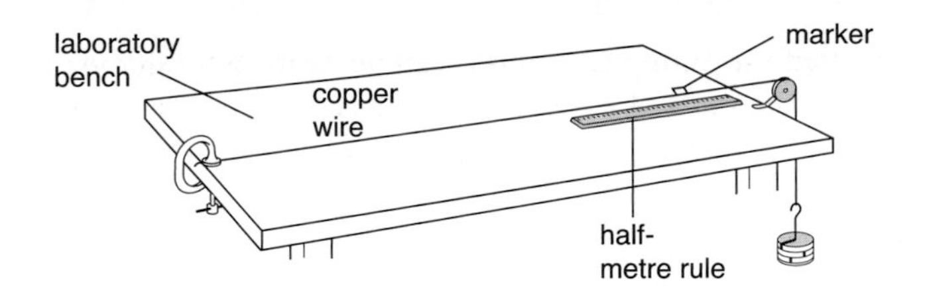

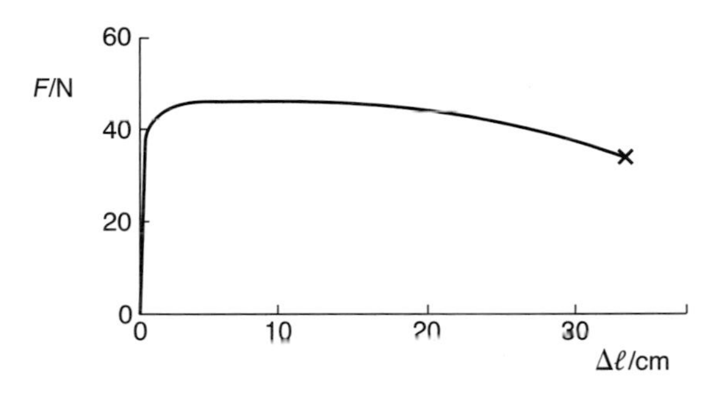

(a) Estimate the stiffness $F/\Delta\ell$ for the initial straight rise of the graph.
(b) The initial length of the wire from the G-clamp to the marker was 1.5 m. What was the strain, expressed as a percentage, at the instant the wire broke?

7.31 If a particular wire is stretched with a steadily increasing force, will the force at which it breaks depend on the original length of the wire? Explain.

7.32 A steel wire has a diameter of 0.36 mm.
(a) What is its cross-sectional area, in m^2?
(b) It is pulled by a force of 3.5 N. What is the tensile stress in the wire?

7.33 What is the tensile stress in
(a) one of the supporting cables of a suspension bridge which has a diameter of 40 mm and which pulls up on the roadway with a force of 30 kN
(b) a nylon fishing line of diameter 0.35 mm which a fish is pulling with a force of 15 N
(c) a tow rope of diameter 6.0 mm which is giving a car of mass 800 kg an acceleration of $0.40\,m\,s^{-2}$ (other horizontal forces on the car being negligible)?

7.34 Express these stresses in $N\,m^{-2}$, giving the number in standard form:
(a) 101 kPa **(b)** 0.27 MPa **(c)** 35 MPa **(d)** 2.8 GPa **(e)** 235 GPa.

7.35 Copy and complete the following table:

	force	cross-sectional area	stress
(a)	6.0 N	$0.10\,mm^2$	
(b)	12 kN	$2.0\,mm^2$	
(c)		$3.4\,cm^2$	6.0×10^6 MPa
(d)		$\pi(0.50\,mm)^2$	6.4 MPa
(e)	0.11 kN		0.22 GPa

7.36 **(a)** Show that the Young modulus E can be expressed as $F\ell/A\Delta\ell$.
(b) Why are the units of E and σ the same?

7.37 Copy and complete the following table:

	stress	strain	Young modulus
(a)	50 MPa	6.0×10^{-4}	
(b)	0.10 GPa	5.0×10^{-2}	
(c)		0.054	0.22 GPa
(d)	0.30 GPa		300 GPa
(e)	1.8 GPa		180 GPa

7.38 The diagram shows force–extension graphs for two wires A and B, made from the same material.

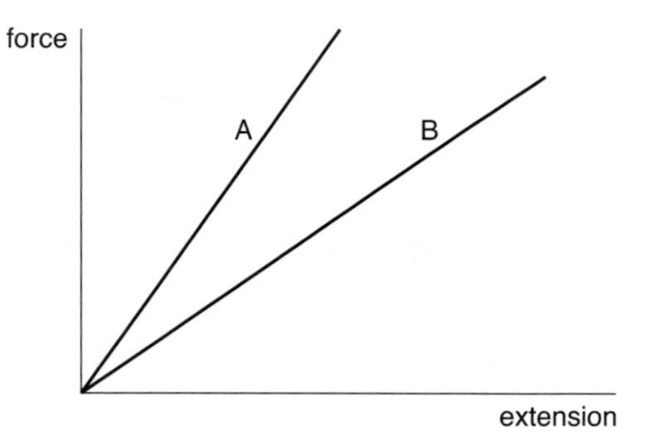

(a) Explain whether it is possible for A, compared with B, to be
(i) thinner and longer
(ii) thinner and shorter
(iii) thicker and shorter
(iv) thicker and longer?
(b) What quantities should be plotted to get the same graph for both wires?

7.39 A copper wire of length 1.2 m and cross-sectional area 0.10 mm² is hung vertically; for copper E = 130 GPa. A steadily increasing force is applied to its lower end to stretch it. When the force has reached a value of 10 N
(a) what is the stress in the wire
(b) what is the strain in the wire
(c) what is the extension of the wire?

7.40 The diagram shows stress–strain graphs for three metals.
(a) Which of the three metals has the greatest Young modulus?
(b) For each metal calculate the Young modulus.

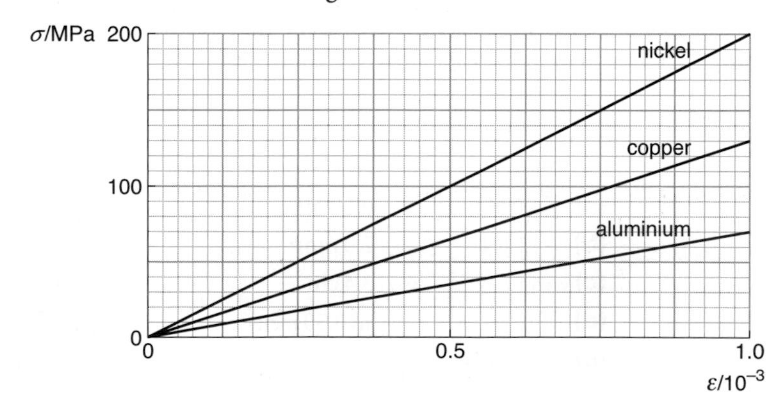

7.41* Human beings have evolved so that their bones are strong enough to withstand the forces met in normal everyday life, e.g. the cross-sectional area of a leg bone is great enough to be able to support the weight of the body and also the increased forces when walking, running or jumping. Imagine a human being whose dimensions were all twice as great as those of a normal human being.

(a) How many times greater would be

- **(i)** its volume and its weight
- **(ii)** the compressive force in a leg bone when it was standing still
- **(iii)** the cross-sectional area of a leg bone
- **(iv)** the compressive stress in a leg bone when it was standing still?

(b) What problems would this human being encounter?

(c) Explain why large animals such as elephants and rhinoceroses have very thick legs.

7.42 The graph shows the behaviour of a long wire made of a ductile material, such as copper, that is successively loaded and unloaded.

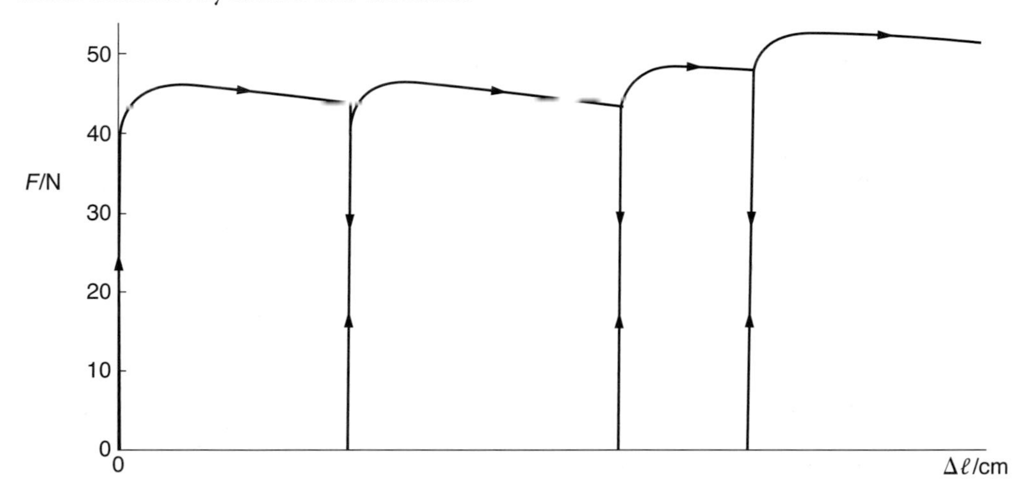

(a) What features of the graph show that copper **(i)** has elastic properties **(ii)** behaves plastically?

(b) How does the stiffness of the wire vary at the beginning and end of each cycle?

(c) How does the graph tell you that the copper becomes work-hardened during successive cycles?

7.43 A lift of mass 3000 kg is supported by a steel cable of diameter 20 mm. The maximum acceleration of the lift is $2.5\,\mathrm{m\,s^{-2}}$. Calculate

(a) the maximum tension in the cable

(b) the maximum stress in the cable

(c) the maximum strain in the cable. [E for steel = 200 GPa]

7.44 A copper wire and a tungsten wire of the same length are hung vertically. The cross-sectional areas of the wires are $0.10\,\mathrm{mm^2}$ and $0.15\,\mathrm{mm^2}$, respectively. Equal steadily increasing forces are applied to each wire. The Young modulus E and ultimate tensile stress σ_u are given in the table:

	E/GPa	σ_u/MPa
copper	130	220
tungsten	410	120

Which of the wires will **(a)** break first **(b)** have the larger extension at that time?

7.45 The diagram shows stress–strain graphs for a glass rod and for a copper wire.

(a) Which material is stronger?

(b) Why are there two graphs for copper wire?

(c) Does copper obey Hooke's law? If so, to what extent?

(d) Explain which material has the larger Young modulus.

(e) What is meant by the term nominal stress?

(f) In **(ii)** why does the graph slope downwards, apparently showing the wire continuing to stretch with decreasing loads?

(g) What feature of the graph for the glass rod shows that glass is brittle?

(h) Make a copy of **(iii)** and add it to a line which shows what you would expect to happen if the load was gradually removed when a strain of 0.0015 had been reached.

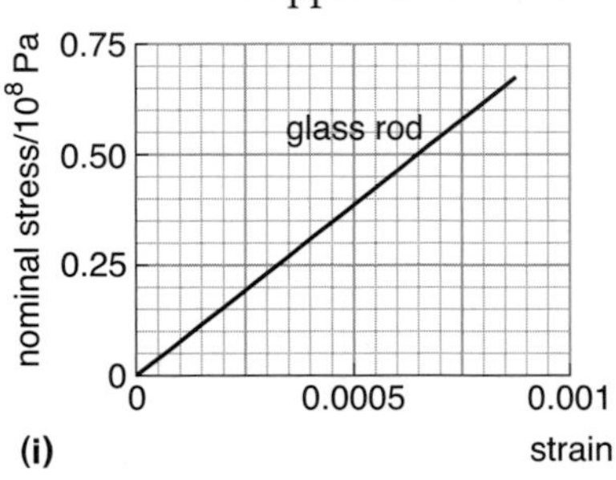

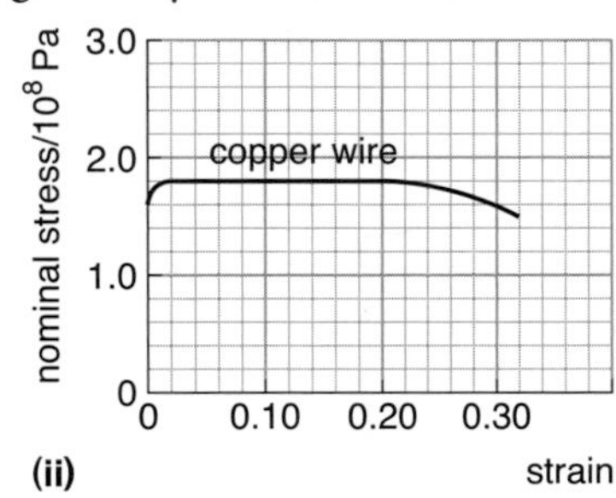

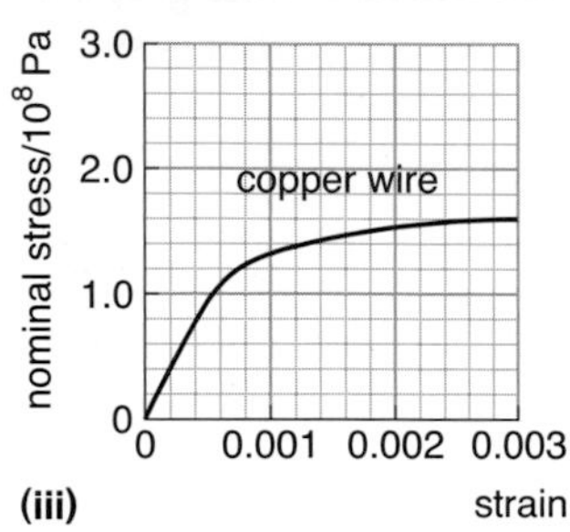

7.46 There is a risk that thermal expansion might cause railway track to buckle if the temperature rises sufficiently. So the track is stretched before it is laid so that it is in tension. If the temperature rises, all that happens is that the tension in the track is reduced.

Suppose a 100 m length of rail is to be laid. The expansivity of the steel is $1.1 \times 10^{-5}\,°C^{-1}$, i.e. the length increases by this fraction of its original length for each Celsius degree rise in temperature.

(a) Show that the expansion of this length for a 25 °C rise in temperature would be 28 mm.

(b) The Young modulus for the steel is 200 GPa. What stress would be needed to push the rail back to its original length?

(c) The cross-sectional area of steel rail is 75 cm². What force would be needed to push the rail back to its original length, so that a later rise in temperature of 25 °C does not put it into compression?

7.47 The diagram shows how a gap can be bridged with a concrete beam. The ultimate tensile stress of concrete is about ten times less than the ultimate compressive stress.

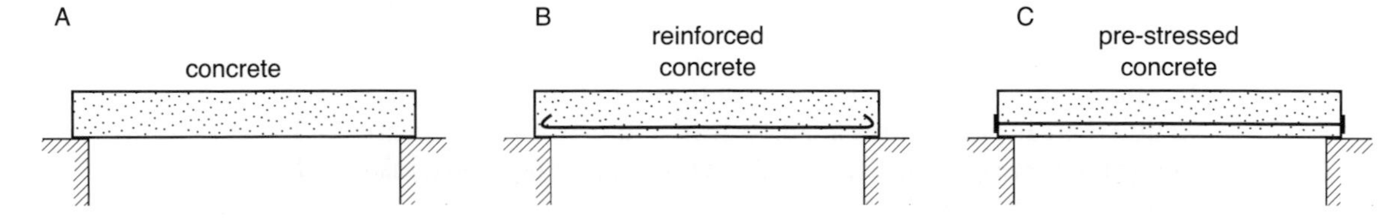

(a) Copy diagram A and mark on it regions that are in tension and regions that are in compression.

(b) Explain why a simple beam like that in A is likely to fail when a heavy load is placed at its centre.

(c) Why is the reinforced concrete beam in diagram B better than that in A at supporting loads?

(d) (i) What is meant by pre-stressed concrete? **(ii)** Explain why diagram C offers a better solution to bridging the gap than either A or B.

7.48 The diagram shows the two bones in a human leg and a graph of compressive or tensile stress against strain for the bones of a 30-year-old person. The bone breaks at point X.

(a) Use the straight part of the graph to calculate the Young modulus for the bone.

(b) What is the maximum compressive or tensile stress for the bone?

(c) What is the maximum strain for the bone?

(d) In fact bones break more often because they are bent than because of simple compression or tension. Draw a diagram of a bent bone and mark on it the regions where tension and compression would occur.

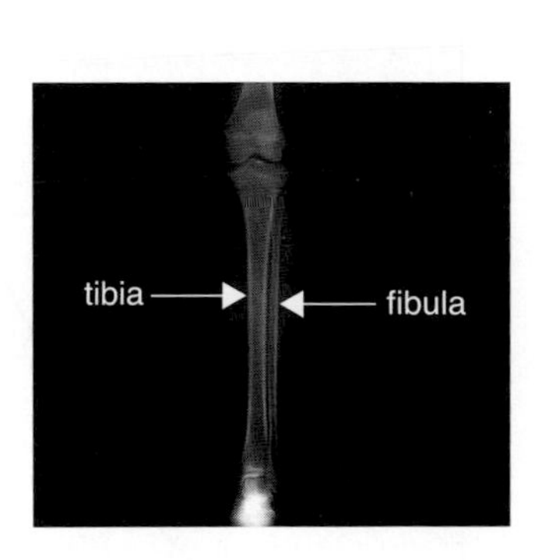

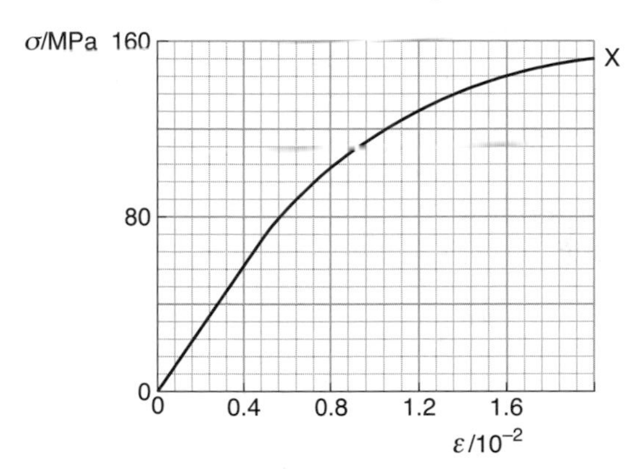

7.49 The diagram shows stress–strain curves for Perspex and polythene up to their breaking points.

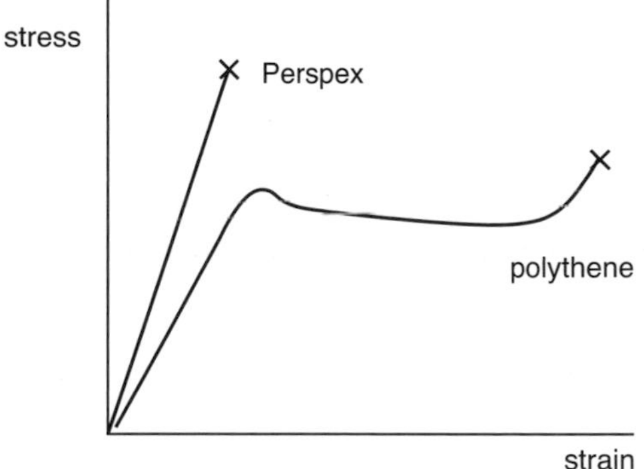

Using the language of materials, describe the differences in the behaviour of Perspex and polythene.

7.3 Energy stored in stretched materials

In this section you will need to

- use the equation $W = \frac{1}{2}kx^2$ to calculate the elastic potential energy W stored in a spring which obeys Hooke's law
- understand that the energy stored in a spring or wire is given by the area beneath the force–extension graph

- use the equation elastic p.e. per unit volume = $\frac{1}{2}$(stress) × (strain) to calculate the elastic p.e. stored per unit volume in a wire which obeys Hooke's law
- understand that the toughness of a material is related to the work that must be done to break it, and that toughness is often a desirable property.

7.50 How much work must be done to stretch a spring of stiffness 30 N m^{-1} by **(a)** 10 cm **(b)** 20 cm **(c)** 30 cm? Assume that the spring obeys Hooke's law.
Explain why your answer to **(b)** is not double your answer to **(a)**.

7.51 The diagram shows a graph of the energy stored W in a spring that obeys Hooke's law against the extension x of the spring.

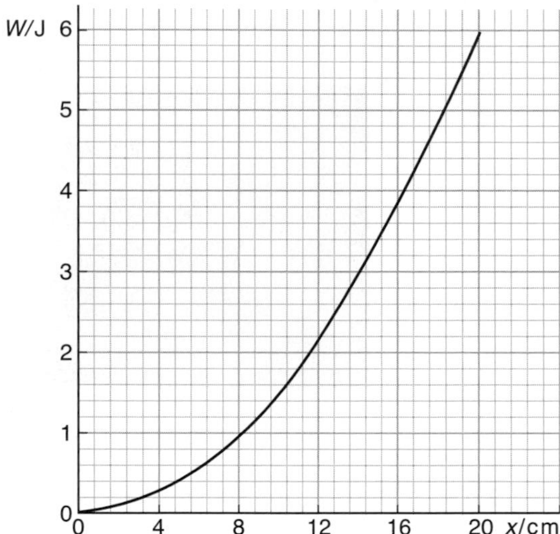

(a) Show that for this spring $k = 300\,\text{N m}^{-1}$.

(b) Sketch a graph of F, the stretching force, against extension x for this spring, for values of x from zero to 20 cm. From your graph calculate, using $W = F_{av}x$, the work done in stretching the spring from **(i)** 0 to 12 cm **(ii)** 0 to 16 cm.

(c) Hence confirm the values of W on the diagram for these two values of the extension.

(d) What is the extra energy stored when the spring is stretched from x = 12 cm to x = 16 cm?

7.52 A steel wire of initial length 1.5 m and diameter 0.50 mm is pulled with a force of 45 N.
(a) What is its extension?
(b) How much energy is stored in it?
[Assume that Hooke's law is obeyed: E for the steel = 200 GPa.]

7.53 Refer to the graph accompanying question 7.27. By considering the average force, estimate the work done when the *extension* of the rubber band increases from
(a) zero to 30 cm
(b) 30 cm to 40 cm.

7.54 The diagram shows a graph produced by stretching a copper wire of length 2.0 m and diameter 0.56 mm. The extension of the wire x is proportional to F up to point A and is elastic up to point B.

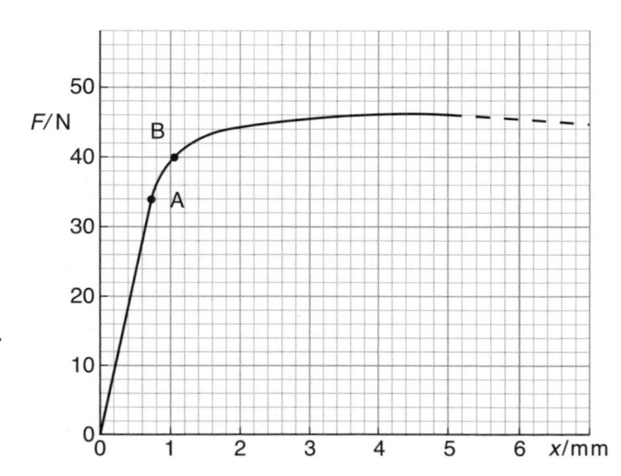

(a) Estimate the work done in stretching the wire up to point B.

(b) Beyond point B the copper wire stretches plastically. Estimate the work done in stretching the wire from 1.0 mm to 5.0 mm extension.
(c) Describe the energy transfers during the elastic and then the plastic deformation of the wire.
(d) What shape would a graph of stress against strain have for this range of force and extension?
(e) Calculate the values of stress and strain at point B on the graph.

7.55 The diagram shows idealised forms of the force–extension curve for specimens (of the same shape) of high tensile steel and mild steel up to the point where each specimen breaks.
(a) Explain which is the stronger material.
(b) By considering, in each case, the area between the graph and the extension axis, find the work which must be done to break each specimen.
(c) Explain which is the tougher material.

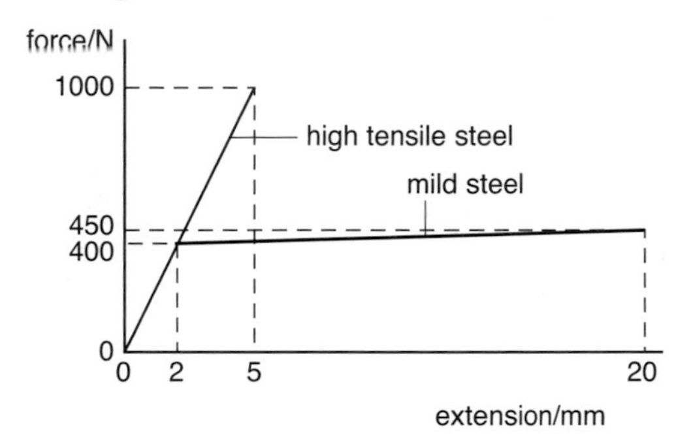

7.56 Consider a wire of length ℓ and cross-sectional area A, whose material has a Young modulus E, which is being pulled by a force F. Its extension is x, and the energy stored, if Hooke's law is obeyed, is given by $W = \frac{1}{2}Fx$.
(a) Express F in terms of E, ℓ, A and x, and hence show that $W = (EAx^2)/2\ell$.
(b) Hence show that the energy stored per unit volume in the wire is equal to $\frac{1}{2}$(stress)×(strain) for a wire which obeys Hooke's law.

7.57 The diagram shows a force–extension graph for a length of rubber cord which is loaded and then unloaded.
(a) Which curve, the lower or the upper, describes the loading?
(b) What name is given to the fact that the two curves do not coincide?
(c) Explain how this effect causes heating in motor car tyres.

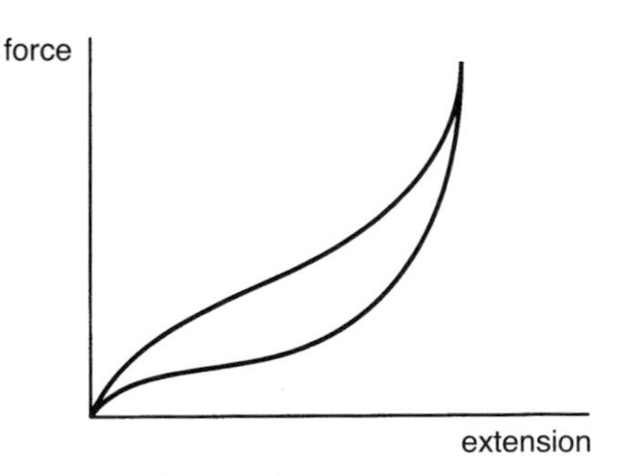

7.58 The diagram on the next page shows a stress–strain graph for a rubber cord stretched to four times its initial length before being relaxed.
(a) Estimate the area beneath **(i)** the loading **(ii)** the unloading curve.
(b) The shaded region of about eight large squares represents the internal energy released during a cycle of stretching and relaxing.
(i) What does one square represent? Express your answer in $\mathrm{J\,m^{-3}}$, and show that this is equivalent to $\mathrm{N\,m^{-2}}$.

(ii) Calculate how much energy per unit volume the shaded area represents.

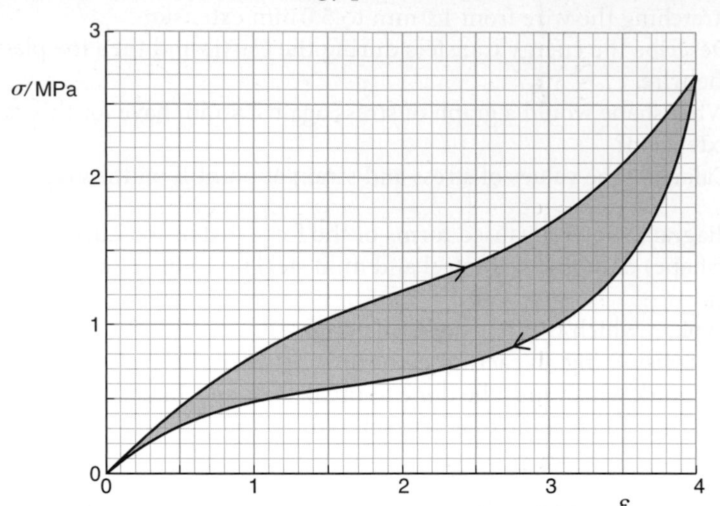

7.59 The diagram shows a stress–strain graph for a specimen of polythene (such as you might find holding together a four-can pack of beer or coke).

(a) Describe what the graph shows about the behaviour of the polythene when it is loaded and unloaded.

(b) What is the approximate value of the Young modulus of the polythene?

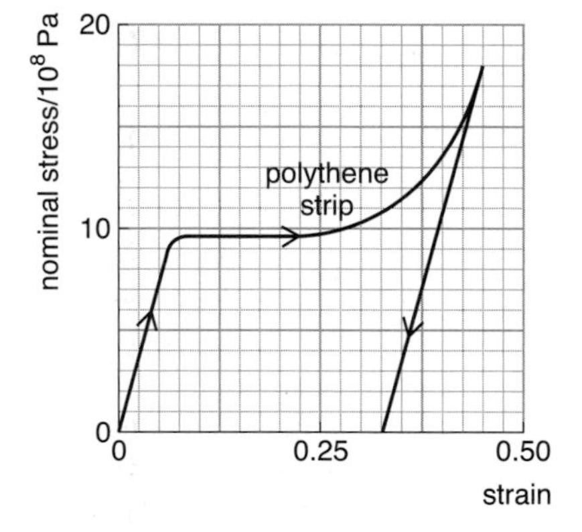

7.60 The diagram shows a stress–strain graph for a thread of spider silk, the material that spiders use to build their cobwebs.

(a) Estimate the Young modulus for this cobweb material up to strains of about 5%.

(b) A wasp flies into a cobweb thread of radius 1.8 μm. This produces a stress of 2.7×10^8 Pa in the thread. **(i)** Is the strain within the 5% calculated in **(a)**? **(ii)** Calculate the resulting tension in the thread.

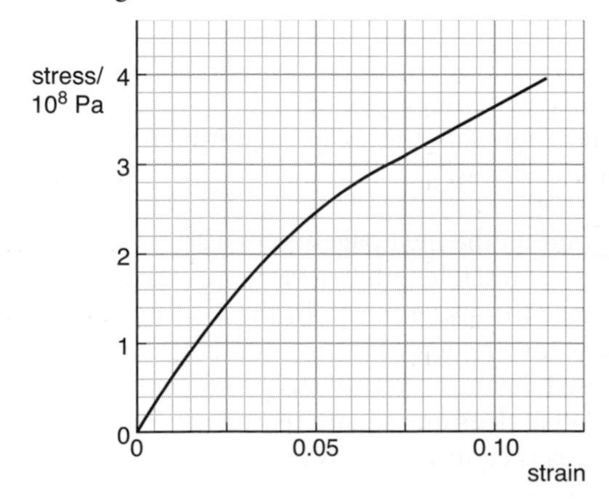

8 Waves

Data: speed of electromagnetic waves in a vacuum $c = 3.00 \times 10^8\,\text{m s}^{-1}$

visible spectrum: from about 4×10^{-7} m to 7×10^{-7} m

speed of sound in air = $340\,\text{m s}^{-1}$

speed of sound in water and sea water = $1500\,\text{m s}^{-1}$

refractive index of crown glass = 1.52

refractive index of water = 1.33

8.1 Wave properties

In this section you will need to

- understand that waves transmit energy from place to place
- use rays or wavefronts to describe the transmission of wave energy
- use the wave equation $c = f\lambda$
- describe waves as transverse or longitudinal
- understand that transverse waves, e.g. all electromagnetic waves, can be plane polarised but that longitudinal waves, e.g. sound, cannot
- remember the order of magnitude of the wavelengths of different parts of the electromagnetic spectrum and remember that all electromagnetic waves travel at $3.00 \times 10^8\,\text{m s}^{-1}$ in a vacuum
- understand that the displacement at any point in the path of a wave is the sum of the displacements affecting the point at that instant
- draw the result of superposition as a sequence of displacement–time graphs
- understand the phrases *in phase* and *in antiphase* for two sinusoidal waves arriving at a point.

8.1 A fishing vessel uses ultrasonic sound pulses to search for shoals of fish.
(a) Draw a diagram to illustrate the principle involved.
(b) Calculate the times for a pulse to return from 35 m and 40 m below the vessel. [Use data.]
(c) Suggest a maximum time for the length of each pulse.

8.2 Light takes 8 minutes 20 seconds to travel from the Sun to the Earth. Calculate the Sun–Earth distance in metres.

8.3 Some ocean waves, which travel in the open sea at $5\,\mathrm{m\,s^{-1}}$, arrive at a beach once every 4 s.
(a) Calculate the distance between wave crests in open sea.
(b) Calculate how long these waves would take to reach the beach if they were formed in a storm 1200 km away. Give your answer in hours.

8.4* A recording station observed that there was an interval of 68 s between the reception of P (push or primary) waves and S (shake or secondary waves) from an underground nuclear test explosion.
The speeds of P and S waves in the Earth's crust are $7800\,\mathrm{m\,s^{-1}}$ and $4200\,\mathrm{m\,s^{-1}}$, respectively. Show that the distance of the test site from the explosion is 620 km.

8.5 Dolphins communicate by emitting ultrasonic waves of frequencies in the range 100 kHz to 250 kHz. What range of wavelengths in water does this represent? [Use data.]

8.6 The diagram shows a 'snapshot' of a transverse wave on a rope. The wave is moving to the right. Several particles P, Q, R, S, T and U on the rope are labelled. The direction of movement of particle R is shown.

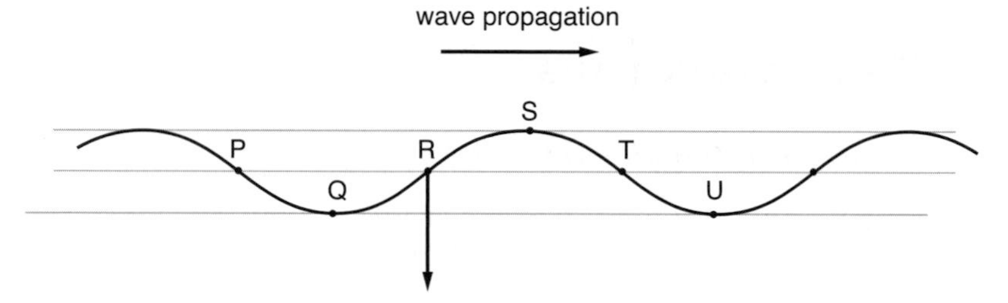

(a) Copy the diagram and draw arrows to show the direction in which the particles P, T and U are moving.
(b) Which pairs of particles are **(i)** in phase, i.e. moving together, and **(ii)** in antiphase, i.e. moving oppositely?

8.7 **(a)** A microwave oven operates at 2.45 GHz (2.45×10^9 Hz). What is the distance between wave crests in the oven? [Use data.]
(b) What frequency of sound would produce waves in air with the same distance between crests? [Use data.]

8.8 The following frequencies and wavelength are used in BBC radio and television broadcasting:

92.6 MHz/3.24 m (FM radio)
0.516 GHz/0.581 m (UHF TV)

Show that the waves carrying these signals travel at the same speed.

8.9 The figure shows a sinusoidal wave travelling to the right along a rope. The dashed line represents the wave at $t = 0$ and the solid line the wave at $t = 25$ ms.
(a) What is **(i)** the amplitude y_0 **(ii)** the wavelength λ of this wave?
(b) What is **(i)** the speed c **(ii)** the frequency f **(iii)** the period T of the wave?

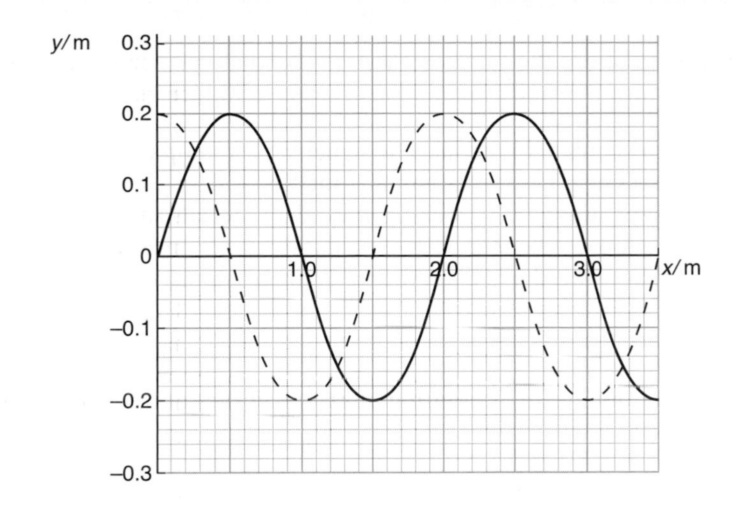

8.10 At large sports meetings the crowd sometimes produces what is called a 'Mexican wave'.

(a) What would be a better name for this manoeuvre?

(b) Describe what the crowd would need to do to produce a wave and suggest values for its frequency and amplitude.

8.11 The speed of ocean waves in deep water can be calculated from the formula

$$v = \sqrt{\frac{g\lambda}{2\pi}}$$

(a) Show that the unit of the right-hand side is equivalent to $\mathrm{m\,s^{-1}}$.

(b) Calculate the wavelength of ocean waves that travel at $6.0\,\mathrm{m\,s^{-1}}$.

8.12 The speed of a transverse wave along a heavy stretched spring is given by $c = \sqrt{(T/\mu)}$ where T is the tension in the spring and μ its mass per unit length.

(a) Show that the unit of the right-hand side is equivalent to $\mathrm{m\,s^{-1}}$.

(b) The speed of a wave pulse on a spring of mass 2.6 kg and stretched length 8.0 m is found to be $5.5\,\mathrm{m\,s^{-1}}$. Calculate the tension in the spring.

8.13 From which parts of the electromagnetic spectrum do waves of the following wavelengths come?

(a) $2 \times 10^{-10}\,\mathrm{m}$ **(b)** $5 \times 10^{-7}\,\mathrm{m}$ **(c)** 0.10 m.

8.14 A student rotates a sheet of Polaroid in front of two different light sources A and B. He finds the transmitted light from A shows no change in intensity, from B is cut off completely twice during each full rotation of the Polaroid sheet.
Describe and explain the state of polarisation of the light from each of the sources A and B.

8.15* Take a piece of Polaroid sheet outside on a sunny day. Look at a part of the blue sky (NOT the Sun) through it and rotate the Polaroid.
Describe what you see and repeat this experiment, describing different parts of the sky. (It is believed that the effect which you see enables certain insects, in particular bees, to navigate when in flight).

8.16 Light reflected from the surface of water is partially plane polarised. Explain what this means and how it enables Polaroid spectacles to reduce glare.

8.17 The diagram shows a TV aerial. The signal is picked up by the quarter wave dipole. The other rods help to aim the aerial at the transmitting station and to give the received signal a greater strength. By looking at local TV aerials and using the data

(a) estimate the wavelength of the broadcast signals and deduce their frequency

(b) explain how the signal from the transmitter is polarised.

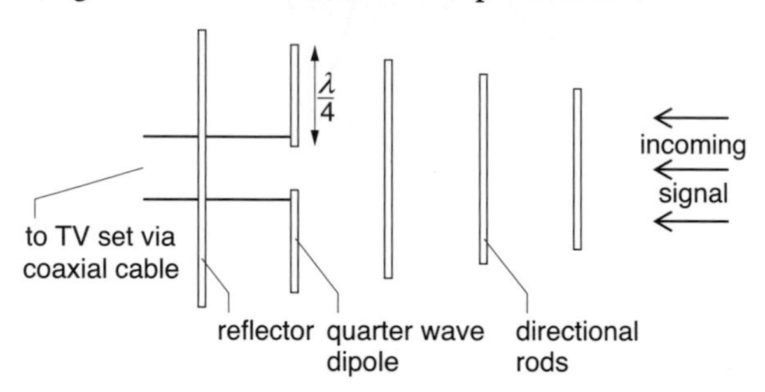

8.18* Both infrared and microwave radiation can be found with wavelengths of 3 mm. Both X-rays and γ-rays can be found with wavelengths of 1 nanometre (1.0×10^{-9} m). Use your textbooks or a search engine to discover why waves of the same wavelength are sometimes described by different names. (Both X-rays and γ-rays are part of the electromagnetic spectrum.)

8.19 A periodic wave of frequency f is travelling along a slinky at a speed v.

(a) What is the distance between points on the slinky that are moving **(i)** in phase, and **(ii)** in antiphase?

(b) Suggest how, in the laboratory and without the use of cameras or other electronic equipment, you would demonstrate your answers to **(a)** to a group of students in a situation where $f = 2.0$ Hz and $v = 0.60\,\mathrm{m\,s^{-1}}$.

8.20 Two speakers connected to the same oscillator are producing sound waves at 600 Hz that are in phase.

(a) What is the result of reversing the connections to one of the speakers?

(b) Predict what a person sitting the same distance from the two speakers will hear when the connections are reversed.

8.21 The ultrasound used in medical imaging uses pulses of waves with frequencies in air in the range 3.0 MHz to 15 MHz. The pulses are of very short duration, containing typically only five oscillations per pulse.

(a) Express the range of wavelength used in millimetres when the waves travel in a medium at $1500\,\mathrm{m\,s^{-1}}$.

(b) What is the length of a typical ultrasonic pulse of frequency 10 MHz in air? [Use data.]

8.22* An ultrasonic A-scan can be used to measure the thickness of an eye lens, as shown in the diagram (the same technique can measure the diameter of a foetal head). A pulse from the ultrasound probe produces the peaks shown on an oscilloscope screen, where the time base is set at $0.20\,\mathrm{div\,\mu s^{-1}}$. Peaks 2 and 3 represent the reflections from the front and back of the eye lens in which the ultrasound travels at $1600\,\mathrm{m\,s^{-1}}$.

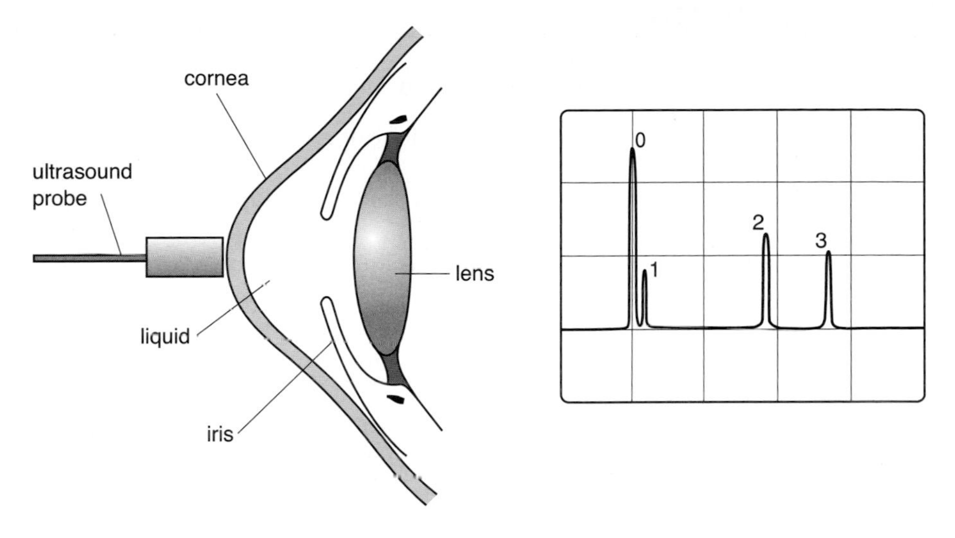

(a) What is the time interval between pulses 2 and 3?
(b) Hence calculate the lens thickness.

8.23* The diagram shows how an ultrasonic B-scan can reveal the details of the foetal face shown in the recorded image.

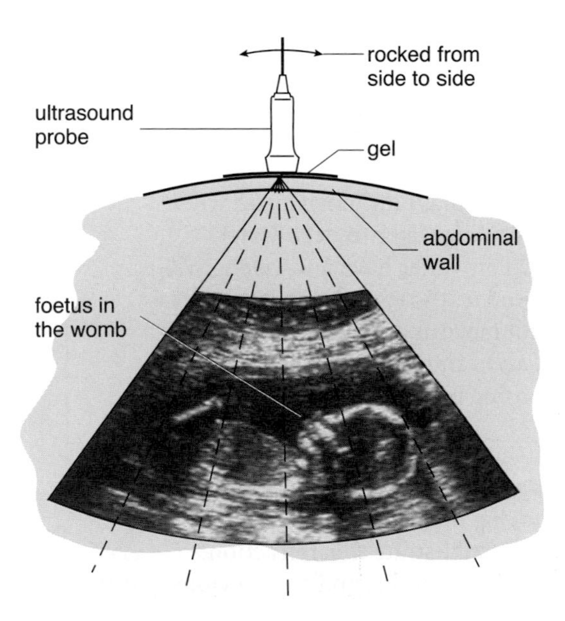

Here the brightness of a spot on the image replaces the height of the oscilloscope trace in an A-scan, such as that in question 8.22, as the ultrasound probe is rocked from side to side.
Using the diagram, explain to a non-scientist how the image is produced.

8.2 Refraction

In this section you will need to

- draw diagrams showing the behaviour of rays and wavefronts as they reflect and refract
- use the law of refraction: that for a ray refracted from medium 1 to medium 2, $n_1 \sin\theta_1 = n_2 \sin\theta_2$
- understand that the ratio of the speeds of light in two media is equal to the inverse ratio of the refractive indexes of the media: $c_1/c_2 = n_2/n_1$
- remember that the refractive index of a given material varies with the wavelength or colour of the light and that this variation is called dispersion
- explain that total internal reflection can occur only when light meets the boundary of a medium with a lower refractive index than the medium in which it is travelling
- use the critical angle condition for total internal reflection: $\sin\theta_c = n_1/n_2$
- understand how light (and infrared radiation) travels along an optical fibre.

8.24 To 'see' plane wavefronts (straight ripples) produced on the surface of water you can use a ripple tank with a vibrating bar.

(a) Calculate the frequency of ripples with a wavelength of 15 mm that move forward at 8.0 cm s^{-1}.

(b) How would you produce ripples with a smaller wavelength?

(c) Sketch the shape of the ripples that would be produced when the vibrating bar is replaced with a single dipper in the shape of a small sphere about a centimetre in diameter.

8.25 The wavefronts in the diagram move faster when they move from shallow water to deep water. The ripples in shallow water have a wavelength λ_1 and move at a speed v_1; those in deep water have a wavelength λ_2 and move at a speed v_2. Explain how λ_1, λ_2, v_1 and v_2 are related, and write down the relationship between them.

8.26 When wavefronts in shallow water meet a boundary with deep water at an angle, the direction in which the waves are moving is changed.

(a) Draw a sketch to illustrate this refraction.

(b) Using the situations in this and the previous question, state what is meant by the refraction of waves.

8.27 A narrow beam of light in air strikes a glass block of refractive index 1.52 at angles of incidence $\theta_a/° = 0, 5, 10, 15, 20, 25$.

(a) Calculate **(i)** the angle of refraction θ_g in each case **(ii)** the angle of deviation $\varphi = \theta_a - \theta_g$ in each case.

(b) Plot a graph of **(i)** θ_a against θ_g **(ii)** θ_a against φ.

(c) Predict the rough shape of your two graphs as θ_a rises from 25° to 90°.

8.28 The diagram shows how a narrow beam of light strikes a layer of oil on the surface of a tank of water at an angle of 58.0°. The refractive index of the oil is 1.28 and of the water 1.33.

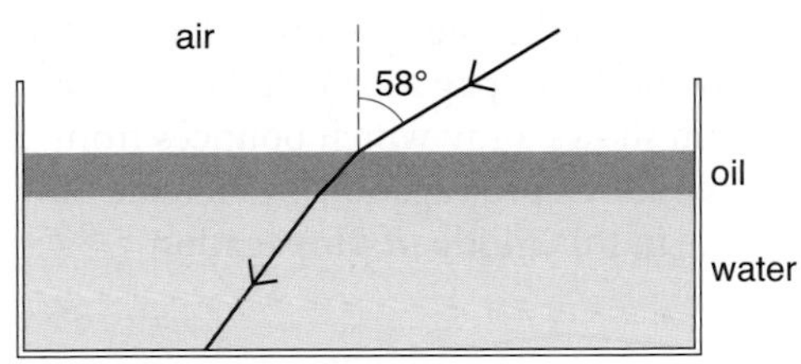

(a) Calculate
  (i) the angle of refraction in the oil, and
  (ii) the angle of refraction in the water.
(b) What would be the angle of refraction in the water if the layer of oil was removed? Comment on your answer.

8.29 The refractive indexes of the glass of a prism for red and blue light are 1.538 and 1.516, respectively. Find the angle between the emerging blue and red light when the white light is incident normally on one face of a prism of angle 40° as shown in the diagram.

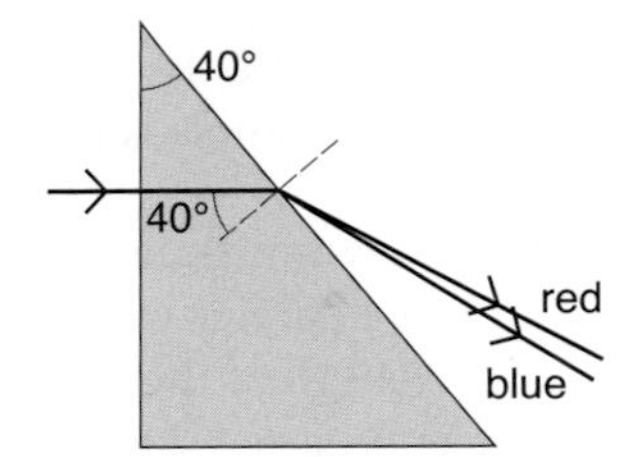

8.30 **(a)** Calculate the wavelength in water for sound waves that have a wavelength of 0.25 m in air. [Use data.]
(b) Draw a scale diagram showing three wavefronts of these sound waves as they travel from air to water. Take the angle of incidence in air to be 45°. [Use data.]

8.31 **(a)** What are the critical angles in air for **(i)** ice $n_i = 1.309$ and **(ii)** common salt $n_s = 1.544$?
(b) The critical angle in air for diamond is 24.42°. What is the refractive index of diamond?

8.32 **(a)** What is the difference between the critical angles of red light and blue light for a glass for which $n_{blue} = 1.639$ and $n_{red} = 1.621$?
(b) Which light has the larger critical angle?
(c) How could you demonstrate that the critical angles were different?

8.33 The output of an LED contains energy in the infrared part of the electromagnetic spectrum. The refractive indexes of a glass fibre for the shortest and longest of these wavelengths are 1.479 and 1.482, respectively.
(a) Calculate the times taken by infrared waves of these two wavelengths to travel 1000 m along the fibre.
(b) What is the time difference?
[Take $c = 2.997 \times 10^8\ \text{m s}^{-1}$.]

8.34 A step-index optical fibre has a core of refractive index 1.48 and cladding of refractive index 1.46. A narrow beam of light in air enters the (flat) end of the core at an angle of 12° to the axis of fibre.
(a) Will the beam be propagated along the fibre or not? Support your answer with a diagram and appropriate calculations.
(b) What is the maximum angle to the axis for propagation to occur?

8.35 An optical fibre core is made of glass of refractive index 1.472 and is surrounded by a cladding of refractive index 1.455.

(a) Calculate the critical angle for the fibre.

(b) How long does it take energy to travel 3000 m along the axis of the fibre (this is called monomode propagation). [Take $c = 2.997 \times 10^8\,\mathrm{m\,s^{-1}}$.]

(c) The diagram shows a ray which bounces from side to side at the critical angle (this is called step-index propagation). Calculate how long energy takes to travel along a 3000 m fibre in this mode of propagation.

8.36* An optical fibre system operates at a wavelength of 1.55 μm and at a bit rate of 700 Mbit $\mathrm{s^{-1}}$.

(a) Where in the electromagnetic spectrum do such waves occur? Why are such wavelengths chosen for 'optical' fibre systems?

(b) Calculate the sampling frequency if there are 16 bits (2 bytes) required to encode each sample, such as might be needed for music transfer.

8.37 The diagram shows the shape of the ripples to the right of a gap in a barrier when the gap width is about three times the wavelength of the ripples.
Make separate sketches of the shape of the ripples passing through the same gap when their wavelength is

(a) about three times greater than

(b) about one-third as much as

the wavelength shown in the diagram

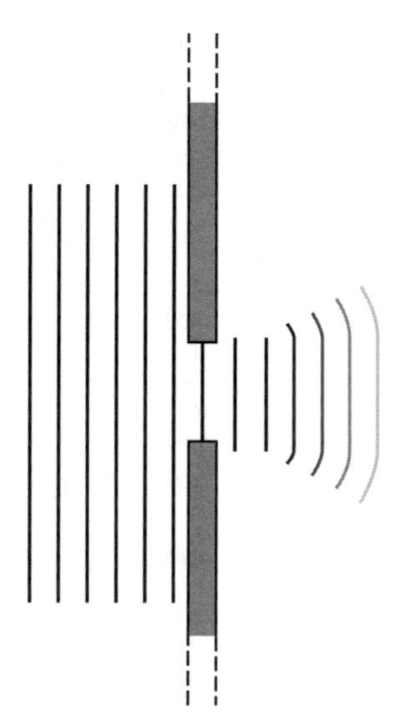

8.38* Your uncle, who has not studied physics, asks the following question:
'I was told that both sound and light travel as waves. So how is it that I can hear a conversation coming from a room with a door open but cannot see the people speaking inside the room?' Explain.

8.3 Two-source interference patterns

In this section you will need to

- remember that, for two sources which are in phase, the result of superposition at a point P is given by: $S_1P - S_2P = n\lambda$ (maximum or constructive interference) and $S_1P - S_2P = n\lambda + \frac{1}{2}\lambda$ (minimum or destructive interference)
- understand that energy is not lost in an interference pattern: it is only redistributed
- use the relationship for light from two adjacent slits:

$$\text{wavelength} = \frac{(\text{slit separation}) \times (\text{fringe width})}{(\text{slit to screen distance})}$$

- understand how to produce two coherent optical sources.

8.39 The photographs show the positions at eight equal time intervals of a stretched rope as two wave pulses moving in opposite directions travel 'through' each other.

(a) Sketch the rope at each stage and add an explanation of what is happening in terms of wave superposition.

(b) Comment on where the energy of the pulses is at each stage. [Hint: The more blurred a piece of the rope is in the photograph, the faster that piece of the rope is moving sideways.]

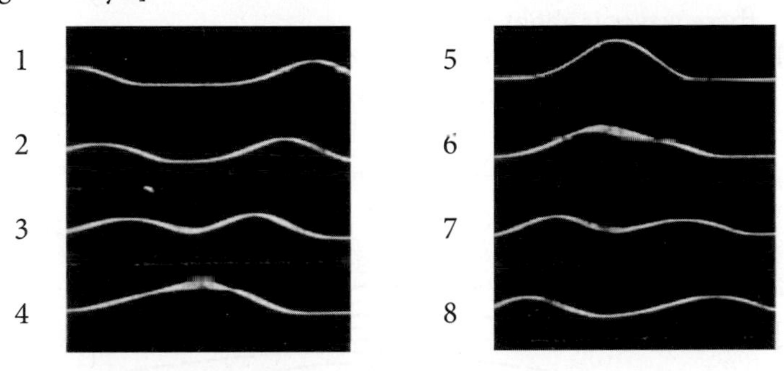

8.40 The diagram shows an experimental arrangement for investigating the superposition of sound waves from two sources S_1 and S_2. The sources are in phase and produce sound of wavelength 80 mm.

(a) When S_1M is 800 mm the trace on the oscilloscope is a maximum. Suggest three possible values for S_2M. [Use data.]

Without moving any of the apparatus the leads to one of the speakers are reversed, i.e. the sources are now in antiphase.

(b) Explain the meanings of *in phase* and *in antiphase* and describe how the trace on the oscilloscope changes when the leads are reversed.

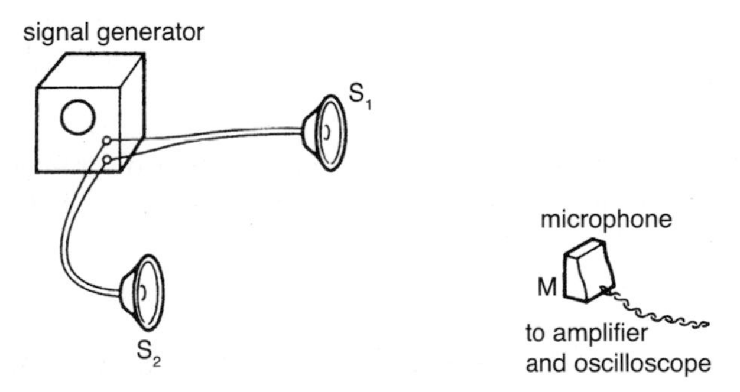

8.41 In another experiment with the apparatus described above, and the speakers in phase, the height of the trace on the oscilloscope was found to be a minimum when $S_1M = 0.80$ m and $S_2M = 1.00$ m.

(a) Why can the wavelength of the sound waves not be found from these measurements alone?

(b) What further observations are needed before the wavelength can be found?

8.42 The only practical method of producing two in-phase (coherent) sources of light from, for example, a monochromatic laser, is to derive two sources from a single source. Explain why this is so and describe one method of achieving the two sources.

8.43 The diagram below shows, to scale, two in-phase sources, S_1 and S_2, e.g. ripple tank dippers. The full lines are lines of constructive superposition (maxima) and the dashed lines are lines of destructive superposition.

(a) Use a ruler to show that on the diagram $S_1P - S_2P \approx 5.0$ mm. Hence explain why the wavelength of these waves is about 5.0 mm.
[You are not expected to measure distances more precisely than ±0.5 mm.]

(b) Use a ruler to show that $S_2Q - S_1Q \approx 2.5$ mm. Hence confirm that the wavelength of these waves is about 5.0 mm.

(c) Choose a point R on one of the outermost full lines. Predict the size of $S_1R - S_2R$ and test your prediction using a ruler.

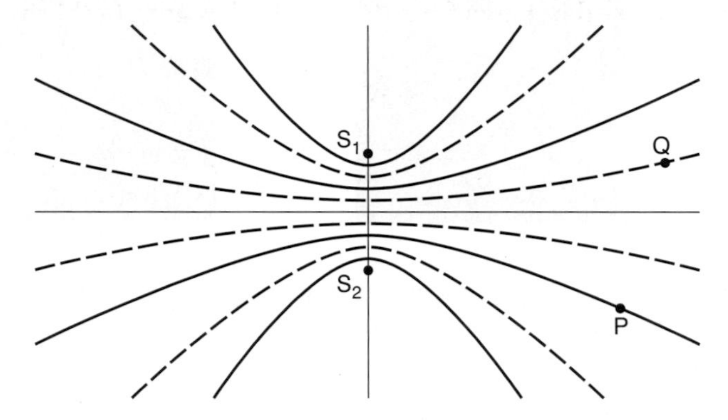

8.44 Take two rectangular glass blocks; clean them and press two flat sides together. Look at the light reflected by the two glass blocks from an overhead fluorescent lamp. If you allow your eyes to focus on them you should see a yellowish pattern that changes shape as you rotate one of the blocks. This is an interference pattern.
Predict how two sources are being produced in your experiment and suggest how the thickness of the gap between the glass blocks might be related to the wavelength of the light from the lamp.

8.45* Between 1802 and 1804 Thomas Young described some of the first experiments producing interference patterns with light. Young used sunlight passing through a tiny hole to make a point source of light and then two narrow slits to produce an interference pattern on a screen in a dark room.

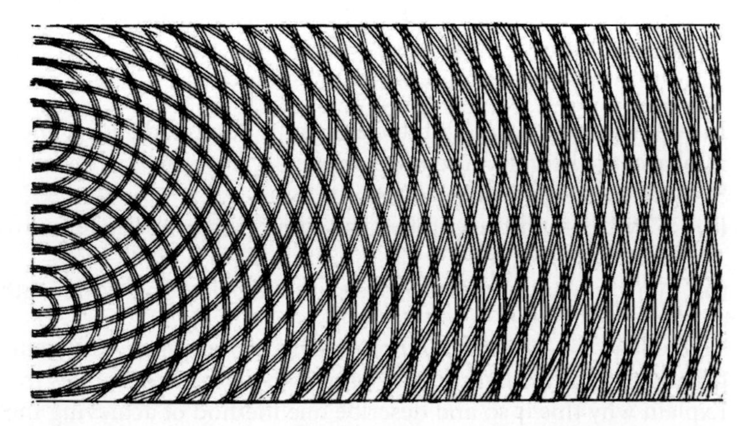

Young's original drawing explaining his wave theory of light is shown in the diagram. (The alternate regions of reinforcement (maxima) and cancellation (minima) can best be seen by holding the page at eye level and looking along the surface from the right.)

(a) Use a search engine to find out about the conflict between Young's wave theory of light and the particle theory of light that Sir Isaac Newton put forward a hundred years earlier.

(b) Discuss whether the modern 'photon' theory of light helps to resolve this conflict.

8.46 The photograph is of interference fringes formed on a screen in a Young's slit experiment. The slit separation was 0.48 mm and the distance from slits to screen was 0.95 m.

(a) By taking measurements from the photograph, which is full size, find a value for the wavelength of the light used and suggest its colour.

(b) In the photograph there is no light energy reaching the minima – the dark lines. Explain where the energy from the two slits has gone in this interference pattern.

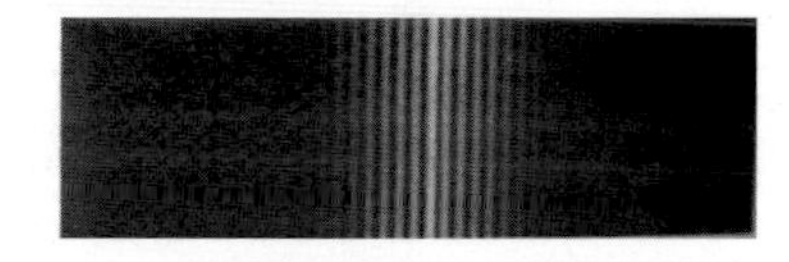

8.47 The eye is more sensitive to wavelengths in the range 500 nm (green) to 600 nm (orange) than to colours in the rest of the visible spectrum.

(a) For a Young's slits arrangement with a slit separation of 0.68 mm, calculate the fringe separation for both of these wavelengths at a distance of 0.80 m from the slits.

(b) Show that the sixth green fringe from the centre of the interference pattern coincides with the fifth orange fringe at this distance.

8.48 A beam of microwaves is directed normally at two narrow slits in a metal sheet. In the diagram, moving away from R, P is the first position where a maximum reading is detected.

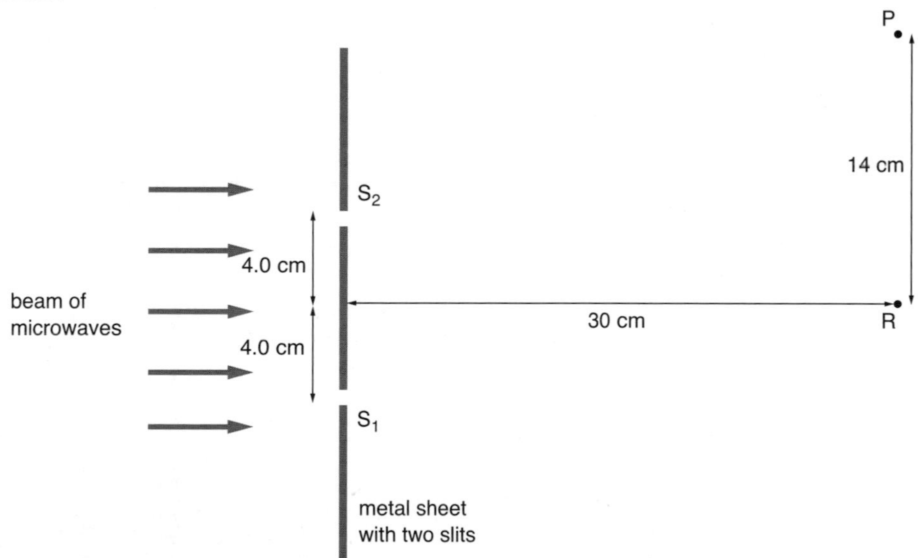

(a) Use Pythagoras' theorem to calculate the two lengths S_1P and S_2P, each from the centre of the slit.

(b) Hence determine the wavelength of the microwaves.

(c) Would you expect the next maximum to be at Q, a distance of 28 cm from R?

8.49 This is a question about the geometry of the two-source Young's slits experiment. (You will need a calculator with at least an 8-digit display.)

(a) In the diagram, which is *not* to scale, sources at S_1 and S_2 send light to a point P. If $a = 0.400\,\text{mm}$, $D = 800.0\,\text{mm}$ and $x = 3.000\,\text{mm}$, calculate

(i) the lengths of S_1P and S_2P using Pythagoras's theorem

(ii) the path difference of $S_2P - S_1P$ between light arriving at P from the two sources.

(b) If the wavelength of the light is known to be $0.60 \times 10^{-6}\,\text{m}$, explain whether P is at a bright (maximum) or a dark (minimum) place in the interference pattern.

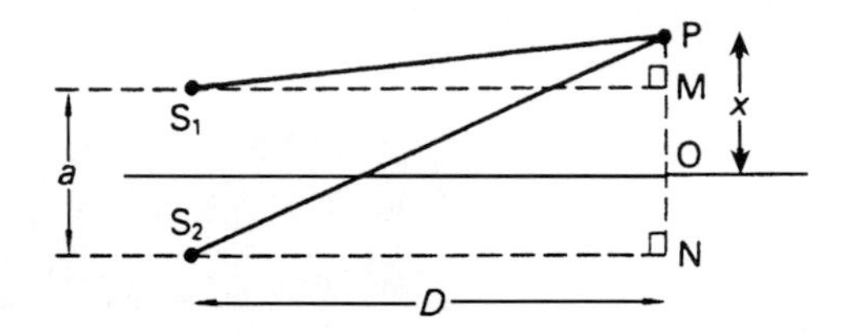

8.50 The frequency used in a microwave oven is, to 2 significant figures, 2.5 MHz. Microwaves from a source at S reach point P in a piece of meat directly and by reflection from the walls of the oven. The direct path is shown, and also one of the reflected paths.

In the diagram (which is about one-fifth scale), SP = 17 cm and SQ = QP = 16 cm.

(a) Calculate the path difference between the two microwaves reaching P and express this path difference as a multiple of the wavelength λ.

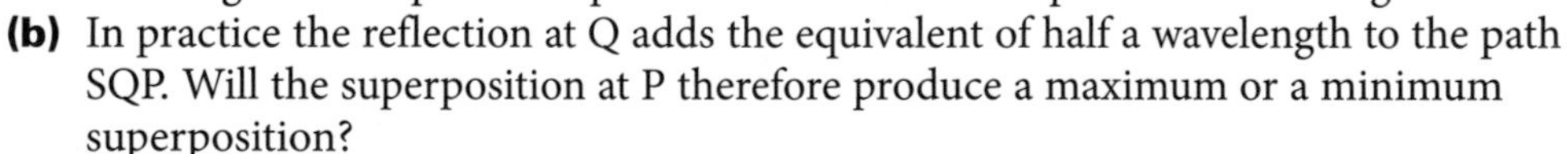

(b) In practice the reflection at Q adds the equivalent of half a wavelength to the path SQP. Will the superposition at P therefore produce a maximum or a minimum superposition?

(c) Suggest why microwave ovens have rotating turntables.

8.51 In order to cut out unwanted reflections and increase the percentage of light that is transmitted through a spectacle lens, a process called blooming is used. This involves coating the lens with a very thin film of magnesium fluoride in which light travels at a speed v, less than the speed of light c.

(a) If $v = 0.69c$, what is the wavelength in the film, of light which has a wavelength of 580 nm in air?

The diagram shows two reflected light rays, R_1 and R_2, which will superpose to the left of the film.

(b) What is the minimum thickness of film which will give rise to R_1 and R_2 being in antiphase?

(c) Suggest why, for this thickness of film, other wavelengths will not be completely cut out.

8.52 Two loudspeakers L_1 and L_2 driven from a common oscillator are arranged as shown in the diagram. A detector is placed at D. It is found that as the *frequency* of the oscillator is gradually changed from 200 Hz to 1000 Hz the detected signal passes through a series of maxima and minima.
Confirm that at frequencies of 170 Hz and 510 Hz, a minimum will be detected at D. [Use data.]

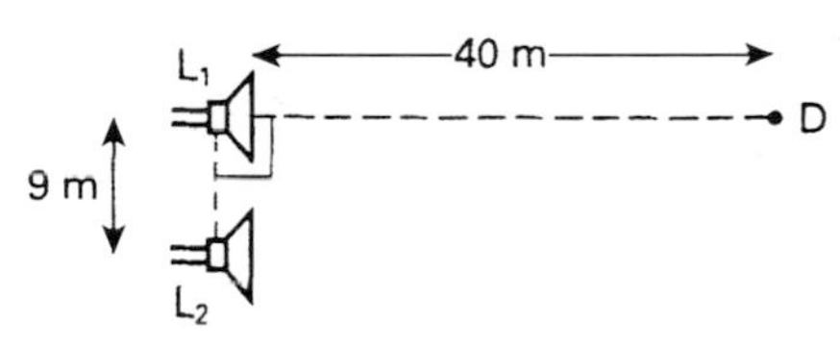

8.4 Stationary wave patterns

In this section you will need to

- explain how two waves travelling in opposite directions produce a stationary wave
- describe how to demonstrate stationary waves on stretched strings and with microwaves
- remember that the distance between adjacent nodes in a stationary wave pattern is $\frac{1}{2}\lambda$
- understand that stationary waves can only be set up at certain frequencies (and with certain wavelengths) on strings and in air columns and that these frequencies are the resonant frequencies of the string or air column.

8.53 **(a)** Sketch three stationary patterns which could be formed on the cord in the diagram. Mark the nodes and antinodes in each sketch.
(b) If in one of your sketches the distance between nodes is 0.55 m when the signal generator frequency is 40 Hz, what is the speed of the waves on the string?

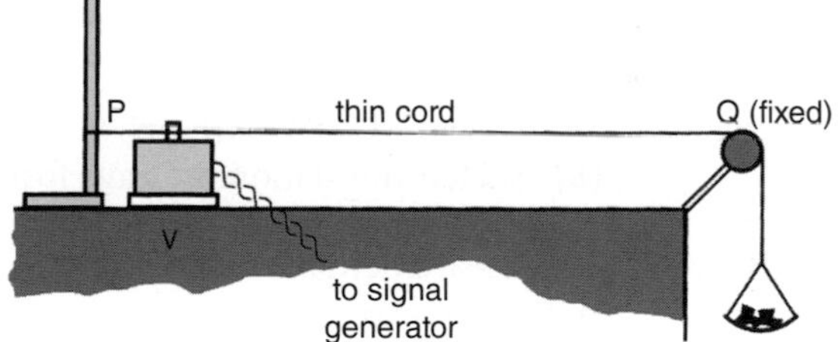

8.54 The natural frequencies f of vibration of the string in the previous question can be found by putting n = 1, 2, 3, etc. in the formula

$$f = \frac{n}{2\ell}\sqrt{\frac{T}{\mu}}$$

where ℓ is the length of the string, T the tension in the string and μ its mass per unit length.
(a) Show that the units of the right-hand side of the formula are s^{-1} or Hz.
(b) A guitar string of mass per unit length $3.8 \times 10^{-4}\,kg\,m^{-3}$ is stretched to a tension of 15 N. Calculate the fundamental frequency of the note that can be played on this string when its length is 'fingered' to be **(i)** 30 cm **(ii)** 45 cm **(iii)** 60 cm.

8.55 A wire with a mass per unit length of $1.6\,\text{g}\,\text{m}^{-1}$ is attached to a fixed block A and pulled by a spring balance B as shown in the diagram. The stretched piece of wire is 1.8 m long and its centre is placed between the poles of a large magnet. An alternating potential difference of frequency 50 Hz is connected across the wire.

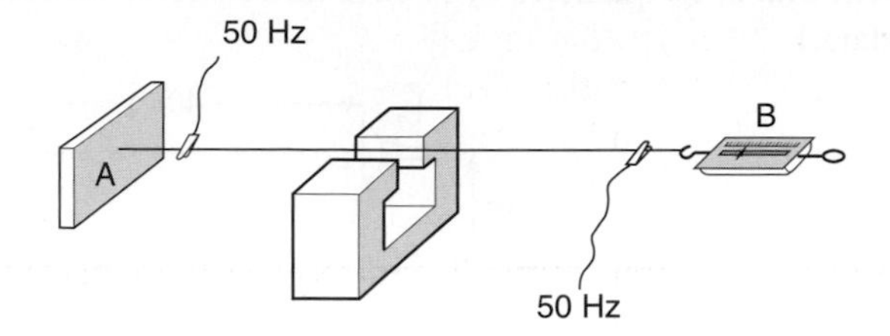

The spring balance is pulled with an increasing force. Show that the wire will oscillate in its fundamental mode, i.e. node–antinode–node, when the tension in the wire is 52 N. [Use the formula in the previous question.]

8.56 Two loudspeakers face each other at a separation of about 30 m. They are connected to the same audio oscillator which is set at 170 Hz.

(a) Describe and explain the variation of sound intensity heard by a person who walks along a line joining the two speakers. [Use data.]

(b) At what speed is he walking if he hears a maximum every 2.5 s?

8.57 The diagram shows a way of representing stationary sound waves in an open tube, e.g. a recorder.

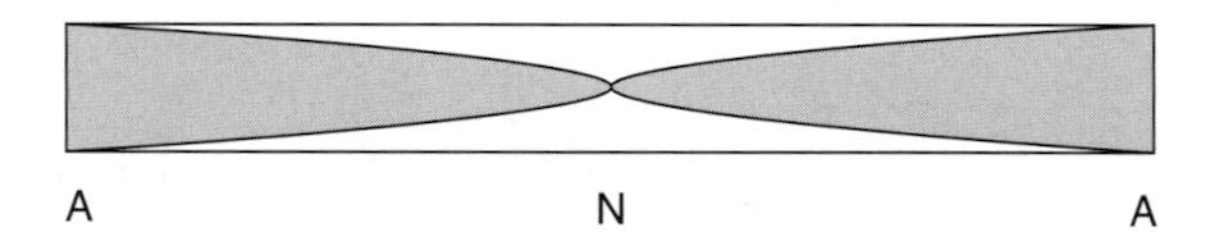

(a) If the length of the tube is 50 cm, calculate the frequency of the stationary wave shown. [Use data.]

(b) Sketch the stationary wave formed, A–N–A–N–A, when this frequency is doubled.

8.58* Use a search engine to find how a stringed musical instrument such as a violin relies on stationary waves.

8.59 Two radio transmitters T_1 and T_2 emit vertically polarised electromagnetic waves at a frequency of 1.44 MHz. A stationary wave pattern is set up along the line T_1T_2.

(a) What is the distance between adjacent nodes in this pattern? [Use data.]

(b) A car is driving along the line between the transmitters (e.g. along the M4) and its radio is receiving the programme carried by the 1.44 MHz waves. Describe what the driver hears if the car is travelling at just over $30\,\text{m}\,\text{s}^{-1}$.

8.60 In the diagram T is a microwave transmitter and P a detecting probe. As P is moved towards the metal reflecting sheet, a series of maxima are found separated by 16 mm.

(a) Deduce, with a full explanation, the wavelength of the microwaves.

(b) As P approaches the metal sheet the intensity of the minima is found to become closer and closer to zero. Explain why this is so.

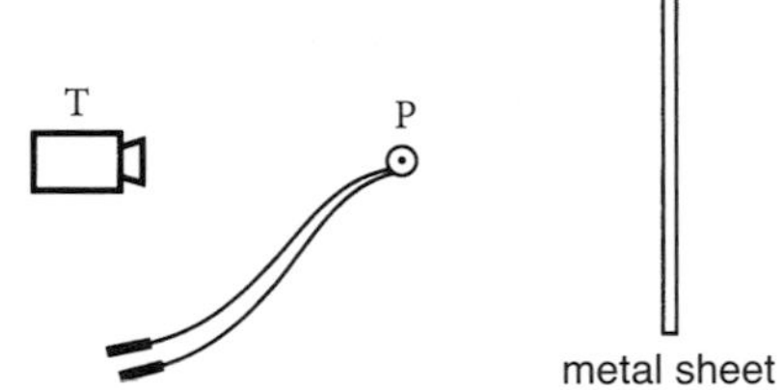

8.61 **(a)** A microwave transmitter T and receiver R are placed side by side facing two sheets of material M and N, as shown in the diagram. It is found that a very small signal is registered by R; what can you deduce about the experimental set up?

(b) When M is moved towards N a series of maxima and minima is registered by R. Explain this and deduce the wavelength of the electromagnetic waves emitted by T if the distance moved by M between the second and seventh minima is 70 mm.

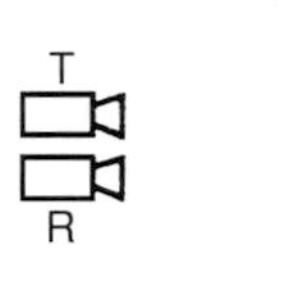

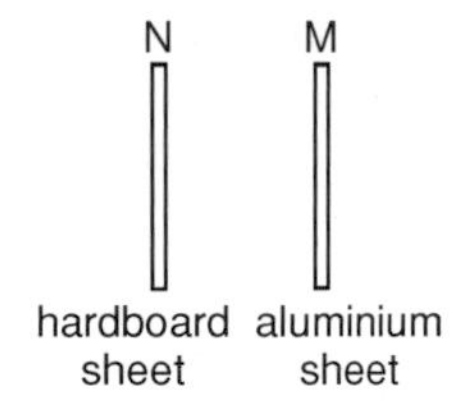

8.5 Diffraction patterns

In this section you will need to

- understand that waves diffract; that is, their energy spreads into the shadow area when part of a wavefront is blocked
- draw graphs of intensity against position for diffraction at a single slit
- use the equation $\sin\theta_m = \lambda/d$ for the first minimum in a single-slit diffraction pattern
- understand that a diffraction grating produces a series of narrow maxima at angles θ_n given by the formula $n\lambda = s\sin\theta_n$
- explain how a diffraction grating can produce spectra.

8.62 Refer to question 8.37.
Describe what is happening to the energy of the water waves as they diffract through the gap between the barriers.

8.63* Take two pencils and hold them vertically and *very close* together (but not touching). Look through the narrow gap between them at a bar or a vertical edge in a window. Describe what you see.
With the pencils still held vertically, what do you see if you look at a horizontal bar or edge in the window?
Explain your observations.

8.64 The graph shows the diffraction pattern produced by a single slit of width 0.30 mm when illuminated by light of wavelength 780 nm (7.8×10^{-7} m).

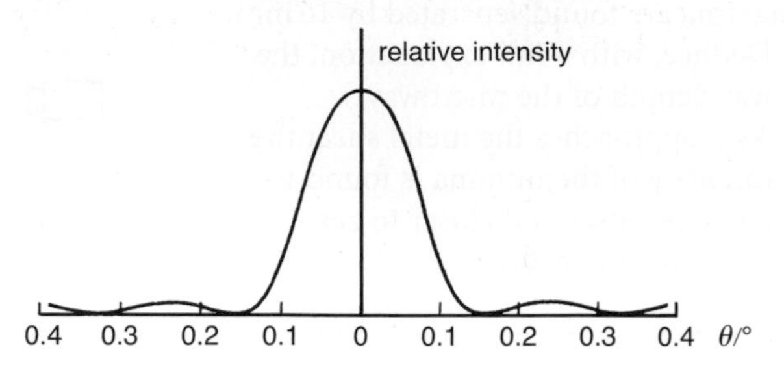

(a) Copy the axes from the diagram and draw a graph to show the appearance of the diffraction pattern when the wavelength of the light is halved.

(b) Describe what will happen to the diffraction pattern if the wavelength is gradually increased.

8.65 Suggest how you might construct a diffraction grating for use with

(a) microwaves of frequency about 10^{10} Hz

(b) sound waves of frequency about 10^{4} Hz. [Use data.]

8.66 A diffraction grating was set up so that parallel light was incident on it normally. For light of wavelength 589 nm the first-order spectral lines are observed in directions making an angle $\theta = 22.0°$ with the straight-through position.

(a) Calculate **(i)** the value of the slit separation s **(ii)** the number of slits per millimetre in the grating.

(b) Calculate the wavelength of light which would give a first-order spectral line at $\theta = 24.3°$.

8.67 A diffraction grating having 5.00×10^{5} lines per metre is illuminated with a parallel beam of white light. The wavelengths of the light at the red and violet ends of the spectrum are 750 nm and 400 nm respectively.

(a) Calculate the angular dispersion between these colours in the first-order spectrum.

(b) Show that the red in the second-order spectrum will overlap the violet in the third-order spectrum.

9 Photons and electrons

Data:

speed of light $c = 3.00 \times 10^{8}\ \mathrm{m\,s^{-1}}$

Planck constant $h = 6.63 \times 10^{-34}\ \mathrm{J\,s}$

electronic charge $e = 1.60 \times 10^{-19}\ \mathrm{C}$

electron volt: $1\ \mathrm{eV} \equiv 1.60 \times 10^{-19}\ \mathrm{J}$

mass of electron $m_e = 9.11 \times 10^{-31}\ \mathrm{kg}$

mass of proton $m_p = 1.67 \times 10^{-27}\ \mathrm{kg}$

9.1 Energy of photons and electrons

In this section you will need to

- use the equation $W = qV$ for the energy transfer when a charge q passes through a p.d. V
- understand that the electron volt (eV) is a unit of energy (though not an SI unit), and remember how it is defined
- understand that the energy E of electromagnetic radiation is quantised
- use the equation $E = hf$ for the energy of a photon, where h is the Planck constant
- use the wave equation $c = f\lambda$
- use the equation $E_k = \frac{1}{2}mv^2$ in solving problems
- use the equation $I = P/4\pi r^2$ to calculate intensity I of radiation.

9.1 What is the energy, in joules, of a photon of **(a)** red light of frequency 4.6×10^{14} Hz **(b)** violet light of frequency 6.9×10^{14} Hz? [Use data.]

9.2 The diagram shows the names given to different ranges of wavelengths in the electromagnetic spectrum. The wavelength scale, from 10^{-12} m to 10^{3} m, is logarithmic. Copy the wavelength scale. Beneath each value of λ add a scale of **(a)** frequency f in Hz **(b)** photon energy E in J.

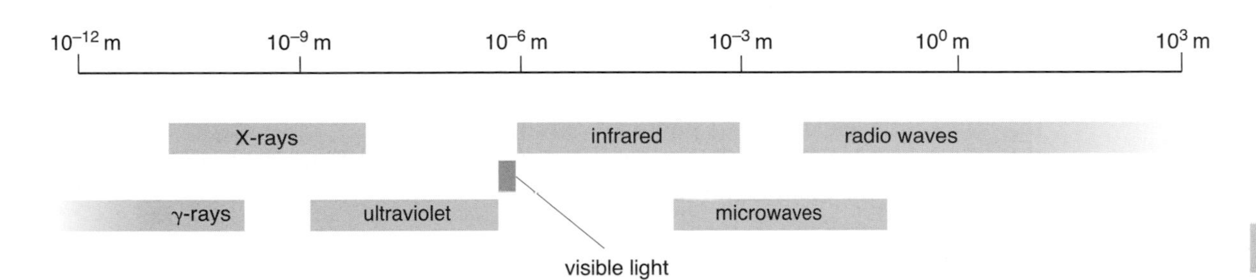

9.3 Calculate the energies, in J, of a photon of
(a) infrared radiation of wavelength 1500 nm
(b) green light of wavelength 546 nm
(c) ultraviolet radiation of wavelength 365 nm
(d) X-radiation of wavelength 154 pm
(e) γ-radiation of wavelength 2.3×10^{-12} m.

9.4 Suppose a photon has an energy of 1.00 eV.
(a) What is **(i)** its frequency **(ii)** its wavelength?
(b) In what part of the electromagnetic spectrum would you find it?
(c) Repeat part **(a)** for a photon with an energy of 2.00 eV.
(d) Repeat part **(b)** for a photon with an energy of 2.00 eV.

9.5* A friend asks you for some evidence that light is made up of individual photons.

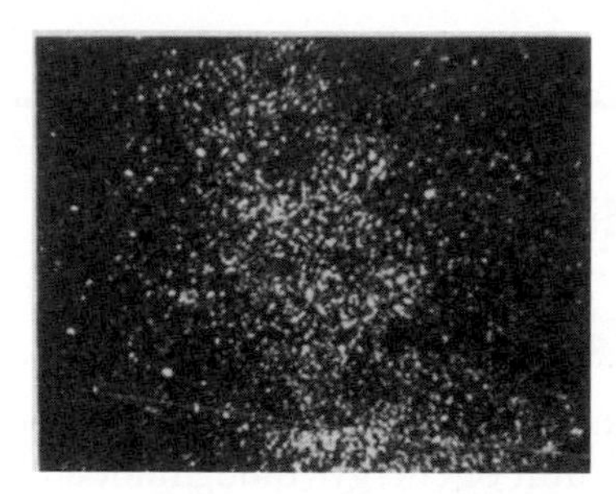

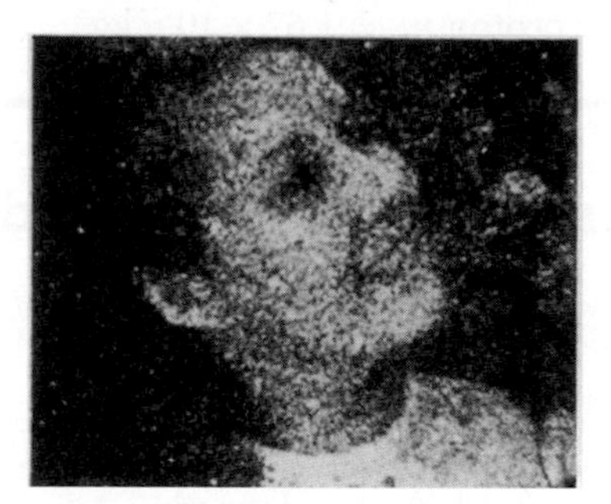

You show him the two photographs of a young woman shown here; adding that the first involved catching only 10 000 photons while the second captured 100 000 photons. He remains sceptical.
Discuss how you would use the photographs to convince him.

9.6 **(a)** In a typical house at night, what objects might be emitting electromagnetic waves of the following wavelengths: **(i)** 5×10^{-7} m **(ii)** 2×10^{-5} m **(iii)** 0.3 m?
(b) What are the energies (in eV) of photons of these wavelengths?

9.7 Copy and complete this table for light from mercury vapour (answers are given for yellow light **(a)**, **(b)**, **(c)** only):

colour	yellow	green	violet	ultraviolet
wavelength/nm	579	546	435	253
frequency/10^{14} Hz	**(a)**			
photon energy/10^{-19} J	**(b)**			
photon energy/eV	**(c)**			

9.8 A laboratory helium–neon laser emits light of wavelength 632.8 nm in a beam of diameter 2.0 mm.
(a) What is the energy of one photon of this light?
(b) If its power is 0.70 mW, how many photons are passing through a 1 mm^2 cross-section of the beam each second?

9.9 A radioactive source is found to emit γ-radiation of wavelength 6.5×10^{-12} m.
(a) What is the frequency of the γ-radiation?
(b) What quantity of energy (in MeV) is carried away from the nucleus by each γ-ray photon?

9.10 The definition of the volt tells us that there is a transfer of energy of 1.0 J when a charge of 1.0 C passes through a p.d. of 1.0 V.
(a) What is the energy transfer when 2.0 C pass through a p.d. of 3.0 V?
(b) What is the energy transfer when an electron passes through a p.d. of 1.0 V? [Use data.]

9.11 What is the energy transfer, in eV, when an electron passes through a p.d. of **(a)** 2.0 V **(b)** 20 V **(c)** 1.0 MV **(d)** 5.0 MV?

9.12 An X-ray has a frequency of 3.20×10^{17} Hz.
(a) Through what p.d. must an electron be accelerated to have the same energy as this X-ray photon?
(b) Show that the speed of such an electron is less that 10% of the speed of light. [Use data.]

9.13 The table shows the relationship between two familiar energy units: the joule (J) and the electron volt (eV), and two energy units no longer used by scientists: the erg and the calorie (cal).
For example, 1 erg ≡ 6.2×10^{11} eV.

	J	erg	cal	eV
1 J =	1	10^{7}	4.2	1.6×10^{-19}
1 erg =	10^{-7}	1	2.4×10^{-8}	6.2×10^{11}
1 cal =	4.2	4.2×10^{7}	1	2.6×10^{19}
1 eV =	1.6×10^{-19}	1.6×10^{-12}	3.8×10^{-20}	1

(a) Write down the equivalence between the joule and the electron volt.
(b) A joule is a $kg\,m^2\,s^{-2}$. Explain why an erg, which is a $g\,cm^2\,s^{-2}$, is equivalent to 10^{-7} J.
[The Calorie referred to in nutrition is equal to 1 kcal.]

9.14 What is the energy transfer, in J, when an electron passes through a p.d. of **(a)** 2.0 V **(b)** 20 V **(c)** 1.0 MV **(d)** 5.0 MV? [Use data.]

9.15 What is the energy transfer, in eV, when
(a) a proton passes through a p.d. of 1.0 V
(b) a doubly-charged ion passes through a p.d. of 1.0 V
(c) a proton passes through a p.d. of 2.0 MV?

9.16 What is the speed of
(a) an electron with a kinetic energy of **(i)** 6.0×10^{-18} J **(ii)** 50 eV [Use data.]
(b) a proton of mass 1.67×10^{-27} kg with a kinetic energy of **(i)** 6.0×10^{-18} J **(ii)** 50 eV?

9.17 In a certain X-ray tube electrons are accelerated through a p.d. of 200 kV. What is their final kinetic energy in **(a)** eV **(b)** J?

9.18 A proton, mass m_p, and a doubly-charged ion of mass $4m_p$ are both accelerated through a p.d V. What are the ratios
(a) (k.e. of proton)/(k.e. of ion)
(b) (speed of proton)/(speed of ion)?

9.19 A lamp with a power output of 150 W, of which 12% is in the visible range, is switched on at night at the top of a high tower.
(a) Calculate the light intensity at distances from the tower of **(i)** 100 m **(ii)** 200 m **(iii)** 300 m.
(b) Did you calculate each value separately? How are the three intensities related?

9.20 A radio-operated garage door opener responds to signals with an intensity greater than 20 μW m^{-2}. For a 250 mW transmitter unit which broadcasts equally in all directions, what is the maximum distance from the garage at which the transmitter will open the door?

9.21 An 11 W lamp hangs from the ceiling in a room. Assume that 10% of the electrical energy is converted to visible radiation.
(a) Calculate an approximate value for the intensity of the visible radiation on a table 1.5 m below the lamp. State any further assumption you make.
(b) The average wavelength of visible radiation is 550 nm. What is the average energy of each photon? [Use data.]
(c) At what rate are photons from the lamp striking an envelope measuring 16 cm by 10 cm that is lying on the table?

9.22 **(a)** What is the kinetic energy **(i)** in eV **(ii)** in J, of an electron accelerated through a p.d. of 100 kV in a tube designed to produce X-rays?
(b) Photons are emitted when the electron hits the metal target in the tube. If the photon emitted from the tube has the whole of this energy, what is **(i)** its frequency **(ii)** its wavelength?

9.2 Energy levels

In this section you will need to

- understand that an atom can exist in a few sharply-defined states of energy because there are only certain 'positions' where its electrons may be
- understand that the energy of an atom may be changed when it is bombarded by electrons or photons, or when it emits photons.

9.23 The diagram shows three energy levels for a sodium atom (at −3.03 eV there are two levels so close together that they cannot be shown separately).
(a) An electron moves from level 1 to the ground state. Calculate the wavelength of the emitted photon. [Use data.]
(b) Calculate the wavelengths of the other two photons that might be emitted as a result of transitions possible in the diagram.

level 2 ——— −1.94 eV

level 1 ——— −3.03 eV

ground state ——— −5.14 eV

(c) Assuming that the photon in **(a)** is in the orange part of the visible spectrum, estimate where the photons in **(b)** lie.

9.24 The diagram shows some of the energy levels for an atom of hydrogen. Photons are emitted when an electron moves down from one level to another.

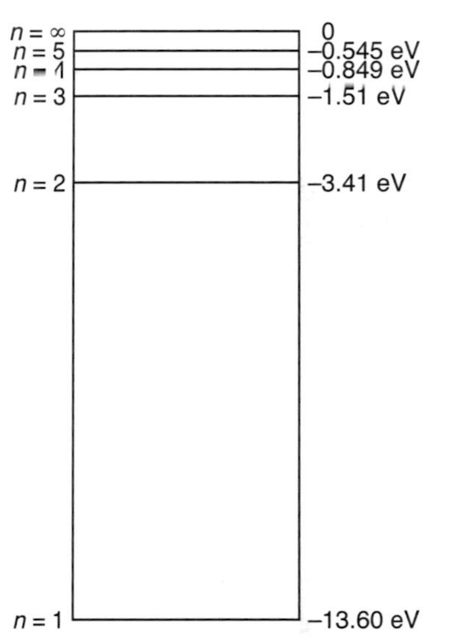

(a) When an electron moves from level 2 to level 1, what is

(i) its loss of energy in eV

(ii) its loss of energy in J

(iii) the frequency of the emitted photon

(iv) the wavelength of the emitted photon

(v) the part of the electromagnetic spectrum in which this radiation occurs?

(b) Repeat part **(a)** for an electron moving from level 3 to level 2.

(c) Repeat part **(a)** for an electron moving from level 4 to level 3.

9.25 The four lowest energy levels for an atom consist of the ground state and three levels above that. How many transitions are possible between these four levels?

9.26 Refer to the diagram for question 9.24.

(a) If the atom is in the ground state, how much energy (in eV) must be given to it to ionise it?

(b) What is the wavelength of the photon which could raise an electron from the −0.849 eV level to the −0.545 eV level?

(c) If an electron returns from the −0.849 eV level to the ground state, what is the wavelength of the photon emitted?

9.27 Suppose an atom has two energy levels E_1 and E_2 above the ground state. Radiation frequencies of f_1 and f_2 correspond to these energies, respectively.

(a) Sketch the energy level diagram of this atom.

(b) What other frequency will be emitted by this atom?

9.28 Refer to question 9.23.

(a) Calculate the frequencies of the three possible photons that can be emitted from these energy levels.

(b) Is there a relationship between your three values? If so, what is it?

9.29 The figure shows an energy level diagram. Sketch a possible line spectrum for the light emitted when electrons make the transitions shown. Label the lines, using the letters shown in the diagram, and indicate on your spectrum diagram which end corresponds to the higher frequency.

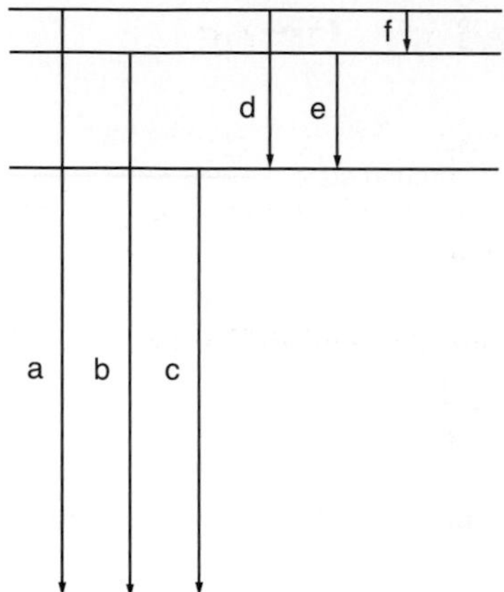

9.30 The ionisation energy of hydrogen is 13.6 eV.

(a) What is the speed of the slowest electron that can ionise a hydrogen atom when it collides with it?

(b) What is the longest wavelength of electromagnetic radiation that could produce ionisation in hydrogen?

9.31 The table gives the frequencies of some of the lines that occur in three groups in the hydrogen spectrum (traditionally the series have the names given at the top of each column):

Lyman $f/10^{14}$ Hz	Balmer $f/10^{14}$ Hz	Paschen $f/10^{14}$ Hz
24.659	4.5665	1.5983
29.226	6.1649	2.3380
30.824	6.9044	2.7399
31.564	7.3084	2.9822

(a) Show that the Balmer series is in the visible part of the spectrum.

(b) Which series is in the infrared part of the spectrum?

(c) Calculate the longest wavelength in each series.

9.32 The diagram shows four lines in the emission spectrum of hydrogen atoms (they are all part of the Balmer series).

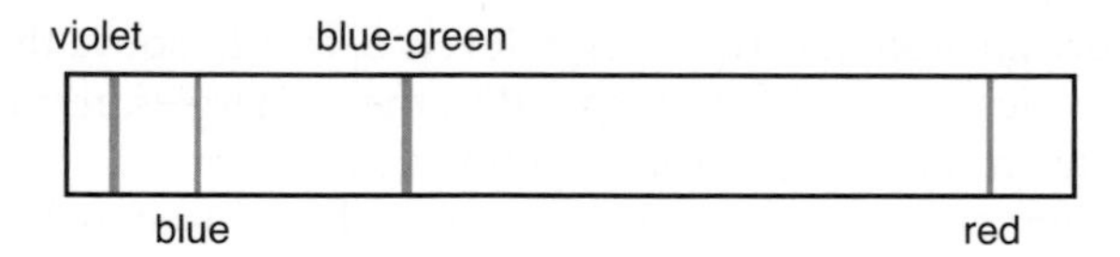

Copy the diagram and explain what the corresponding absorption spectrum would look like when white light is passed through atomic hydrogen.

9.33 The diagram in the previous question covers from 400 nm to 700 nm linearly. Sketch an energy level diagram that illustrates the production of the violet, blue and red lines. (Ignore the blue-green line.)

9.3 Waves and particles

In this section you will need to

- understand that we cannot describe light in terms of other phenomena (i.e. it is not 'like' water waves or bullets)
- understand that just as light has particle properties, so electrons (and other particles) have wave properties, and diffraction experiments are evidence for this
- use the de Broglie equation for momentum $p = h/\lambda$ to calculate the wavelength associated with a particle
- understand how quantisation explains the photoelectric effect
- use the photoelectric equation $hf = \varphi + \frac{1}{2}m(v_{max})^2$ where φ is the work function of a metal
- understand that the maximum k.e. $\frac{1}{2}m(v_{max})^2$ may be equated to the electrical potential energy eV_s, where V_s is the stopping potential to give $hf = \varphi + eV_s$.

9.34 What type of experimental evidence suggests that photons, e.g. in a light beam, have particle-like properties?

9.35 What type of experimental evidence suggests that the particles which we call electrons have wave-like properties?

9.36 On some days a dairy delivers milk to its customers continuously in a pipe but on other days it delivers it in packets on a conveyor belt. Each system delivers, on average, one pint every 10 minutes.

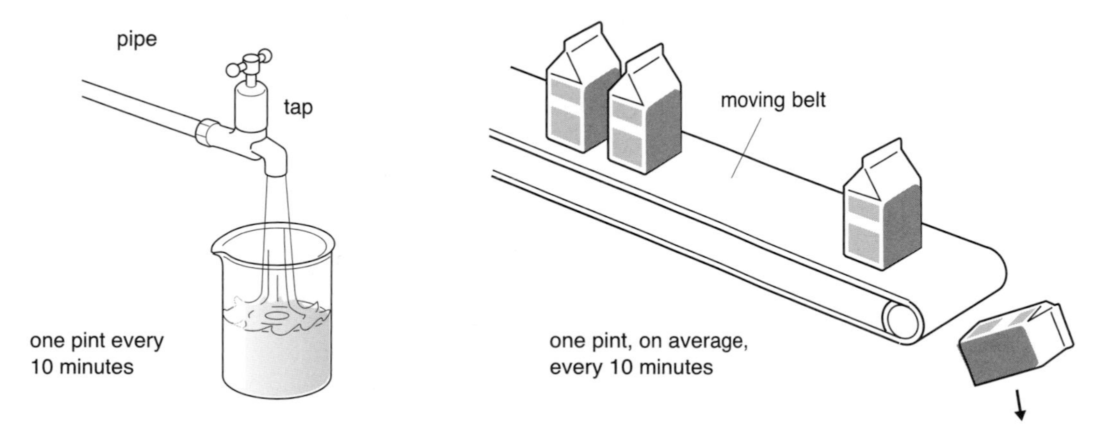

(a) Describe the different experiences of a customer who can be there for only 20 minutes but who wishes to buy two pints of milk.

(b) What has this got to do with the ideas of waves and particles?

9.37 **(a)** What are the energies (in eV) of the following types of photon:
(i) a radio wave of wavelength 1500 m
(ii) infrared radiation of wavelength 5.0×10^{-5} m
(iii) a gamma ray of wavelength 2.0×10^{-12} m? [Use data.]
(b) Which type would you think of as being most like a wave, and which most like a particle?
(c) Does any of this radiation behave entirely like a wave or entirely like a particle?

9.38 **(a)** What are, to 1 significant figure, the de Broglie wavelengths associated with the following moving particles:
(i) an electron at a speed of 20 m s^{-1}
(ii) a uranium atom of mass 4.0×10^{-25} kg at a speed of 1.6×10^{4} m s^{-1}
(iii) a raindrop of mass 1.5×10^{-13} kg falling at 8.0 m s^{-1}? [Use data.]
(b) Which of these would you think behaves most like a wave, and which most like a particle?
(c) Do any of these particles behave entirely like a wave or entirely like a particle?

9.39 Use the de Broglie equation to calculate the wavelength associated with each of the particles listed below. [Use data.]
(a) an electron of energy 10 keV
(b) a proton of energy 10 keV
(c) an α-particle of energy 10 keV.

9.40 How would you distinguish, experimentally, between a photon and an electron if each has the same de Broglie wavelength associated with it?

9.41 The emission of photoelectrons from a metal surface illuminated by short-wavelength radiation can take place instantaneously.
Explain why this experimental result
(a) supports a particle theory of radiation
(b) cannot be explained by a wave theory of radiation.

9.42 The diagram shows a coulombmeter with a negatively charged zinc plate connected to one of its terminals.

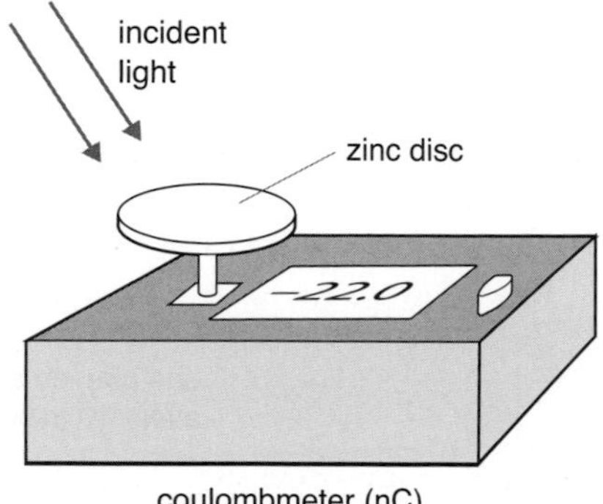

Incident light with a frequency below a certain threshold frequency does not discharge the zinc disc. Explain how this experiment supports the idea that light is made of ‘particle-like’ bundles of energy.

9.43 The diagram shows, side by side, photographs of the diffraction patterns taken by passing (on the left) X-rays and (on the right) electrons through the *same* very thin aluminium foil. The X-rays have a wavelength of about 2×10^{-11} m.

(a) Suggest what wavelength the experiment tells you is associated with the electron beam.

(b) Calculate an approximate value for the p.d. through which the electrons have been accelerated.

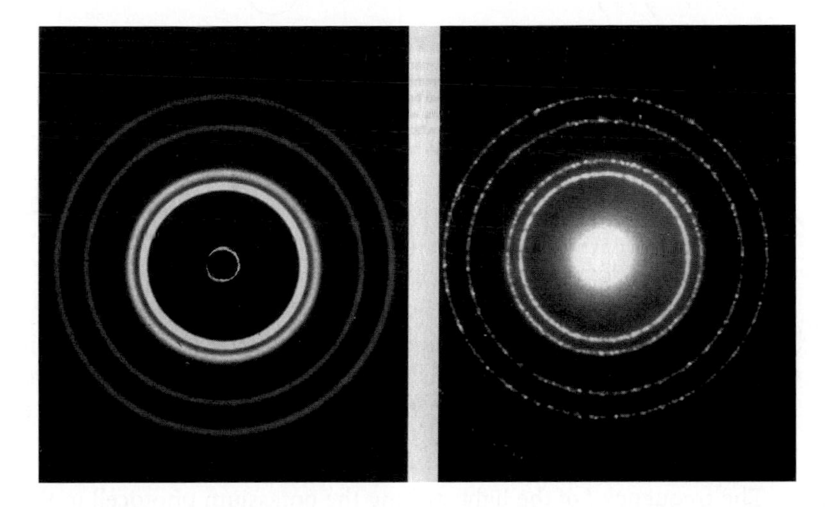

9.44 How does the quantum theory explain that

(a) electrons are emitted by a particular metal only when radiation of less than a certain wavelength falls on it

(b) the rate of emission of electrons is proportional to the intensity of the radiation

(c) the maximum speed of the emitted electrons is independent of the intensity of the radiation?

9.45 The work functions of sodium and zinc are 3.7×10^{-19} J and 6.8×10^{-19} J respectively.

(a) Explain what is meant by the work function of a metal. Express these two work functions in electron volts.

(b) Will either metal emit electrons when illuminated by light of frequency 8.0×10^{14} Hz? Support your answer with suitable calculations. [Use data.]

9.46 The minimum frequency of light which will cause photoelectric emission from a lithium surface is 7.0×10^{14} Hz.

(a) Calculate the work function of lithium.

If the surface is lit by light of frequency 8.0×10^{14} Hz, calculate

(b) **(i)** the maximum energy of the electrons emitted **(ii)** the maximum speed of these electrons.

9.47 The work function of a freshly cleaned copper surface is 4.16 eV. Calculate

(a) the minimum frequency of the radiation which will cause emission of electrons, and state whether this radiation is visible

(b) the maximum energy of the electrons emitted when the surface is exposed to radiation of frequency 1.20×10^{15} Hz.

9.48 The diagram show a photoelectric cell connected to a variable power supply and a picoammeter. A voltmeter registers the p.d. across the photocell.

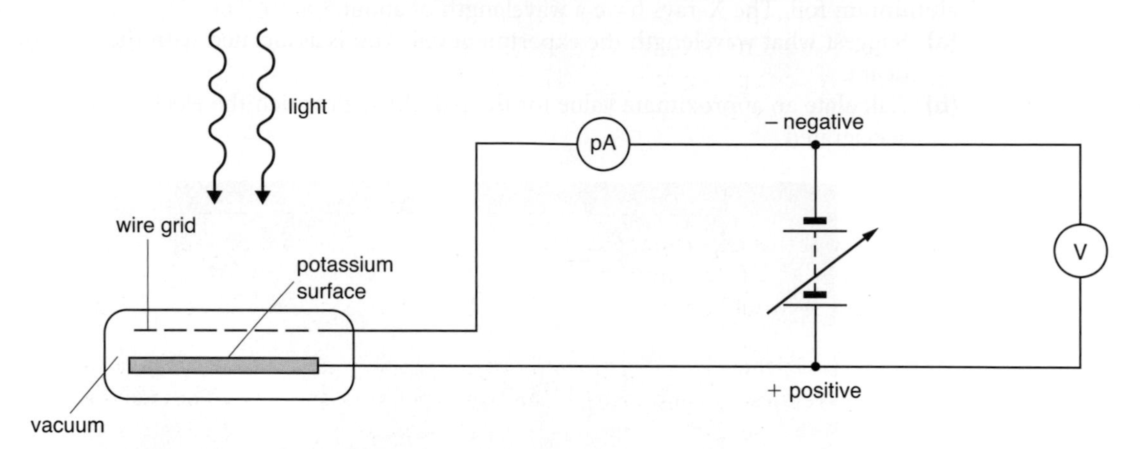

(a) When ultraviolet light is shone through the grid onto the potassium, electrons are emitted with a range of kinetic energies. Why do the electrons not all have the same k.e.?

(b) The circuit can be used to measure the maximum energy of these electrons. Describe how this is done.

9.49 The frequency f of the light striking the potassium photocell in the previous question is varied and the p.d. needed to reduce the current in the circuit to zero, the stopping potential V_s, is measured.

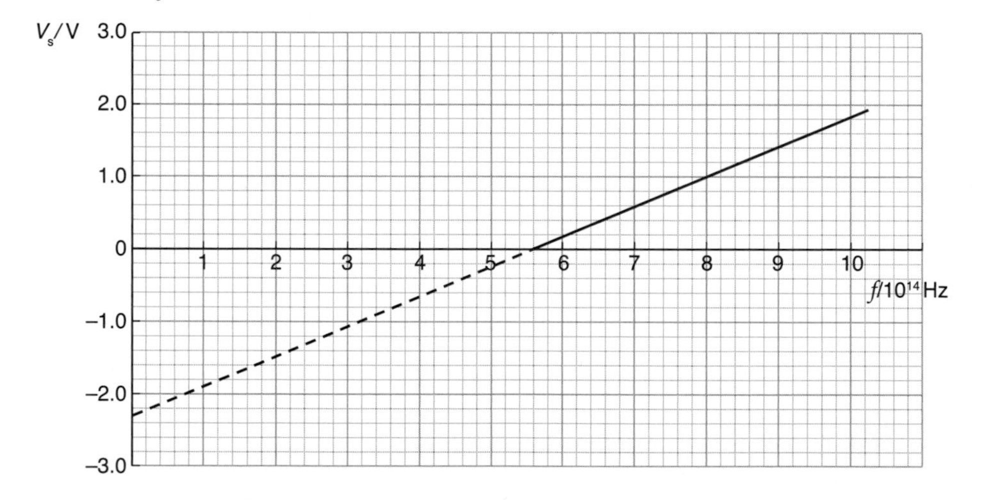

The graph shows the resulting relationship between V_s and f. [The graph line is dashed where no experimental results can be obtained.]

(a) Rearrange the equation between V_s and f to read 'V_s =' and measure the gradient of the graph.

(b) Hence calculate a value for the Planck constant h.

(c) From the graph deduce the work function of potassium.

(d) Describe in words what the line on this graph would look like for caesium, which has a work function of 2.1 eV.

9.50 In an electron diffraction tube the electron beam passes through a very thin crystalline foil. The beams diffracted by the crystals form circles on the end face of the tube. The diameter d of one prominent particular circle is measured for a range of values of the accelerating p.d. V, and the following values are obtained:

V/kV	1.5	2.0	3.0	4.5	6.0
d/mm	68	58	48	40	35

The theory of this experiment leads us to expect that d should be proportional to $1/\sqrt{V}$.

(a) Plot a graph of d against $1/\sqrt{V}$, and comment on the result.

(b) What p.d. would be required to give a diffraction circle of diameter 25 mm?

9.51* It is said that in the 1920s physicists thought of light as being carried by waves on Mondays, Wednesdays and Fridays, but by particles on Tuesdays, Thursdays and Saturdays (most people worked a 6-day week in those days).
Describe why there was such a problem. Is it now resolved?

10 Practising calculations

10.1 Numbers

In this section you will need to

- understand what is meant by expressing a number to a certain number of significant figures
- understand that the number of significant figures in a result should not be more than the smallest number of significant figures in the data
- understand what is meant by expressing a number in standard form
- use a calculator to add, subtract, multiply and divide numbers expressed in standard form
- use a calculator to raise numbers to powers, and to take square and cube roots of numbers
- remember the meaning of the following prefixes: p, n, μ, m, k, M, G, and know what they stand for: i.e., pico-, nano-, micro-, milli-, kilo-, mega-, giga-.

10.1 Express the following numbers in standard form, to two significant figures: **(a)** 0.0342 **(b)** 0.005291 **(c)** 0.145 **(d)** 153.2 **(e)** 674.

10.2 The numbers in these calculations have been given to different numbers of significant figures. In each case what should be the number of significant figures in the answer? **(a)** $5.72 \times 11.1 \times 1.1$ **(b)** $6.35 \times (2.35)^2 \times 0.1152$ **(c)** $67 \times 2.46 \times 0.010$.

10.3 *Without using a calculator*, find the value of **(a)** $10^3 \times 10^6$ **(b)** $10^2 \times 10^4$ **(c)** $10^6 \times 10^3$ **(d)** $10^{-3} \times 10^{-6}$ **(e)** $10^6 \div 10^3$ **(f)** $10^4 \div 10^{-2}$ **(g)** $(10^6 \times 10^{-3})/10^{-4}$ **(h)** $1 \div 10^3$ **(i)** $1 \div 10^{-6}$.

10.4 *Using a calculator*, find the value of **(a)** $10^3 \times 10^{-5}$ **(b)** $10^2 \div 10^6$ **(c)** $10^{-3} \times 10^5$ **(d)** $1 \div 10^{-3}$ **(e)** $10^{-3} \div 10^5$.

10.5 Giving your answers to two significant figures, multiply together each of the following pairs of numbers: **(a)** 1.34×10^4, 1.05×10^2 **(b)** 4.78×10^5, 1.34×10^{-2} **(c)** 9.11×10^{-31}, 7.432×10^{-12}.

10.6 Refer to the previous question. Giving your answers to two significant figures, divide the first number in each pair by the second number.

10.7 Do the following calculations, leaving your answers in standard form, to the appropriate number of significant figures: **(a)** $\pi(2.35)^2$ **(b)** $4\pi^2(1.63 \times 10^{-3})^2$ **(c)** $\sqrt{(25/\pi)}$.

10.8 Express the following percentages as fractions to two significant figures (e.g. 50% = 0.50): **(a)** 20% **(b)** 2.0% **(c)** 0.040% **(d)** 300%.

10.9 What is the percentage increase in a quantity if it is **(a)** doubled **(b)** trebled?

10.10 If $x = 60$, what does x become if there is a percentage increase in it of **(a)** 50% **(b)** 100% **(c)** 120% **(d)** 500%?

10.11 Express the following fractions as percentages: **(a)** 0.10 **(b)** 0.87 **(c)** 2.3 **(d)** 0.0020.

10.12 **(a)** If 30% of x is 37.2, what is x?
(b) If 18% of x is 43.9, what is x?
(c) If 75% of x is 0.972, what is x?

10.13 Measurements of the speeds of 10 cars on a motorway gave the following results:
$v/\mathrm{m\,s^{-1}} = 30.1, 25.6, 28.9, 34.5, 32.7, 33.1, 29.6, 29.9, 38.2, 23.1$.
Which of these were within 5% of the $31.3\,\mathrm{m\,s^{-1}}$ speed limit?

10.14 Giving your answers to two significant figures, do the following calculations:
(a) $\sqrt{(0.242)}$ **(b)** $\sqrt{(0.00476)}$ **(c)** the reciprocal of 43.98 **(d)** the reciprocal of 0.0824
(e) $\sqrt[3]{2.45}$ **(f)** $(9.73)^3$ **(g)** $(0.431)^{1/2}$ **(h)** $\pi(0.142)^2$ **(i)** $4.96^{3/2}$ **(j)** $(1 - 0.15^2)^{1/2}$.

10.2 Units and equations

In this section you will need to

- remember that the base units in the Système International (SI) are the metre, kilogram, second, ampere and kelvin
- understand that quantities need to be expressed in base units when used in equations for calculations
- rearrange equations so that a different quantity is the subject of the equation
- use equations to show how a derived unit is related to the base units
- know how to check that an equation is homogeneous with regard to units.

10.15 The base units in the SI system are the metre, the kilogram and the second. Express the following quantities in the appropriate base units, in standard form to two significant figures:
(a) 6.34 cm **(b)** 12 mm **(c)** 832 km **(d)** 546 nm **(e)** 53.4 g **(f)** 500 t
(g) 123 mg **(h)** 2.3 μg **(i)** 30 minutes **(j)** 23 ms **(k)** 24 hours **(l)** 45 litres.

10.16 Express the following areas and volumes in the appropriate base units:
(a) $1.6\,\mathrm{cm^2}$ **(b)** $5.3\,\mathrm{mm^2}$ **(c)** $0.017\,\mathrm{cm^2}$ **(d)** $7.8\,\mathrm{cm^3}$ **(e)** $34\,\mathrm{mm^3}$.

10.17 Calculate the areas of circles with these diameters, giving your answer in $\mathrm{m^2}$ and to two significant figures: **(a)** 2.5 m **(b)** 54 cm **(c)** 2.6 mm.

10.18 Write down the following quantities as numbers in standard form (to two significant figures) together with the appropriate unit without any prefix (such as p or M):
(a) 470 pF **(b)** 1.5 kV **(c)** 50 MW **(d)** 40 ns.

10.19 The force F exerted on an area A by a pressure p is given by the equation $F = pA$. Calculate F, giving your answer to the appropriate number of significant figures, when $p = 1.01 \times 10^5\,\mathrm{N\,m^{-2}}$ and $A = 1.25\,\mathrm{m^2}$.

10.20 The volume V of a cylinder is given by the equation $V = \pi r^2 h$. Calculate V, giving your answer in the base unit and to the appropriate number of significant figures, when r = 3.54 cm and h = 0.25 m.

10.21 Rearrange these equations to make the quantity given in the bottom row the subject of the formula:

(a)	**(b)**	**(c)**	**(d)**	**(e)**	**(f)**	**(g)**
$x = vt$	$V = bdh$	$A = \pi r^2$	$V = \pi r^2 h$	$\rho = m/V$	$v = u + at$	$v^2 = u^2 + 2ax$
v	h	r	h	V	a	a

(h)	**(i)**	**(j)**	**(k)**	**(l)**	**(m)**	**(n)**
$\eta = 1 - (T_2/T_1)$	$P = I^2R$	$P = V^2/R$	$T = 1/f$	$T = 2\pi\sqrt{(\ell/g)}$	$V = E - Ir$	$V = \frac{4}{3}\pi r^3$
T_1	I	R	f	ℓ	r	r

10.22 Use the equation $c = f\lambda$ to calculate f when $c = 3.00 \times 10^8\,\text{m s}^{-1}$ and λ = 546 nm.

10.23 Use the equation $A = \pi r^2$ to calculate r when $A = 12\,\text{cm}^2$.

10.24 Use the equation $V = \frac{4}{3}\pi r^3$ to calculate r when $V = 3.2 \times 10^{-3}\,\text{m}^3$.

10.25 Use the equation $E = F\ell/eA$ to calculate e when F = 12 N, ℓ = 1.5 m, $A = 0.010\,\text{mm}^2$ and $E = 200\,\text{GN m}^{-2}$.

10.26 Use the equation $GM = 4\pi^2 r^3/T^2$ to calculate T when $G = 6.67 \times 10^{-11}\,\text{N m}^2\,\text{kg}^{-2}$, $M = 5.97 \times 10^{24}\,\text{kg}$ and $r = 6.37 \times 10^6\,\text{m}$.

10.27 What are the base units, in the SI system of units, of **(a)** length **(b)** mass **(c)** time **(d)** electric current **(e)** temperature?

10.28 The unit of force F is the newton (N). Use the equation $F = ma$, where m is mass and a is acceleration, to express the newton in terms of the base units.

10.29 The unit of pressure p is the pascal (Pa). Use the equation $p = F/A$, where F is force and A is area, to express the pascal in terms of the base units.

10.30 The unit of energy W is the joule (J). Use the equation $W = Fs$, where F is force and s is distance, to express the joule in terms of the base units.

10.31 The unit of power P is the watt (W). Use the equation $P = W/t$, where W is energy and t is time, to express the watt in terms of the base units.

10.32 The unit of electric charge q is the coulomb (C). Use the equation $I = q/t$, where I is electric current and t is time, to express the coulomb in terms of the base units.

10.33 The unit of potential difference V is the volt (V). Use the equation $V = W/q$, where W is energy transferred and q is electric charge, to express the volt in terms of the base units.

10.34 The unit of frequency f is the hertz (Hz). Use the equation $T = 1/f$, where T is the time period, to express the hertz in terms of the base units.

10.35 The unit of electrical capacitance C is the farad (F). Use the equation $C = q/V$, where q is electric charge and V is potential difference, to express the farad in terms of the base units.

10.36 Simplify the following expressions, performing the calculation on the units as well as the numbers. What type of quantity is each expression? (In answering parts of the question it may help you to know that N m can be written as J, and $\mathrm{N\,m^{-2}}$ as Pa.)

(a) $20\,\mathrm{m\,s^{-1}} + (5\,\mathrm{m\,s^{-2}})(2.0\,\mathrm{s})$

(b) $\frac{1}{2}(20\,\mathrm{N\,m^{-1}})(0.01\,\mathrm{m})^2$

(c) $(1.29\,\mathrm{kg\,m^{-3}})(9.81\,\mathrm{N\,kg^{-1}})(2.3\,\mathrm{m})$

(d) $\frac{4}{3}\pi(2.63\,\mathrm{m})^3$

10.37 Calculate the following and express the results by means of numbers in standard form with a single unit:

(a) $470\,\mathrm{pF} \times 1.5\,\mathrm{kV}$ given that $1\,\mathrm{F} - 1\,\mathrm{C\,V^{-1}}$

(b) $50\,\mathrm{MW} \times 40\,\mathrm{ns}$ given that $1\,\mathrm{W} = 1\,\mathrm{J\,s^{-1}}$

(c) your answer to **(b)** divided by your answer to **(a)** given that $1\,\mathrm{V} = 1\,\mathrm{J\,C^{-1}}$.

10.38 Check the homogeneity of the following equations with respect to units:

(a) $W = mgh$ where W = change of gravitational potential energy, m = mass, h = change of height

(b) $W = \sigma\varepsilon$ where W = change of elastic potential energy per unit volume, σ = stress and ε = strain

(c) $p = \rho gh$ where p = change of pressure, ρ = density of fluid and h = change of height

(d) $T = 2\pi\sqrt{(\ell/g)}$ where T is the time period of a pendulum, ℓ is its length and g is the gravitational acceleration at the Earth's surface

(e) $\Delta E = c^2\Delta m$ where ΔE is the energy equivalent of a mass Δm and c is the speed of light.

10.39 In an old scientific textbook the intensity of the solar radiation arriving at the Earth (before absorption, etc. by the atmosphere) is quoted as 2.0 cal per square centimetre per minute. Express this quantity in the appropriate SI unit, given that 1.0 cal = 4.2 J.

10.40 Tyre pressures for road vehicles are often still given in British units, e.g. $30\,\mathrm{lb/in^2}$ (p.s.i.). Express this quantity in the appropriate SI unit, given that 1.00 lb = 4.45 N and 1.00 inch = 2.54 cm.

10.41 Land areas are often still given in acres. One acre is an area which has a width of 1 chain and a length of 1 furlong. 1 chain = 22 yards; 1 furlong = 220 yards. The metric unit of land area is the hectare (ha): one hectare is an area which is 100 m square. Given that 1 yard = 0.914 metre, express 1 hectare in acres.

10.42 The cost of electrical energy is quoted as 6.51p per kilowatt-hour. Work out the cost of cooking a large meal, if this requires the use of three cooking rings for 2.5 hours; the rings are each rated at 2.8 kW.

10.3 Graphs and relationships

In this section you will need to

- understand what is meant by saying that one quantity is proportional to another
- understand what is meant by saying that there is a linear relationship between two quantities
- understand that an equation of the form $y = mx + c$ represents a linear relationship and that when y is plotted against x the result is a straight line
- understand that when an equation is not of the form $y = mx + c$ it may still be possible to obtain a linear graph by a suitable choice of quantities to plot

- measure or calculate the gradient of a straight-line graph
- measure or calculate the gradient at a point on a non-linear graph by drawing a tangent to the curve
- understand that the area between a graph and the horizontal axis may represent a quantity
- calculate the quantity represented by the area between a graph and the horizontal axis.

10.43 State the meaning of: **(a)** $t > 10\,\text{s}$ **(b)** $20\,\text{s} < t \leq 40\,\text{s}$ **(c)** $A \propto r^2$ **(d)** $\Delta x = 0.35\,\text{m}$.

10.44 Write the statements in the previous question as algebraic relationships using only mathematical symbols:

(a) The resistance R of a conductor is directly proportional to its length ℓ, and inversely proportional to its cross-sectional area A.

(b) The centripetal force F on a body moving in a circular path with constant speed is directly proportional to the square of its speed v, and inversely proportional to the radius r of its path.

10.45 What is **(a)** Δv if v changes from $5.3\,\text{m}\,\text{s}^{-1}$ to $6.7\,\text{m}\,\text{s}^{-1}$ **(b)** Δt if t increases from 2.34 s to 3.98 s **(c)** ΔV if V changes from $65\,\text{cm}^3$ to $18\,\text{cm}^3$?

10.46 Write the following statement out in full without mathematical symbols, where V refers to the potential difference across a thermistor, I refers to the current in it, and T is its temperature: $\Delta V = -2.0\,\text{V}$, when $\Delta T = +10\,\text{K}$, if $\Delta I = 0$.

10.47 Answer yes or no to the following:

	relationship	constant quantity	are these quantities proportional?
(a)	$F = pA$	p	F, A
(b)	$F = pA$	A	F, p
(c)	$W = Fs$	F	W, s
(d)	$W = Fs$	s	W, F
(e)	$P = W/t$	t	P, W
(f)	$P = W/t$	W	P, t
(g)	$P = W/t$	W	$P, 1/t$
(h)	$E = Fx^2$	x	E, F
(i)	$E = Fx^2$	F	E, x
(j)	$E = Fx^2$	F	E, x^2

10.48 The diagram shows three graphs for varying quantities.

(a) Which of the graphs show a linear relationship between y and x?

(b) Which of the graphs show that y is proportional to x?

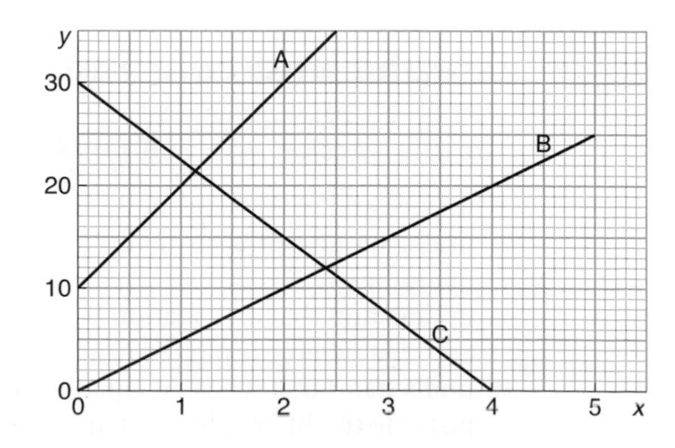

10.49 The diagram shows three graphs for varying quantities.

(a) Which of the graphs show a linear relationship between y and x?

(b) Which of the graphs show that y is proportional to x?

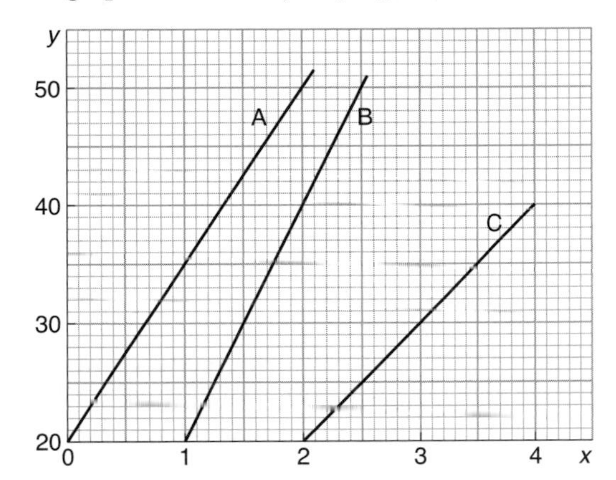

10.50 Calculate the gradients of these graphs. If you can, simplify the unit to express it in its simplest form.

s/m 20 10 0 | 0 5 10 | t/s

(a)

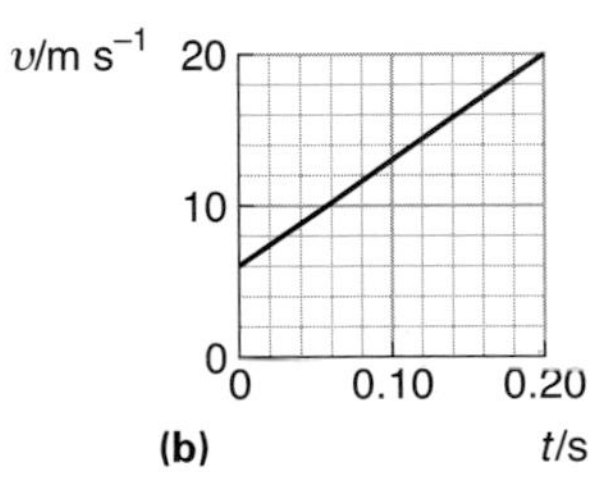

(b)

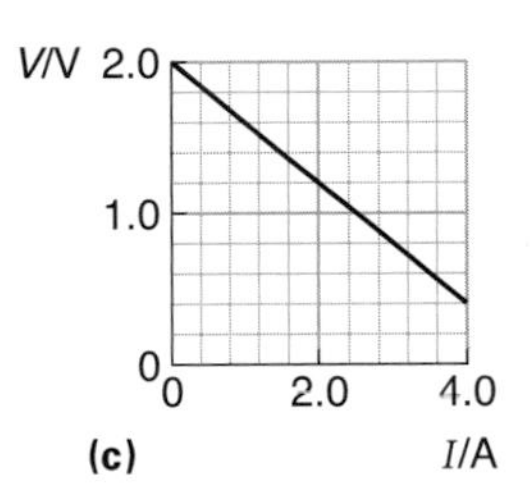

(c)

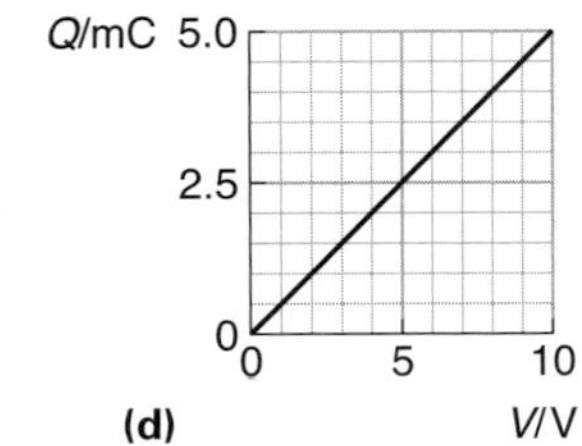

(d)

g/m s⁻² 10 5.0 0 | 0 1.0 2.0 3.0 | 1/r²/10⁻¹⁴ m⁻²

(e)

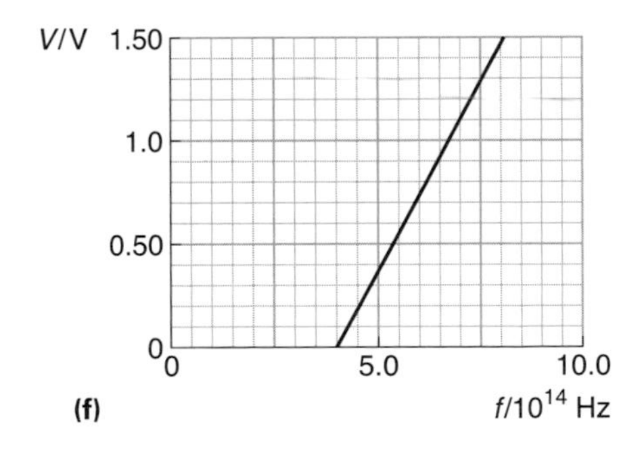

(f)

10.51 Calculate the quantity represented by the areas between these graphs and the horizontal axis. If you can, simplify the unit to express it in its simplest form.

υ/m s⁻¹ 20 10 0 | 0 5.0 10 | t/s

(a)

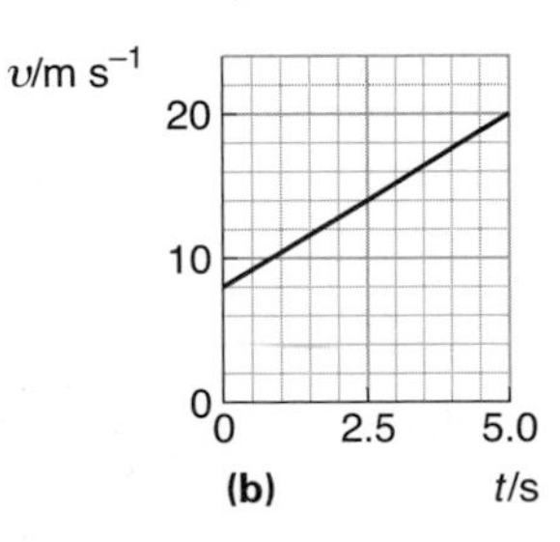

(b)

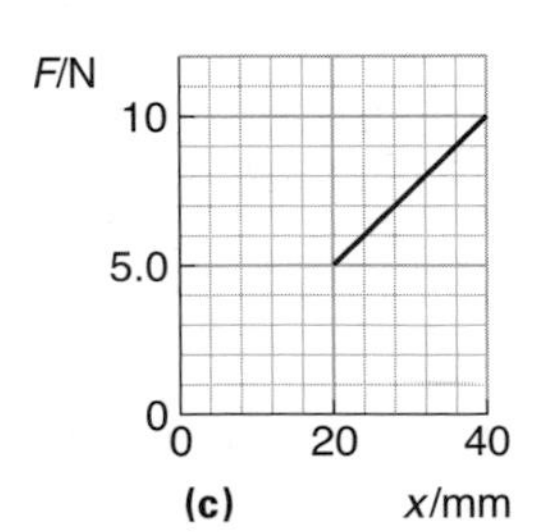

(c)

10.52 Calculate the quantity represented by the areas between these graphs and the x-axis. [Hint: first find the quantity represented by the small square and then estimate the number of these between the graph and the horizontal axis.] If you can, simplify the unit to express it in its simplest form.

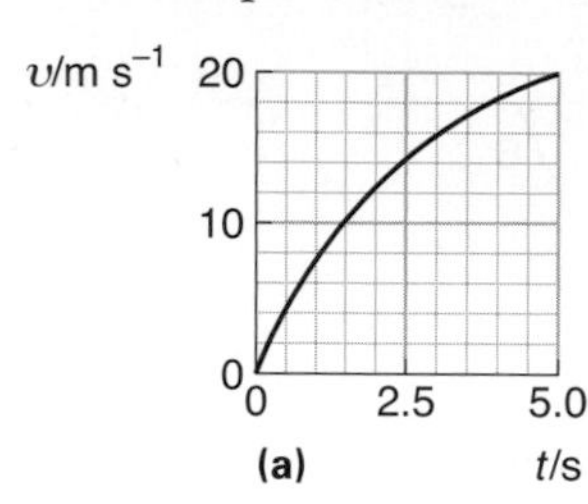

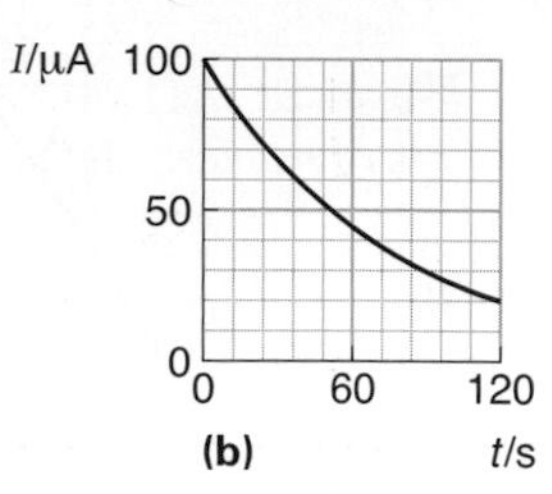

10.53 The table gives some relationships between quantities: in the next columns the two varying quantities are shown. All the other quantities are constant. Copy the table and in each case complete the columns to show what quantities you would plot on the y-axis and x-axis if you wanted to obtain a straight-line graph, and what would be the gradient.

	relationship	varying quantities	plot on x-axis	plot on y-axis	gradient
(a)	$W = \frac{1}{2}kx^2$	W, x			
(b)	$E = V/d$	E, d			
(c)	$F = Gm_1m_2/r^2$	F, r			
(d)	$c = f\lambda$	f, λ			
(e)	$C = \varepsilon_r\varepsilon_0 A/d$	C, A			
(f)	$C = \varepsilon_r\varepsilon_0 A/d$	C, d			
(g)	$T = 2\pi\sqrt{(\ell/g)}$	T, ℓ			
(h)	$T = 2\pi\sqrt{(m/k)}$	T, k			
(i)	$eV_s = hf - \varphi$	f, V_s			

10.54 The extensions x of a copper wire are measured for different values of force F, and the following results are obtained:

F/N	10	20	30	40	50
x/mm	1.5	3.1	4.6	6.2	7.7

(a) Plot a graph of F (on the y-axis) against x (on the x-axis).
(b) Is there a linear relationship between F and x?
(c) Is F proportional to x?
(d) Calculate the gradient of the graph, remembering to give the units.
(e) The gradient is equal to EA/ℓ, where A is the cross-sectional area of the wire, 0.10 mm^2, and ℓ is the length of the wire, 2.0 m. Calculate the value of E, the Young modulus of copper.

10.55 The potential difference V across an electric cell is measured for different values of the current I taken from it, and the following results are obtained:

I/A	0	0.20	0.40	0.60	0.80	1.00
V/V	1.52	1.40	1.28	1.16	1.04	0.92

(a) Plot a graph of V (on the y-axis) against I.

(b) Is there a linear relationship between V and I?
(c) Is V proportional to I?
(d) What is the gradient of the graph?
(e) The equation connecting V and I for this cell is in the form $V = mI + c$. What are the values of the quantities m and c? (Give their units as well as the numbers.)

10.56 The following table gives a series of values of the wavelength λ of sound in air measured for different values of the frequency f.

f/kHz	0.20	0.40	0.60	1.00	1.50	2.00
λ/m	1.71	0.86	0.57	0.34	0.23	0.17

(a) Plot graphs of **(i)** f against λ **(ii)** f against $1/\lambda$, with f on the vertical axis in each case.
(b) What can you deduce from these graphs about the relationship between f and λ?
(c) Measure the gradient of whichever graph is a straight line. [Remember that $1\,\text{Hz} = 1\,\text{s}^{-1}$.]

10.4 Sines, cosines and tangents

In this section you will need to

- understand that the ratios of the sides of right-angled triangles are called sine, cosine and tangent
- remember which ratio is referred to by the terms sine, cosine and tangent
- remember that the sines and cosines of angles such as 0°, 30°, 60°, 90°, 180° have simple values
- understand how to use, in a right-angled triangle, an angle and a side to calculate the other sides
- understand how to find the resolved part of a vector quantity
- understand how Pythagoras's theorem can be used to calculate the third side of a right-angled triangle.

10.57 If $\theta = 25.3°$, what, to three significant figures, is **(a)** $\sin\theta$ **(b)** $\cos\theta$ **(c)** $\tan\theta$?

10.58 If $\theta = 0°$, what is **(a)** $\sin\theta$ **(b)** $\cos\theta$ **(c)** $\tan\theta$?

10.59 If $\theta = 90°$, what is **(a)** $\sin\theta$ **(b)** $\cos\theta$?

10.60 Why, if you use your calculator to find $\tan 90°$, does the display read 'E' or 'Error'?

10.61 **(a)** In this triangle what is $\sin\theta$?
(b) Use the $\sin^{-1}$ function on your calculator to find θ to four significant figures.

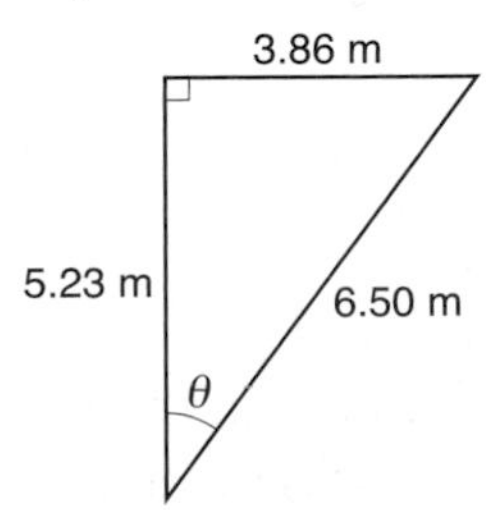

10.62 Refer to the previous question. Also find θ by first calculating
(a) $\cos\theta$ and using the $\cos^{-1}$ function on your calculator
(b) $\tan\theta$ and using the $\tan^{-1}$ function on your calculator.

10.63 In the triangle shown, what are the lengths of **(a)** AB **(b)** BC?

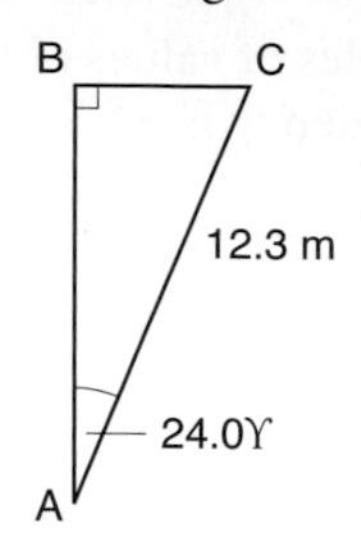

10.64 In the diagram what is the perpendicular distance from O to the line AB?

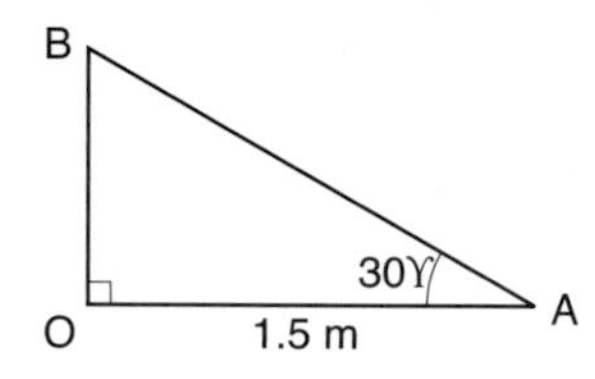

10.65 The diagram shows various vectors and two perpendicular directions x and y into which the vectors are to be resolved.
(a) In each case what is the other angle which makes up the right angle?
(b) Write down the two resolved parts in each case, using only the angle between the vector and the direction in which you are resolving. Leave your answer in the form $X\cos\theta$. You are not asked to do the calculations.

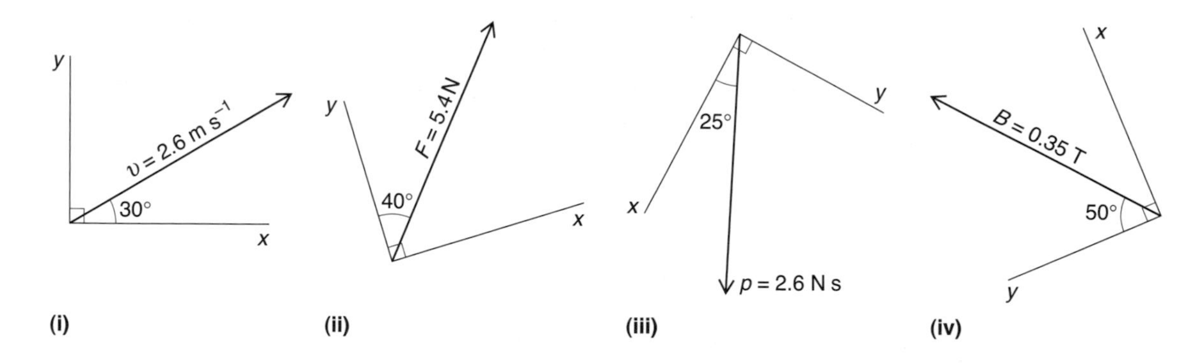

10.66 Use Pythagoras' theorem to check whether these could be the sides of right-angled triangles:
(a) 3 m, 4 m, 5 m **(b)** 5 m, 12 m, 13 m
(c) 11 m, 22 m, 25 m **(d)** 9 m, 40 m, 41 m
(e) 15 m, 36 m, 39 m.

10.67 Find the third side of a right-angled triangle whose two short sides are
(a) 6.32 m and 4.17 m **(b)** 5.21 m and 1.87 m.

10.68 Find the third side of a right-angled triangle if **(a)** its shortest side is 3.46 cm and its longest side is 7.21 cm **(b)** its shortest side is 0.542 cm and its longest side is 0.825 cm.

10.69 The diagram shows two forces of 24 N and 7.0 N acting at a point. Their resultant is given by the diagonal of the rectangle. Find **(a)** the size of the resultant of these forces **(b)** the angle which it makes with the 24 N force.

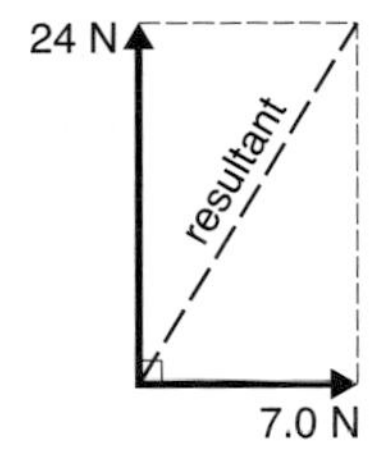

10.70 The diagram shows an air liner flying at a speed of 350 m s^{-1} relative to the air, in a cross wind of 50.0 m s^{-1}. Its velocity relative to the ground is given by the diagonal of the rectangle. Find the size and direction of this velocity.

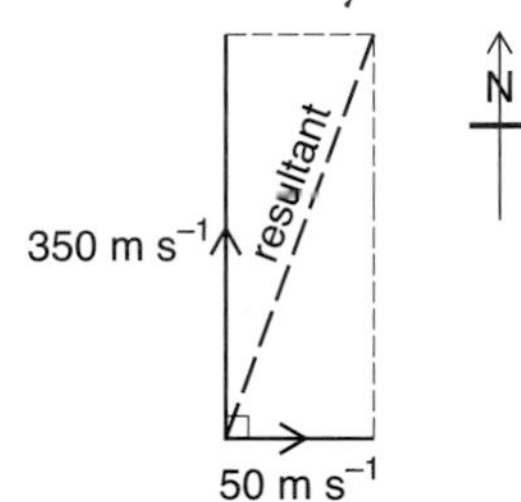

10.5 Angles

In this section you will need to

- understand that angles may be measured in radians as well as in degrees
- remember that π radians = 180°
- understand the meaning of the word 'subtend' in a phrase like 'the arc subtends an angle of 30° at the centre of the circle'
- understand that some calculations are simpler when the radian measure of angle is used
- understand that it is possible to make the approximation that $\sin\theta$ and $\tan\theta$ are equal to the angle θ measured in radians if the angle is small and answers are required to three or fewer significant figures.

10.71 Express the following angles in degrees, to three significant figures: **(a)** 0.10 rad **(b)** 1.00 rad **(c)** 2.00 rad **(d)** 2π rad.

10.72 Express the following angles in radians, as fractions or multiples of π rad: **(a)** 90° **(b)** 60° **(c)** 120°.

10.73 The diagram shows a segment of a circle of radius 1.50 m. The angle at the centre is 1.00 rad.
(a) What is the length of the arc?
(b) If the angle at the centre is increased to 2.00 rad, what is the new length of the arc?
(c) Write down an equation relating s, the arc length, to the radius r and the angle θ measured in radians.

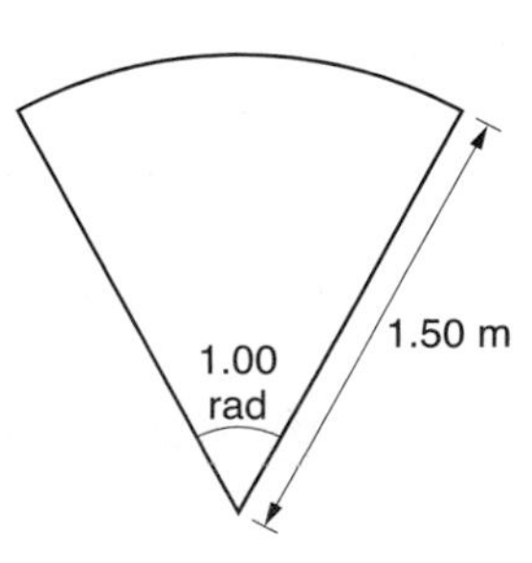

10.74 An arc of a circle of radius 2.50 m subtends an angle of 60.0° at the centre. What is **(a)** 60.0° in radians **(b)** the length of the arc?

10.75 A motor car tyre of radius 0.270 m is punctured in close succession by two nails on the road. The angle subtended at the centre of the wheel by the part of its circumference between the two nails is 140°. What was the shortest possible distance on the road between the nails?

10.76 A pole of height 2.156 m is placed 500.0 m away on a distant hillside. A man looks at the pole; his eyes are level with the foot of the pole.

(a) Calculate, by these two methods, the angle subtended at his eye:

- **(i)** Use trigonometry to work out the tangent of the angle, and hence the angle in degrees. Convert this angle to radians, giving your answer to four significant figures.
- **(ii)** Assume that the pole is the arc of a circle of radius 550.0 m, and use the equation $\theta = s/r$ to calculate the angle.

(b) The second method is approximate – but are the answers the same to four significant figures?

(c) Which method is easier?

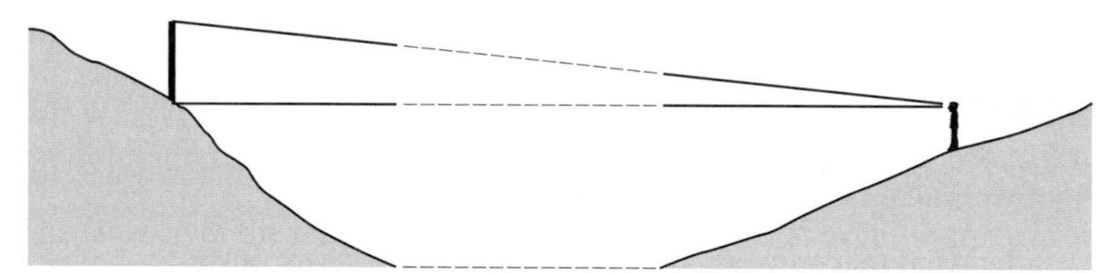

10.77 The top of a tree is seen at an elevation of 9.0° above its base when observed from a point at a distance of 120 m from it. Estimate the height of the tree, without using sine, cosine or tangent functions on your calculator.

10.78 The visible diameter of the planet Venus is 1.22×10^7 m, and its closest distance of approach to the Earth is 4.10×10^{11} m. Calculate its maximum angular size as seen from the Earth **(a)** in radians **(b)** in minutes of angle.

10.79 **(a)** A unit of distance used in astronomy is the parsec. This is defined as the distance from the solar system at which the radius of the Earth's orbit round the Sun (1.50×10^{11} m) subtends an angle of 1 second. (60 seconds = 1 minute; 60 minutes = 1 degree.) Calculate the size of the parsec in metres.

(b) The nearest star to the solar system is α-Centauri at a distance of 1.32 parsec. What angle does the radius of the Earth's orbit subtend at α-Centauri?

10.80 In order to measure the angular width of the scene photographed through a camera lens a boy takes a picture of a horizontal metre scale set at right angles to the lens axis and at a distance of 1.2 m from it. The developed photograph shows 0.92 m of the scale.

(a) Calculate the angular width of the scene **(i)** assuming that 0.92 m is the length of a circular arc of radius 1.2 m **(ii)** taking this distance more exactly as the shortest distance between these edges.

(b) What is the percentage error in using the approximate method of calculation?

10.6 Sinusoidal oscillations

In this section you will need to

- remember the shape of the sine and cosine functions for an angle θ as θ increases from 0° to 360° (or from 0 to 2π radians), and understand that they repeat indefinitely for larger angles
- remember what is meant by the amplitude, frequency and period of a sinusoidally varying quantity
- understand that an expression of the form $x = x_0 \cos(2\pi ft)$ represents a quantity x which varies sinusoidally with time; its amplitude is x_0 and frequency f
- remember that the maximum rate of change of x is given by $2\pi fx_0$, and that when x is a displacement, this will give the maximum velocity
- remember that in a similar way the maximum rate of change of velocity, i.e. the maximum acceleration, is given by $2\pi f(2\pi fx_0)$ or $(2\pi f)^2 x_0$.

10.81 **(a)** Sketch a graph to show how $\sin\theta$ varies with θ for angles from 0° to 360°.
(b) Also label the angle axis with the same angles in radians.
(c) What are the maximum and minimum values of $\sin\theta$?
(d) What are the angles when $\sin\theta$ varies most rapidly?

10.82 Repeat the previous question for $\cos\theta$.

10.83 In the expression $(1.5\,\text{m})\cos(2\pi ft)$ the frequency f is measured in Hz (which is the same as the unit s^{-1}) and t is measured in seconds. The angle is expressed in radians. Calculate the value of the expression where $f = 50\,\text{Hz}$ and $t = 12\,\text{ms}$. [Hint: Adjust your calculator so that it deals with angles in radians and not degrees, before finding the cosine.]

10.84 Repeat the previous question for $f = 100\,\text{Hz}$ and $t = 5.0\,\text{ms}$.

10.85 **(a)** Plot a graph to show how x varies with t, where $x = x_0\cos(2\pi ft)$, $x_0 = 1.2\,\text{m}$ and $f = 0.25\,\text{Hz}$, using the following values of t in s: 0, 0.25, 0.50, 1.00, 1.50, 1.75, 2.00. [Hint: The angle $2\pi ft$ will be in radians, so set your calculator to work in radians before calculating the cosines.]
(b) What is the period of this sinusoidal variation?
(c) If x represents displacement from the equilibrium position, what does the gradient of the graph represent?
(d) At what time(s) does the graph have a maximum gradient?
(e) Measure the gradient of the graph at $t = 1.00\,\text{s}$, and verify that its size is equal to $2\pi fx_0$.
(f) What would be the maximum speed for the motion for which
(i) $x_0 = 1.2\,\text{m}$ and $f = 0.50\,\text{Hz}$ **(ii)** $x_0 = 0.60\,\text{m}$ and $f = 0.25\,\text{Hz}$?

10.86 When a tuning fork is set in vibration the displacement x of the end of one of the prongs from its equilibrium position at time t is given by $x = x_0\cos(2\pi ft)$.
(a) The frequency f of a particular fork is 440 Hz. Calculate the maximum speed v_0 of the ends of its prongs, if the maximum displacement x_0 is 0.50 mm.
(b) If the speed v of the ends of the prongs also varies sinusoidally, calculate the maximum acceleration of the prongs.

10.87 At a certain point in a radio set tuned to the BBC long-wave transmitter there is a sinusoidal alternating potential V of peak value 2.5 V and frequency 200 kHz. Calculate
(a) the maximum rate of change of V
(b) the period of the oscillation
(c) the value of V at a time 0.20 μs after it changes direction.

10.88 The tides of the sea move up and down a certain beach at a frequency of approximately twice per day.
(a) What is this frequency in Hz? [1 Hz = 1 s^{-1}.]
(b) The amplitude of movement of the tide is 15 m (measured along the surface of the beach) on either side of the mean tide-line. The beach can be assumed to have a constant slope. What is the maximum rate at which the tide moves up the beach, assuming that the variation is sinusoidal?

10.7 Exponentially varying quantities

In this section you will need to

- understand what is meant by the logarithm of a number, and remember that the base of logarithms may be 10 ($\log_{10} x$ or $\lg x$) or e ($\log_e x$ or $\ln x$)
- understand that an exponentially varying quantity is one whose rate of increase is proportional to the quantity itself
- understand that exponential increases and decreases occur naturally and in many man-made situations
- understand that the general form of an exponential change is given by the equation $X = X_0 e^{-kt}$
- 'take logarithms' of both sides of an equation such as $X = X_0 e^{-kt}$ to give $\ln X = -kt + \ln X_0$
- understand that ln is the symbol for the natural logarithm of a number.

10.89 Use a calculator to find, to three significant figures **(a)** lg 25 **(b)** lg 250 **(c)** lg 2500.

10.90 Use a calculator to find, to three significant figures **(a)** ln 25 **(b)** ln 250 **(c)** ln 2500.

10.91 Look at your answers to the last two questions. What pattern do you see in the answers?

10.92 What is, to three significant figures **(a)** lg 10 **(b)** lg 100 **(c)** lg 1000?

10.93 What is, to three significant figures **(a)** $\ln e$ **(b)** $\ln e^2$ **(c)** $\ln e^3$?

10.94 What is the logarithm to base 10, to three significant figures, of x if **(a)** $x = 10^2$ **(b)** $x = 10^3$ **(c)** $x = 10^{2.5}$?

10.95 What is the logarithm to base e, to three significant figures, of x if **(a)** $x = e^2$ **(b)** $x = e^3$ **(c)** $x = e^{2.5}$?

10.96 Calculate the values of $e^{-\lambda t}$ where **(a)** $\lambda = 0.675\ s^{-1}$ and $t = 4.0$ s **(b)** $\lambda = 2.45\ min^{-1}$ and $t = 50$ s **(c)** $\lambda = 6.54 \times 10^{-5}\ s^{-1}$ and $t = 2.0$ hours.

10.97 It is thought that the current I in a circuit component varies with the applied p.d. V according to the equation $I = kV^n$, where k and n are both unknown quantities, at least for some values of V. The following measurements of V and I are made.

V/V	0	25	50	75	100	125	150	175	200
I/mA	0	3.2	9.1	16.9	25.6	30.1	32.0	32.3	32.3

(a) Take logarithms to base e of both sides of the equation and plot $\ln(I/\text{mA})$ (on the y-axis) against $\ln(V/\text{V})$ (on the x-axis).
(b) Find the value of n from the straight part of the graph.
(c) Verify that $\log_e x = 2.30 \times \log_{10} x$, where x is any number. What difference would there have been in your graph, or the value for n, if you had taken logarithms to base 10?

10.98 The number of people in a certain group is 1000, and is increasing at a rate of 20 each year. If the increase is exponential, what is the rate of increase when the number of people in the group is **(a)** 2000 **(b)** 5000?

10.99 Why might you expect the rate of growth of the following to be exponential initially? In each case, why would the increase stop being exponential after a time?
(a) the number of bacteria in some food
(b) the number of people ill with flu
(c) the number of owners of Blu-ray players.

10.100 An exponentially decreasing quantity has the value 1000 at 12 noon. Its value at 1 p.m. is 800.
(a) What will be its value at **(i)** 2 p.m. **(ii)** 3 p.m.?
(b) What was its value at 11 a.m.?

10.101 A quantity X decreases according to the equation $X = X_0 e^{-at}$, where $X_0 = 400$ and $a = 0.500\,\text{s}^{-1}$.
(a) What is X when **(i)** $t = 2.00$ s **(ii)** $t = 3.00$ s **(iii)** $t = 4.00$ s?
(b) After what time will $X = 200$?
(c) After what time will X be 1% of its initial value?

10.102 A specimen of radioactive material contains 2.0×10^9 undecayed atoms at $t = 0$. This number N decreases exponentially. When $t = 10$ s, $N = 1.6 \times 10^9$.
(a) Use this information to draw a graph to show how N decreases with time t for $t = 0$ to $t = 40$ s.
(b) Call the initial number N_0. At what time does $N = \frac{1}{2} N_0$? What name is given to this time?
(c) At what time does $N = \frac{1}{4} N_0$?
(d) At what time does $N = \frac{1}{e} N_0$?

10.103 **(a)** Calculate the values of N, where $N = 10.0 e^{-0.200t}$ for $t = 0, 1.00, 2.00, 3.00, 4.00, 5.00$.
(b) Plot a graph of $\ln N$ against t.
(c) How would you describe the relationship between $\ln N$ and t?

10.104 The current in one kind of circuit containing a resistor and a capacitor varies with time t according to the equation $I = I_0 e^{-t/RC}$, where R and C are the resistance and the capacitance. The initial current $I_0 = 100$ mA.
(a) What is the unit of the quantity RC? (You do not need to know the units of R and C to be able to answer this.)

(b) Suppose $R = 150\,\text{k}\Omega$ and $C = 2.2\,\mu\text{F}$. Calculate RC.
(c) After what time will the current in the circuit have fallen to **(i)** $1/\text{e}$ **(ii)** $1/\text{e}^2$ **(iii)** $1/\text{e}^3$ of its initial value?
(d) Find the time taken for the current to become 50 mA.
(e) How long will it take for the current to become **(i)** 25 mA **(ii)** 12.5 mA?

10.105 The equation $N = N_0\text{e}^{-\lambda t}$ describes how the number N of undecayed nuclei varies with time t. N_0 is the initial number of undecayed nuclei and λ is the decay constant.
(a) What feature of the equation tells you that the number N decreases with time?
(b) What feature of the equation tells you how rapidly N decreases?
(c) In a particular case the following readings of N were taken for different values of t:

t/h	0	1.0	2.0	3.0	4.0
N	8532	5548	3601	2348	1523

(c) Take logarithms, to base e, of both sides of the equation and plot a graph of $\ln N$ against t.
(d) Use your graph to calculate the value of λ.
(e) The half-life $t_{1/2}$ is related to λ by the equation $\lambda t_{1/2} = \ln 2$. Calculate $t_{1/2}$ and use your graph to verify this result.
(f) Take *any* two times which are 1.6 h apart (e.g. $t = 1.0$ h and $t = 2.6$ h) and find the ratio of the final activity to the initial activity.

10.106 **(a)** Explain why if radioactive decay is a random process and each nucleus has the same chance of decaying in a certain time, the number of undecayed nuclei must decrease exponentially.
(b) If the number of undecayed nuclei decreases exponentially, why must the *rate* of decrease also decrease exponentially?

10.107 The activity of a radioactive source is $4200\,\text{s}^{-1}$ when it is first measured. The half-life for this source is 2.8 h. After 2.8 h its activity will be $2100\,\text{s}^{-1}$.
(a) What is its activity after two half-lives?
(b) Continue this process to find the number of half-lives which will need to elapse for the activity to fall to less than 1% of its initial value.
(c) Will this be true for all radioactive sources?
(d) Why would there be no point in trying to continue the process to find out, for example, how many half-lives would elapse for the activity to fall to 0.1% of its initial value?

10.108 A certain bacterial population number N grows exponentially in the right environment according to the relation $N = N_0\text{e}^{bt}$ where N_0 is the number of bacteria when $t = 0$. In an environment of tinned meat at room temperature the population doubles in 5.0 hours.
(a) What is the value of b for this population?
(b) If a tin of meat is polluted with a single live bacterium, what will the bacterial population become after one week?

10.109 The pressure in the atmosphere decreases with height above the Earth's surface according to the equation $p = p_0\text{e}^{-kh}$, where $k = g_0\rho_0/p_0$.
(a) Calculate k, if $g_0 = 9.81\,\text{m}\,\text{s}^{-2}$, $\rho_0 = 1.29\,\text{kg}\,\text{m}^{-3}$, $p = 101\,\text{kPa}$, giving the units, if any, of k.
(b) Why would you expect the unit of k to be what you have found?
(c) Calculate the value of p at heights of **(i)** 10 km **(ii)** 20 km.
(d) What do you notice about the sizes of p_0 and your answers to **(c)**?

10.110 Plot the exponential curve $z = z_0 e^{bt}$ for values of t at 2 s intervals from 0 up to 10 s, taking $b = 0.20\,s^{-1}$ and $z_0 = 2.0$.
Measure the gradient of this curve at a number of points by drawing tangents at these points; and plot another graph of the gradient against t. Test whether the value of b deduced from this graph agrees with that used above.

10.111 The excess pressure p in a leaky gas cylinder is measured at 9 a.m. every morning at times t (in days) with the following results:

t/d	0	1	2	3	4	5
p/MPa	6.0	3.8	2.4	1.5	0.96	0.63

(a) Plot a graph of $\ln(p/\text{MPa})$ against t.
(b) Do you consider that the measurements justify stating that the pressure excess in the cylinder is decaying exponentially?
(c) If so, the relationship is of the form $p = p_0 e^{-at}$, where a is the time constant of the decay process. Calculate the time constant.
(d) After what time would you expect the pressure to have fallen to half its initial value?

10.112 In a particular electrical circuit the p.d. V across a component increases with time t according to the equation $V = V_0(1 - e^{-t/RC})$.
(a) The quantity RC has the numerical value of 2.00. What is its unit?
(b) If V_0 is 5.00 V what is V after a time **(i)** $t = 1.00$ s **(ii)** $t = 2.00$ s?
(c) After what time is $V = 3.50$ V?

10.113 The number N of bacteria in a population increases with time t as shown in the table:

t/days	0	1	2	3	4
N	25	185	1365	10086	74524

If you wished to represent this variation on a graph, what difficulty would you meet, and how would you overcome it?

10.114 A logarithmic scale on one axis of a graph has these values, spaced at equal intervals: 1.00×10^7, 1.00×10^8, 1.00×10^9, 1.00×10^{10}. Make a copy of this scale for yourself. What is the value on the scale half-way between 1.00×10^8 and 1.00×10^9? Is it **(i)** 5.00×10^8 **(ii)** $1.00 \times 10^{8.5}$ **(iii)** 3.16×10^8? (More than one of these alternatives may be correct.) Explain how you made your choice.

10.115 Refer to the previous question: What is the value on the scale at a point which is between 1.00×10^7 and 1.00×10^8, and **(a)** halfway between 1.00×10^7 and 1.00×10^8 **(b)** one-tenth of the distance up from 1.00×10^7 to 1.00×10^8?

11 Thermal physics

Data: $g = 9.81\,\text{N kg}^{-1}$

$0\,°\text{C} \equiv 273\,\text{K}$ (to 3 s.f.)

1 kilowatt-hour $\equiv 3.6\,\text{MJ}$

density of water $= 1000\,\text{kg m}^{-3}$

specific heat capacity (s.h.c.) of water $= 4200\,\text{J kg}^{-1}\,\text{K}^{-1}$

specific latent heat of fusion of ice $= 0.334\,\text{MJ kg}^{-1}$

specific latent heat of vaporisation of water $= 2.26\,\text{MJ kg}^{-1}$

molar gas constant $R = 8.31\,\text{J mol}^{-1}\,\text{K}^{-1}$

Boltzmann constant $k = 1.38 \times 10^{-23}\,\text{J K}^{-1}$

Avogadro constant $N_A = 6.02 \times 10^{23}\,\text{mol}^{-1}$

11.1 Thermal energy

In this section you will need to

- understand what is meant by absolute zero
- understand that $0\,\text{K} = -273.16\,°\text{C}$ (though $0\,\text{K}$ may be taken as equal to $-273\,°\text{C}$ for most purposes)
- understand what is meant by internal energy
- understand the difference between heating and working
- understand the first law of thermodynamics
- use the equation $\Delta U = \Delta Q + \Delta W$ where ΔU is the change of internal energy of a body when energy ΔW is supplied to the body by working and energy ΔQ is supplied to the body by heating
- understand that the internal energy of a gas may be changed by either heating or working, and understand the difference between these methods
- understand the principle of a heat engine
- understand that the thermal efficiency η of a heat engine is defined by the equation

$$\eta = \frac{\text{work done}}{\text{energy supplied by heating}}$$

- understand that the maximum thermal efficiency of a heat engine is governed only by the temperatures between which the heat engine works and use the equation maximum thermal efficiency $\eta_{max} = (T_1 - T_2)/T_1$
- understand that the actual (or overall) efficiency of a heat engine is less than the thermal efficiency because of unintentional energy transfers to the surroundings

- understand that a heat pump is a heat engine working in reverse, and that a refrigerator is one example of this
- understand that the heat pump coefficient of performance η_r is given by $\eta_r = Q_c/W$.

11.1 **(a)** What are **(i)** the kelvin temperatures corresponding to 37 °C and −196 °C and the Celsius temperatures corresponding to 4 K and 234 K?
(b) Use the internet to find out whether your answers correspond to the melting or boiling points of any elements.

11.2 What is meant by absolute zero?

11.3 **(a)** Explain what is meant by internal energy.
(b) Does a block of ice have any internal energy?
(c) Could a lump of iron at 20 °C have more internal energy than another lump of iron at 80 °C? Explain your answer.

11.4 A hot copper block and a cold copper block are placed together in good thermal contact. They soon reach the same temperature. Explain whether the blocks necessarily then have the same internal energy.

11.5 What is meant by the internal energy of a body?
If a car is stopped by being braked, in what sense is the increase of energy of the molecules of the brake drums, tyres, road, etc., different from the original kinetic energy of the car?

11.6 Write down a statement of the first law of thermodynamics, using the symbols ΔU, ΔQ and ΔW, and explain what each symbol means.

11.7 Would you describe the following energy exchanges as heating or working? State in each case **(i)** the body losing the energy and the kind of energy lost, and **(ii)** the body gaining the energy and the kind of energy gained.
(a) A man sandpapers a block of wood: its temperature rises.
(b) A night storage heater cools down during the day.
(c) A tennis ball is dropped and after several bounces comes to rest.
(d) A girl pumps air into a bicycle tyre: the pump and air become hotter.

11.8 When a car's brakes are applied frictional forces do 0.20 MJ of work. Because they are hot they lose 0.080 MJ of energy to the surroundings. What are **(a)** ΔW **(b)** ΔQ **(c)** ΔU for this process?

11.9 A battery drives an electric current through a resistor.
(a) Initially the resistor is warming up. Are ΔU, ΔQ and ΔW then positive, negative or zero for the resistor?
(b) Are ΔU, ΔQ and ΔW positive, negative or zero when the resistor reaches a steady temperature?
(c) Describe the nature of the work done, i.e. explain what ΔW represents.

11.10 The diagram shows some gas contained in a cylinder fitted with a piston. Initially the piston is fixed in position, and the gas is heated. On a second occasion, with the gas in the same initial state, the gas undergoes the same amount of heating, but with the piston free to move.

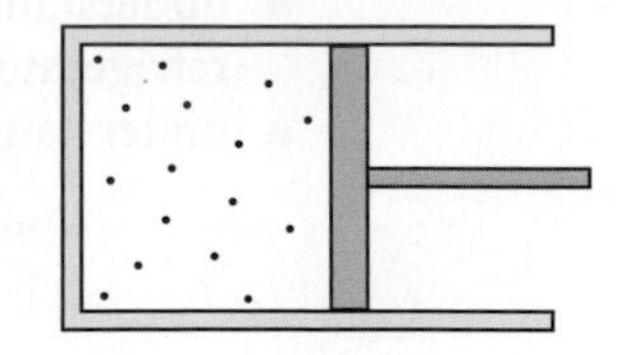

(a) Explain whether the gain of internal energy will be the same on this second occasion.

(b) Will the temperature rise be the same?

11.11 What is the maximum thermal efficiency of a heat engine working between temperatures of **(a)** 300 K and 550 K **(b)** 300 K and 650 K **(c)** 300 K and 750 K?

11.12 The diagram shows a schematic diagram for a heat engine. Energy Q_h is taken from the hot reservoir and energy Q_c is delivered to the cold reservoir (e.g. in a power station the hot reservoir is the furnace and the cold reservoir is the surroundings) and as a result work W is done by the heat engine.

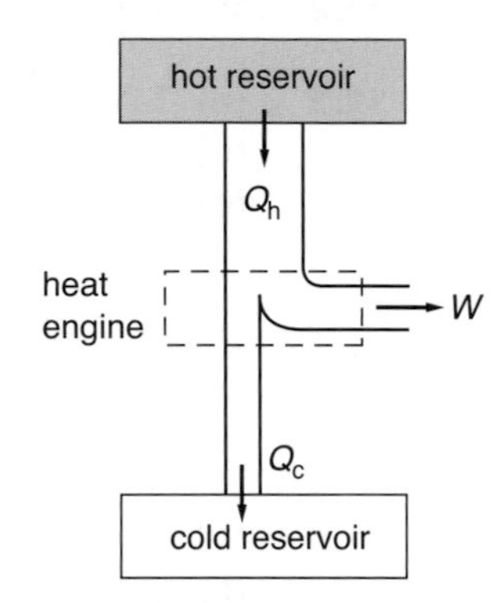

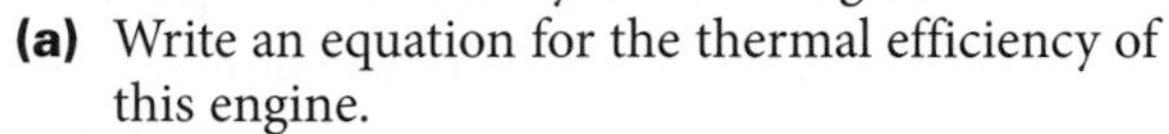

(a) Write an equation for the thermal efficiency of this engine.

(b) Calculate the thermal efficiency if Q_c = 2.4 GJ and Q_h = 3.9 GJ.

11.13 The schematic diagram shows the flow of energy through a heat pump such as a domestic refrigerator. During a typical day it requires 2.0 MJ of energy to run it, and during that time it pumps 3.0 MJ from the contents of the refrigerator to the kitchen.

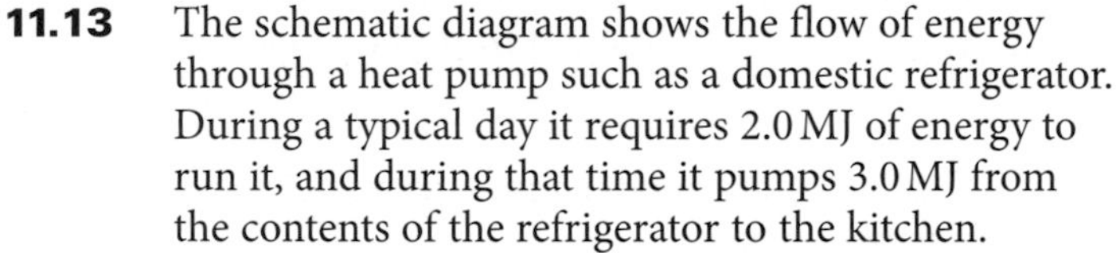

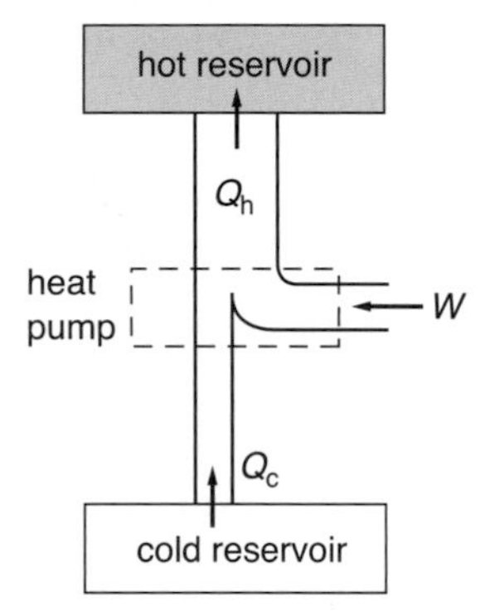

(a) For a refrigerator, what is the 'hot reservoir' and what is the 'cold reservoir'?

(b) How much energy is delivered to the kitchen during the day?

(c) What is the coefficient of performance for this refrigerator?

11.14* Your friend says 'You're telling me that you can use a power of 1.0 kW and deliver a power of 3.0 kW to a room? I don't believe it – that's something for nothing.' Explain to your friend how this can be done with a heat pump.

11.15* The coefficient of performance of a heat pump is defined by the equation η_r = (energy transferred from cold reservoir)/(work done) or $\eta_r = Q_c/W$ where W is the work which must be done to transfer energy Q_c from the cold reservoir to the hot reservoir.

(a) Show that it is possible to write $\eta_r = Q_c/(Q_h - Q_c)$ where Q_h is the energy delivered to the hot reservoir.

Heat pumps are in use to warm large buildings such as hospitals and shopping centres. Suppose that a hospital needs an average power of 120 kW for heating, and the cost of electrical energy is 20p per kW h.

(b) What is the cost per day of using conventional electrical heating?

(c) Suppose a heat pump with a coefficient of performance of 2.0 is used to deliver a power of 120 kW to the buildings. What power must be supplied to the heat pump?

(d) Where does the rest of the energy come from?
(e) What is the cost saving per day?

11.16 A thermocouple consists of two junctions between two different metals or alloys, as shown in the diagram. A difference of temperature between the junctions can produce an e.m.f. and drive a current round a circuit. This is a heat engine.

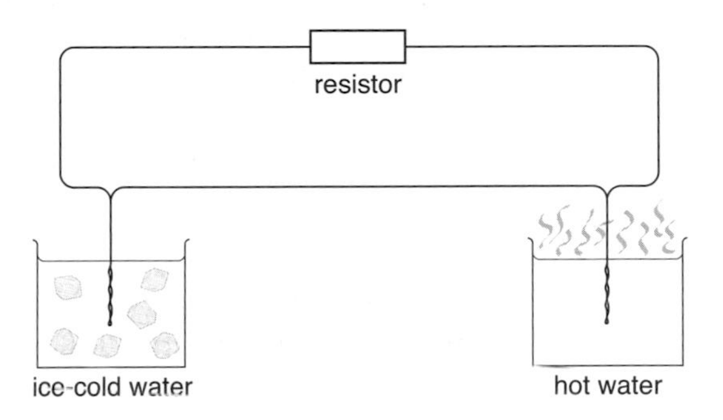

(a) What are the hot and cold reservoirs?
(b) Where is the work done?

11.2 Heating solids and liquids

In this section you will need to

- use the equation $\Delta Q = cm(\Delta\theta)$, which defines specific heat capacity (s.h.c.) c
- understand that the rate of loss of energy from a body depends on the temperature difference between itself and the surroundings, and on its surface area and the nature of the surface
- understand that a heated body must eventually reach a steady equilibrium temperature when it is losing exactly as much energy as it is gaining
- use the equation $\Delta Q = cm(\Delta\theta)$ in the form: rate of heating = cm(rate of change of temperature)
- understand the significance of water having a particularly high s.h.c.
- use the equation $\Delta Q = m\ell$, which defines specific latent heat ℓ
- explain in molecular terms why energy is needed to melt a solid or evaporate a liquid
- understand the importance of evaporation in regulating human body temperature.

11.17 In a steel-making furnace 5.0 tonnes of iron have to be raised from a temperature of 20 °C to the melting point of iron (1537 °C). Find how much energy (in GJ) is needed to do this. [s.h.c. of iron = 420 J kg^{-1} K^{-1}.]

11.18 Make a rough calculation of the cost of using electrical energy to heat water for a bath, if about 0.3 m^3 of water have to be heated from 5 °C to 35 °C. Assume that the cost of 3.6 MJ of electrical energy is about 20p. [Use data.]

11.19 The bit of a soldering iron is made of copper and has a mass of 3.3 g. If the power of its electrical heater is 45 W, how long will it take to raise its temperature from 15 °C to 370 °C, assuming that there are no energy losses to the surroundings? [s.h.c. of copper = 385 J kg^{-1} K^{-1}.]

11.20 A washing machine has a heater of power 2.5 kW. Six litres (6.0 kg) of water enters at 12 °C and has to be heated to 60 °C for one of the cycles. Assuming all the energy is used to heat the water, calculate how long this heating will take. [Use data.]

11.21 An electric shower has a maximum power of 7.0 kW.

(a) If water enters at 10 °C and leaves it at a temperature of 41 °C, what is the maximum rate of flow of water? [Use data.]

(b) If you were content with an output temperature of 37 °C, what rate of flow could you have?

11.22* It is sometimes said that the cost of running an upright freezer is greater than the cost of running a chest freezer because each time the door of an upright freezer is opened cold air falls out and warm air from the room enters and has to be cooled down. Consider an upright freezer of capacity 0.20 m^3 and discuss whether there is any truth in this statement. The temperature inside a freezer may be assumed to be about −18 °C, and the s.h.c. of air (under these conditions) is 600 J kg^{-1} K^{-1}. The density of air is 1.3 kg m^{-3}.

11.23 A small electrical heater was used to heat 65 g of water in a thin plastic cup. The graph shows the temperature θ of the water against time t.

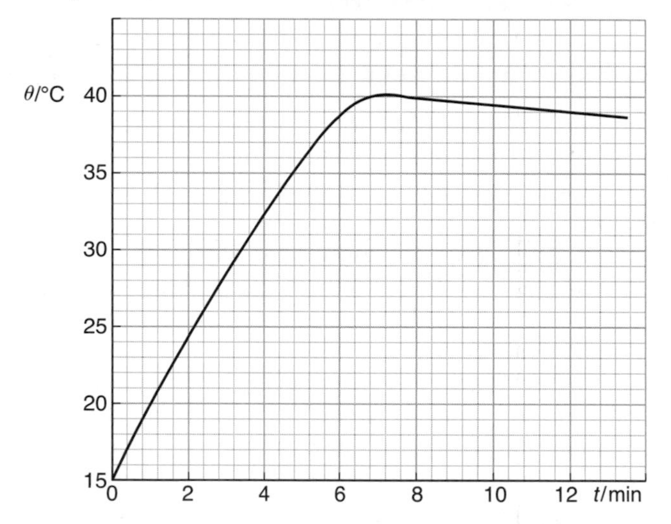

(a) The heater is switched off after 6 minutes. Explain why the temperature of the water continues to rise after the heater is switched off.

(b) Estimate the initial rate of rise of temperature of the water. Hence calculate the power of the heater. [Use data.]

(c) What is the rate of loss of heat energy from the water between 8 and 12 minutes?

11.24 Energy is transferred to a block of lead by repeatedly striking it with a hammer. The head of the hammer has a mass of 0.31 kg and it strikes the block with a speed of 15 m s^{-1}. The lead block has a mass of 1.2 kg and is struck 10 times.

(a) What is the kinetic energy of the head of the hammer as it hits the lead block?

(b) Estimate the temperature rise of the block and discuss whether in reality its rise in temperature would be greater or smaller.
[The s.h.c. of lead is 130 J kg^{-1} K^{-1}.]

11.25 How long would it take a 2.0 kW heater to warm the air in a room which measures 4 m by 3 m by 2.5 m from 5 °C to 20 °C?
[The s.h.c. of air under these conditions is about 1000 J kg^{-1} K^{-1}. The density of air is 1.3 kg m^{-3}.]
Give at least two reasons why in practice it takes much longer (perhaps an hour) for the heater to warm such a room.

11.26* A gas-fired power station needs to get rid of energy at a rate of 900 MW and does so by warming up water in a river that flows past it. The river is 30 m wide, 3.0 m deep and flows at an average speed of $1.2\,\mathrm{m\,s^{-1}}$.

(a) How much water flows past the power station every second? [Use data.]

(b) How much warmer is the river downstream of the power station? [Use data.]

(c) Is your answer to **(b)** one that is acceptable environmentally?

11.27 A squash ball of mass 46 g is struck so that it hits a wall at a speed of $40\,\mathrm{m\,s^{-1}}$; it rebounds with a speed of $25\,\mathrm{m\,s^{-1}}$.

(a) What is its rise in temperature? [S.h.c. of rubber = $1600\,\mathrm{J\,kg^{-1}\,K^{-1}}$.]

(b) Why is it unnecessary to know its mass?

(c) What will happen to its temperature if the players continue to hit it against the wall?

11.28* Water has a very high s.h.c. Discuss what effect this has on the climate of different countries at the same latitude in Western and Eastern Europe.

11.29 Some tiny spheres of lead (lead shot) of total mass 380 g are trapped in a long plastic tube of length 80 cm. At one end there is a thermocouple thermometer connected by long leads to a data logger.

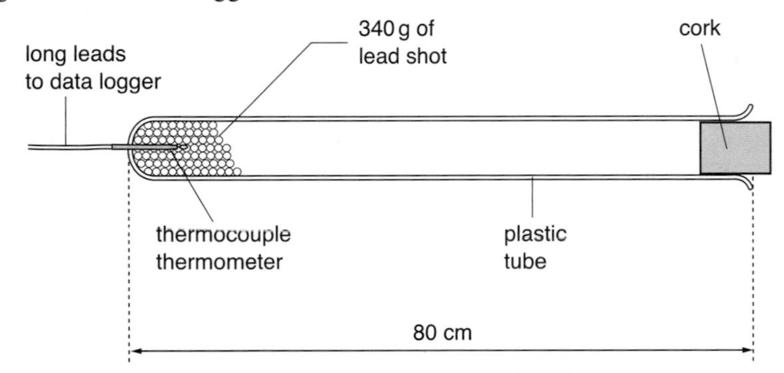

The plastic tube is held vertically and then rapidly inverted 50 times, allowing all the lead shot to fall to the bottom after each inversion. The data logger registers a rise in temperature of 2.4 K. *Estimate* the s.h.c. of lead.

11.30* A storage heater consisting of a set of heavy blocks is made of a material with a density of $3.2 \times 10^3\,\mathrm{kg\,m^{-3}}$ and s.h.c. $820\,\mathrm{J\,kg^{-1}\,K^{-1}}$. The blocks are warmed by a 2.0 kW electrical heater for 7 hours during the night when electrical energy is relatively cheap.

(a) Assuming that an overnight temperature rise of 80 K is needed, show that the volume of the blocks can be calculated to be $0.24\,\mathrm{m^3}$, and explain why this result is only an estimate.

(b) Suggest values for the dimensions of the blocks to produce a volume of $0.2\,\mathrm{m^3}$.

(c) Sketch a temperature–time graph for the blocks over a 24-hour period, and discuss any disadvantages of this form of space heating.

11.31* An immersion heater is fitted to a hot water tank in a house. In the summer months hot water is needed only for washing. Dad says it is better to keep the heater running all the time because if the water is allowed to cool down there will be extra energy needed to warm it up again. Clare says it is best to have the heater switched on only when it is needed. Explain whether you think Dad or Clare is right.

11.32 How much energy must
(a) be given to 2.0 litres of water at 100 °C to evaporate it
(b) be taken from 0.50 kg of water at 0 °C to freeze it?
[Use data.]

11.33 How long will it take
(a) a 1000 W heater to evaporate 1.0 kg of water that is already at 100 °C, its normal boiling point
(b) a refrigerator to freeze 0.40 kg of water that is already at 0 °C, its normal freezing point, if it can remove energy at an average rate of 75 W?
[Use data.]

11.34 Using the values of specific heat capacity and specific latent heat given at the start of this chapter, draw a graph, with labelled axes, to show how the temperature varies with time when a block of ice, of mass 2.0 kg, is placed in a sealed container and has an immersion heater of power 200 W placed in it. The ice is initially at −10 °C; continue the graph until the temperature of the water vapour is 120 °C.
[S.h.c. of water vapour under these conditions = 1400 J kg^{-1} K^{-1}.]

11.35 The mass of liquid nitrogen in an open beaker is found to have decreased by 46.3 g in 10 minutes. If the specific latent heat of vaporisation of nitrogen at its boiling point is 1.99×10^5 J kg^{-1}, at what rate were the surroundings heating the beaker?

11.36 A coffee machine in a café passes steam at 100 °C into 0.18 kg of cold coffee (s.h.c. the same as that of water) to warm it. If the initial temperature of the coffee is 14 °C, what mass of steam must be supplied to raise the temperature of the coffee to 85 °C?

11.37 The graph shows how the temperature θ varies with time t for 0.25 kg of a substance cooling in surroundings that were at a steady temperature.

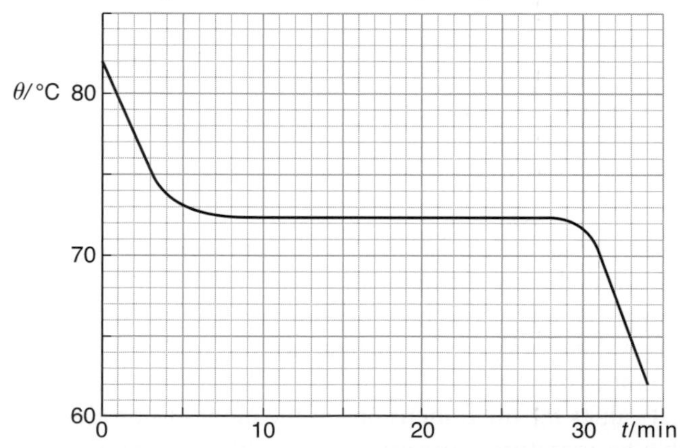

(a) Explain why the temperature remained constant at 72.3 °C for about 23 minutes.
(b) If the s.h.c. of the substance in the liquid phase is 950 J kg^{-1} K^{-1}, estimate its s.h.c. in the solid phase.
The rate of loss of heat of the substance at its freezing point is about 10 W.
(c) Explain why this value is consistent with the data on the graph.
(d) Estimate the specific latent heat of fusion of this substance.

11.38 A thermocouple probe connected to a multimeter gave a reading of 20.5 °C for room temperature on a particular day. Explain the following observations:

(a) When the probe was placed in some ethanol in a watch glass the recorded temperature fell to 14.8 °C.
(b) When the probe was removed from the ethanol and placed in the air again the temperature fell further to 6.8 °C, but after a few seconds the temperature began to rise again, eventually reaching 20.5 °C.

11.39* Explain why snow and ice lie on the ground for some days after the air temperature has risen above 0 °C.

11.40 A female marathon runner of mass 60 kg generates internal energy at the beginning of a marathon at the rate of 800 W. Assuming that she loses no energy, and that the average s.h.c. of her body is the same as that of water, her temperature will rise at 0.19 K min^{-1}.

(a) If she loses energy through conduction, convection and radiation at a rate of 300 W, at what rate will her temperature rise?
(b) Ideally her temperature during the race should remain constant. Evaporation (from skin and through exhaled air from her lungs) is an additional mechanism by which she can lose energy. At what rate, in g min^{-1}, must she evaporate water in order to keep her temperature constant? [Use data.]
(c) When she stops running she generates internal energy at a rate of 100 W, but continues to lose energy at 800 W. Explain this photograph of such a runner just after the end of a marathon.

11.41* In 2010 one electricity company charged 20p during the day for each kW h of energy; in the same area a gas company charged 4.0p.
(a) Assuming that the use of electrical energy for heating homes is 100% efficient, but the use of gas for heating is only 75% efficient, how many times more expensive is it to use electrical energy rather than gas for heating water?
(b) What is the underlying cause of this difference?

11.42* Discuss the effects on everyday life if, over a few thousand years, the latent heat of vaporisation of water gradually fell to one-tenth of its present value.

11.3 The ideal gas

In this section you will need to

- understand that the temperature used in gas laws is the kelvin temperature, and that 0 °C = 273 K
- remember Boyle's law (pV = constant, if the temperature and amount of substance are constant)
- understand that amount of substance n is measured in moles
- understand the similarity of the equations $pV = NkT$ and $pV = nRT$ and understand that R is the molar gas constant and k (the Boltzmann constant) can be thought of as the molecular gas constant
- remember that the number of particles in a mole is called the Avogadro constant N_A so that the number of molecules N is given by $N = nN_A$

- interpret $pV \propto nT$ in molecular terms
- remember the assumptions made in the kinetic theory of gases
- understand what is meant by the root mean square (r.m.s.) speed of a collection of molecules
- understand that the average kinetic energy E of a molecule of a gas is given by $E = \frac{3}{2}kT$, where k is the Boltzmann constant (i.e. it is proportional to the kelvin temperature of the gas)
- understand that the internal energy U of an ideal gas is given by $U = N(\frac{3}{2}kT)$
- remember that the work done ΔW by a gas at constant pressure is given by $\Delta W = p(\Delta V)$ where p is the pressure and ΔV the change of volume.

11.43 Express the following volumes in m^3: **(a)** 1.7 litres **(b)** 6.5 cm^3 **(c)** 3.3 mm^3.

11.44 Some gas occupies a volume of $6.0 \times 10^{-3}\,m^3$ and exerts a pressure of 80 kPa at a temperature of 20 °C. What pressure does it exert if, separately

(a) the temperature is raised to 40 °C
(b) the volume is halved
(c) the temperature is raised to 586 K
(d) the volume becomes $2.5 \times 10^{-3}\,m^3$
(e) the volume becomes $12 \times 10^{-3}\,m^3$ and the temperature becomes 57 °C?

11.45* In an experiment the pressure of a gas and its volume were measured at constant temperature and the following readings were obtained:

p/kPa	102	143	178	200	233
V/cm^3	40.5	28.7	23.4	20.7	17.8

(a) Calculate values of the product pV to show that the gas is behaving like an ideal gas.
(b) Sketch a graph of p against $1/V$ to confirm that pV is constant.

11.46 The diagram shows a graph of p against V, and a graph of p against $1/V$, for a fixed mass of gas kept at constant temperature. Copy the graphs and on each sketch a graph for the following separate changes: **(a)** a lower temperature and **(b)** an increased mass of gas. Label your graphs to make it clear which is which.

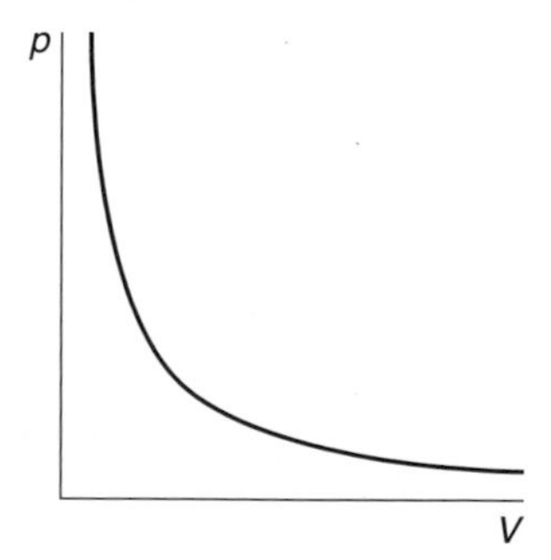

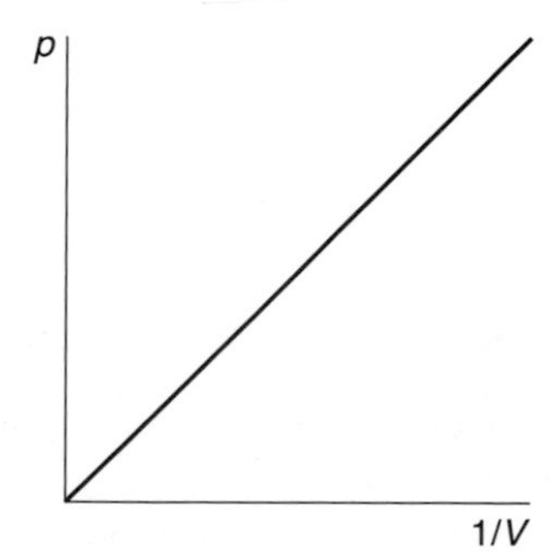

11.47 A cylinder of volume 0.20 m^3 contains gas at a pressure of 200 kPa and a temperature of 17 °C. Use data to calculate **(a)** how many molecules of gas there are in the cylinder and **(b)** how many moles of gas this represents.

11.48 Show that the unit of each of the quantities pV, nRT and NkT is the joule (J).

11.49 At the start of a journey the pressure of the air in a car tyre is 276 kPa and the temperature is 12.0 °C. After being driven the pressure is 303 kPa. Assuming that the volume of the air remains constant, what is its temperature now?

11.50 On a day when atmospheric pressure is 105 kPa, air is pushed into a new vehicle tyre until the pressure of the air in it is 360 kPa. If the volume of the inside of the tyre is 0.150 m^3, what volume of air at atmospheric pressure is pushed in, assuming that the volume of the tyre, and the temperature of the air, remain constant?

11.51* A student put half a mole of helium into a flask of volume 0.016 m^3. He measured the pressure of the trapped gas over a range of temperatures and plotted the graph shown.

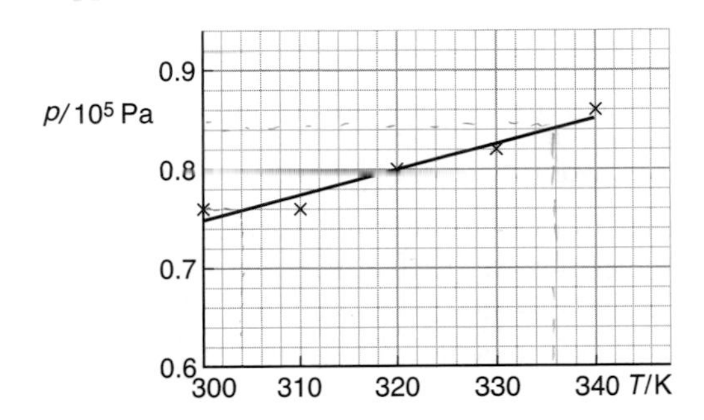

(a) The student used his graph to find a value for the molar gas constant *R*. What value did he deduce?

(b) Suggest why he did not simply take a single set of readings to calculate a value for *R*.

11.52 The volume of one cylinder in a diesel engine is 360 cm^3 and the cylinder contains a mixture of fuel vapour and air at a temperature of 320 K and a pressure of 101 kPa. The volume of the mixture is then reduced to 20 cm^3 and at the same time the temperature rises to 1000 K.

(a) Calculate the new pressure in the cylinder.

(b) What assumption have you made?

11.53 The diagram shows two graphs of p against $1/V$ for a gas.

(a) What is the gradient of graph B?

(b) If the temperature of the gas was 290 K, what was the amount of gas?

Graph A is for the same amount of gas at a different temperature.

(c) What was the new temperature?

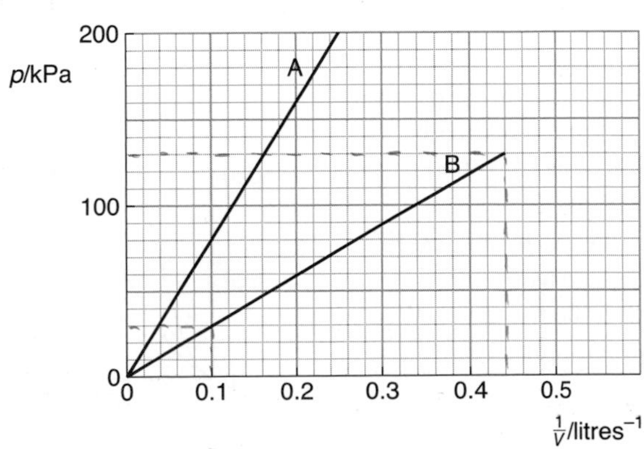

11.54 The graph shows the variation of pressure with volume for a gas at two temperatures 300 K and 600 K. Use information from the graph to calculate
(a) the number of moles of gas present
(b) the pressure of the gas at a
(c) the volume of the gas at b.
[Use data.]

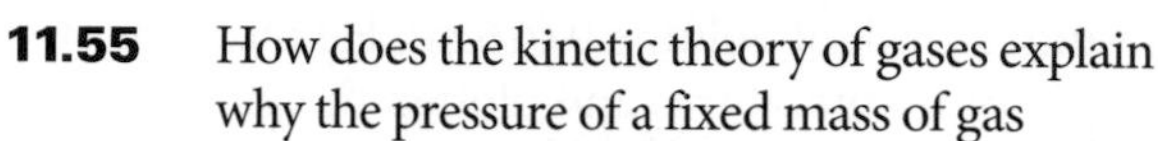

11.55 How does the kinetic theory of gases explain why the pressure of a fixed mass of gas
(a) at constant volume increases as its temperature rises
(b) at constant temperature is inversely proportional to its volume?

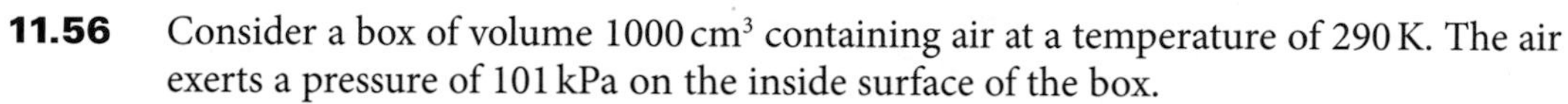

11.56 Consider a box of volume 1000 cm^3 containing air at a temperature of 290 K. The air exerts a pressure of 101 kPa on the inside surface of the box.
(a) How many molecules are there in the box? [Use data.]
(b) What is the space 'available' to each molecule?
(c) Taking a typical air molecule to be a sphere of radius 1.5×10^{-10} m, show that its volume is 1.4×10^{-29} m^3.
(d) How many times larger than this is its available space?

11.57 The density of argon gas is 1.61 kg m^{-3} at a pressure of 100 kPa.
(a) What is the r.m.s. speed of argon molecules under these conditions?
(b) What would be the r.m.s. speed if the pressure were halved, the temperature remaining the same?

11.58 The measured speeds of 10 vehicles on a motorway are, in m s^{-1}, 42, 32, 28, 40, 33, 32, 35, 34, 32, 25. Calculate **(a)** their mean speed **(b)** their r.m.s. speed.

11.59 What is the temperature of a gas if its molecules have an average k.e. of **(a)** 6.21×10^{-21} J **(b)** double that amount? [Use data.]

11.60 The graph shows the distribution of molecular speeds v for 1 000 000 oxygen molecules at two temperatures 300 K and 600 K.
(a) Explain which graph corresponds to 300 K and which to 600 K.
(b) Estimate the number of oxygen molecules that have speeds within ±0.5 m s^{-1} of 750 m s^{-1} at **(i)** 300 K **(ii)** 600 K.
(c) Is the area beneath the 300 K curve the same as the area beneath the 600 K curve? What does this area represent?

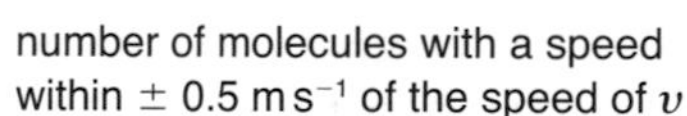

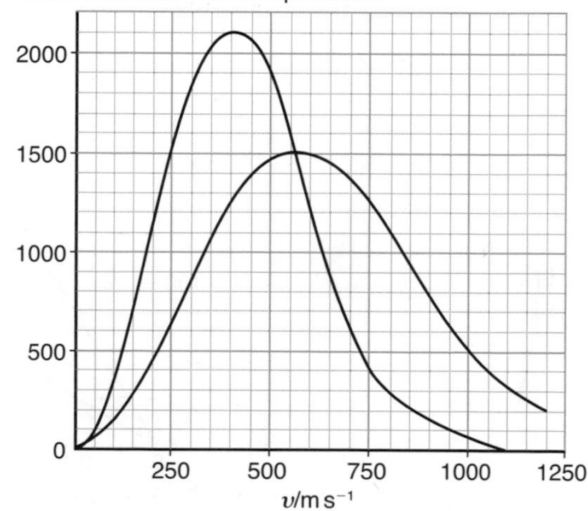

12 Linear momentum

Data: $g = 9.81\,\text{N}\,\text{kg}^{-1}$

12.1 Conservation of linear momentum

In this section you will need to

- remember that linear momentum p, defined as mass times velocity, is a vector quantity
- understand that linear momentum is conserved in collisions and explosions provided no external forces act during the interaction
- understand that kinetic energy is conserved only in elastic collisions
- draw sketches when applying the principle of conservation of linear momentum which show what is happening immediately before and after an interaction
- describe experiments to demonstrate the validity of the principle of conservation of linear momentum.

It is helpful to draw sketches showing 'before and after' when solving many of these problems.

12.1 What is the momentum of
(a) a boy of mass 50 kg running round a track at a constant speed of $3.0\,\text{m}\,\text{s}^{-1}$ at the moment when he is **(i)** moving to the north **(ii)** moving to the south
(b) a car of mass 800 kg moving east at a speed of $25\,\text{m}\,\text{s}^{-1}$
(c) an oil tanker of mass 250 000 tonnes which is moving west at a speed of $20\,\text{m}\,\text{s}^{-1}$?

12.2 Explain why, when a stationary object explodes into two pieces,
(a) the pieces move off in opposite directions
(b) the piece with the smaller mass moves off with the greater speed.

12.3 The unit for momentum, $\text{kg}\,\text{m}\,\text{s}^{-1}$, can also be written as N s, i.e. newton times second. Explain why these units are equivalent.

12.4 Calculate the unknown velocity v in each of the following collisions.

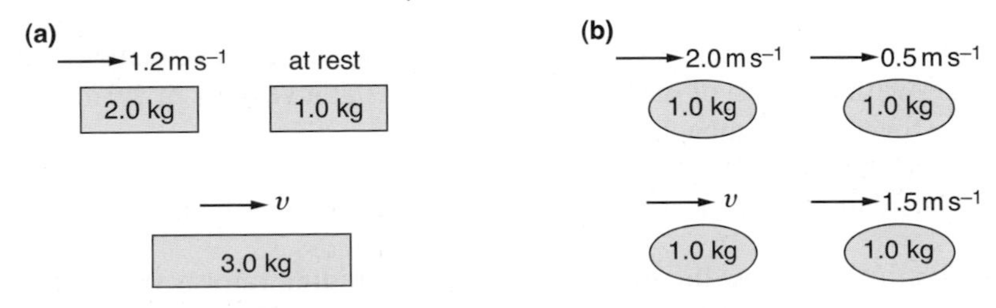

12.5 A woman of mass 60 kg steps out of a canoe of mass 40 kg onto the river bank.
(a) If she steps out at a speed of $2.0\,\text{m}\,\text{s}^{-1}$, what happens to the canoe?
(b) Can she step out without the canoe recoiling?

12.6 Two air-track gliders are at rest with a spring compressed between them. A thread tied to each prevents them moving apart. When the thread is burned, one glider moves away with a speed of 0.32 m s^{-1}, and the other moves in the opposite direction with a speed of 0.45 m s^{-1}.

(a) If the first glider has a mass of 0.40 kg, what is the mass of the other?

(b) Explain how you would measure the speeds of the gliders as they move apart.

12.7 In the diagram two trolleys are at rest and are then 'exploded' by the release of a spring-loaded piston or plunger at the front of one of them. Each trolley has a mass of 1.0 kg, but one carries a block of unknown mass m. The result of the explosion is also shown.

(a) Show that m = 2.0 kg.

(b) How much energy does the spring release in this explosion?

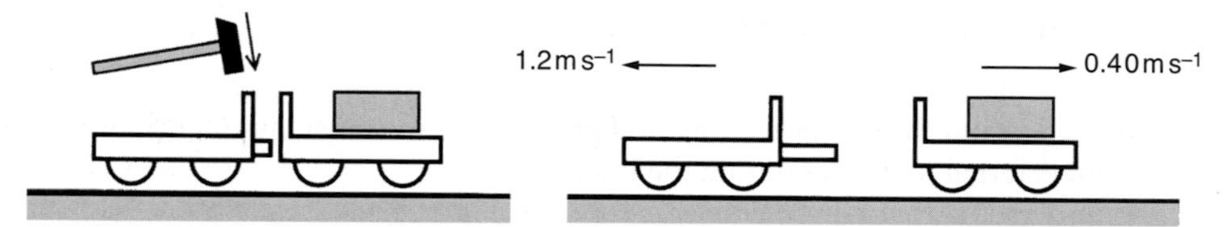

12.8 A rugby player of mass 70 kg, running south at 6.0 m s^{-1}, tackles another player whose mass is 85 kg and who is running directly towards him at a speed of 4.0 m s^{-1}. If in the tackle they cling together, what will their common velocity be immediately after the tackle?

12.9 A supermarket trolley loaded with 9.5 kg of goods was rolled at 2.0 m s^{-1} towards a stationary stack of two identical empty trolleys. The linked trolleys move off initially at 1.0 m s^{-1}. Calculate the mass of a single supermarket trolley.

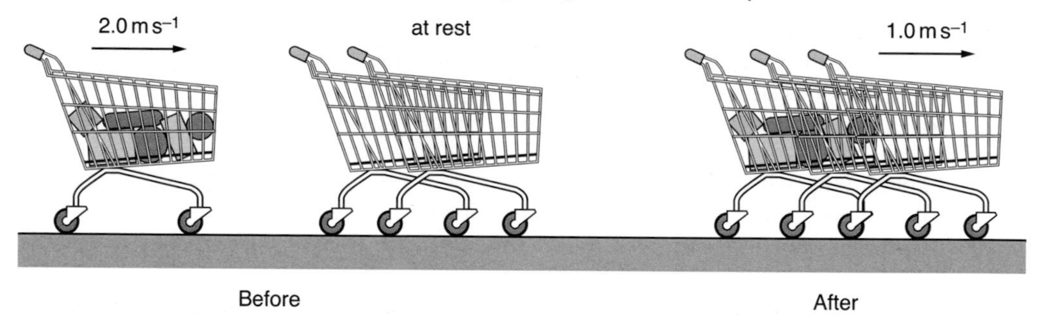

12.10* When a bullet hits a 'baddie' in the chest in a TV western, he sometimes slumps forward. As a physics student you know he ought to jerk backwards. Explain why.

12.11 A railway wagon of mass 15 tonnes (15 000 kg) which has a velocity of 5.0 m s^{-1} north collides with a wagon of mass 10 tonnes which has a velocity of 2.0 m s^{-1} south. They couple together on impact.

(a) Find their velocity after the collision.

(b) Calculate the k.e. transformed to other forms of energy in the collision.

12.12 The graph shows how the momentum of two colliding railway trucks varies with time. Truck A has a mass of 20 tonnes (20 000 kg) and truck B has a mass of 30 tonnes.

(a) What is the *change* of momentum of **(i)** truck A **(ii)** truck B?

(b) Calculate **(i)** the initial velocities **(ii)** the final velocities of the two trucks.

(c) Calculate the *total* momentum of the two trucks at t/s = 0.4, 0.8, 1.2 and comment on your answers.

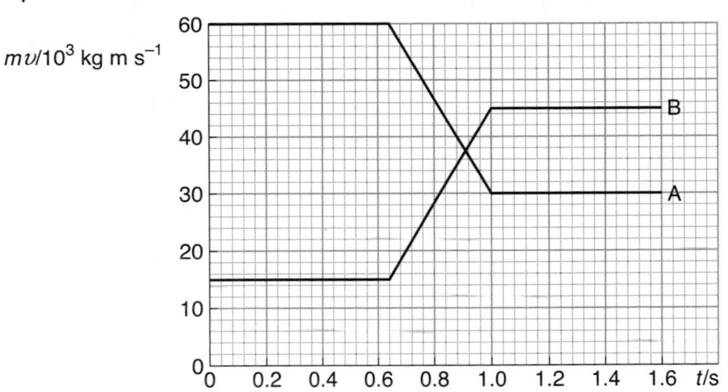

12.13 **(a)** Body A of mass 3.0 kg has a velocity of +4.0 m s^{-1} and collides head-on with a body B, which has a mass of 2.0 kg and a velocity of −2.0 m s^{-1}. After the collision the velocity of B is found to be +3.0 m s^{-1}. Find the velocity of A.

(b) Sketch a graph to show how the momentum of the two bodies varies with time before and after the collision.

12.14 A pair of skaters such as those shown in question 2.5 are together moving across the ice at a speed of 6.0 m s^{-1}. The man has a mass of 80 kg and the woman a mass of 60 kg. They push each other apart along their line of motion so that after they separate the man is moving in the same direction at 4.0 m s^{-1}.

(a) What is the woman skater's new velocity?

(b) How much energy is transferred to k.e. as the skaters push each other apart?

12.15* The diagram shows the result of a helium ion of mass $4m$ bouncing back from a head-on collision with a stationary oxygen nucleus of mass $16m$. Their velocities are shown above the particles.

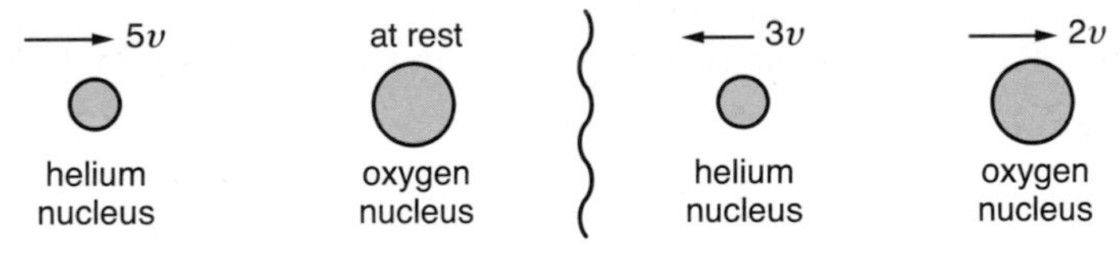

(a) Show that both linear momentum and kinetic energy are conserved in this collision.

(b) If T_0 and T are the initial and final kinetic energies of the helium ion, what is the ratio T/T_0?

(c) In general, in a head-on elastic collision between a stationary ion of mass m_s and a rebounding ion of mass m_r, the k.e. T of the rebounding ion is a fraction k of its initial k.e. T_0 given by

$$k = (m_s - m_r)^2 \div (m_s + m_r)^2$$

Use this equation to calculate k for the helium ion striking an oxygen nucleus.

(d) What would be the value of k for a helium ion striking a carbon nucleus? (The mass of the carbon nucleus is $12m$.)

(e) For a helium ion striking a surface with nuclei of mass xm, k is found to be 0.751. What is the value of x, and what element might the surface consist of?

(f) Use a search engine to find out about Rutherford back-scattering.

12.16 The nucleus of lanthanum-140 decays by α-particle emission to caesium-136. The speed of the α-particle is $1.40 \times 10^7\,m\,s^{-1}$.

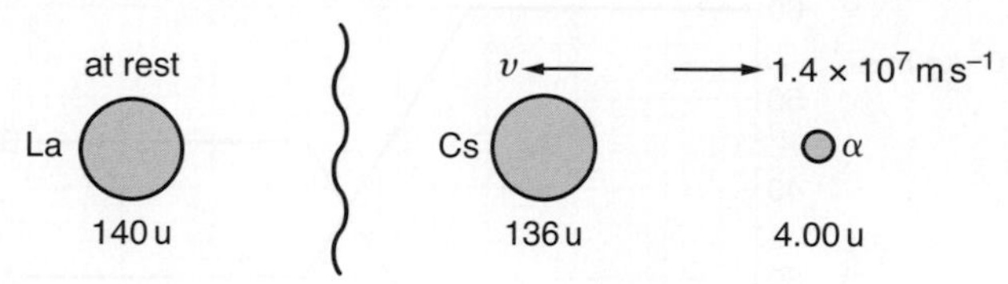

The diagram, in which all the masses are expressed in unified atomic mass units u, illustrates the decay.

(a) Calculate the recoil speed v of the caesium nucleus.

(b) What percentage of the total kinetic energy released in this decay is carried by the caesium nucleus?

12.17 A Frisbee of mass 114 g is caught in a tree. To dislodge it you toss a 230-g lump of clay vertically upwards. The clay and the Frisbee stick together and rise to a maximum height of 1.5 m above the Frisbee's initial position. Show that the speed of the clay as it hits the Frisbee was $8.1\,m\,s^{-1}$.

12.18 Discuss how momentum is conserved when

(a) a train accelerates from rest

(b) a ball falls to the ground and bounces up again.

12.19 A bullet of mass 16 g is fired into a block of wood of mass 4.0 kg which is supported by vertical threads. After the bullet has become embedded in the block, the block swings and rises (vertically) 50 mm.

(a) Draw an energy flow or Sankey diagram for this event.

(b) Calculate the speed of the block and bullet immediately after the collision.

(c) What was the speed of the bullet before the collision?

12.20 Two identical pendulum bobs are suspended from strings of equal lengths, and bob P is released from a height h as shown in the diagram.
When bob P hits bob Q, the two stick together. Calculate how high the combination rises.

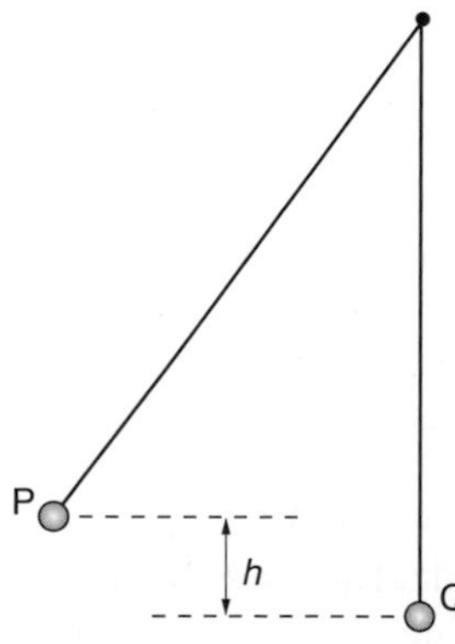

12.21* A Newton's cradle consists of five steel balls suspended by threads so that the balls lie in a horizontal line, almost touching one another. When one ball is pulled back and released so that it strikes head-on the row of the remaining four balls, the ball stops on impact, and just one ball at the far end moves off with the velocity which the first ball had, the others now being stationary. Clearly both linear momentum and energy are conserved. Suppose now that two balls are pulled back together and then released. Show that it is not possible for one ball to move off from the other end of the row without contradicting one of the conservation principles.

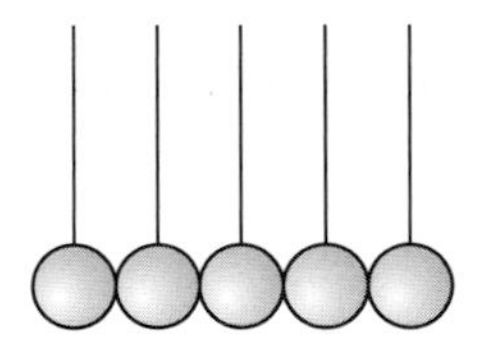

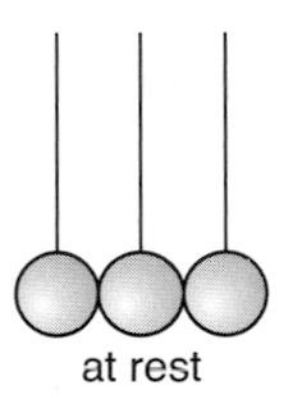

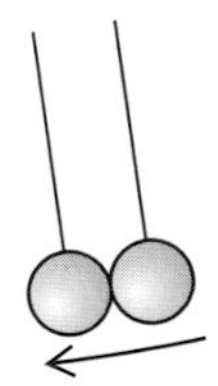

12.22* The white snooker ball, of mass m, is cued to strike a red ball at rest, also of mass m. The diagrams show **(i)** the momentum vector mu_w before the collision and **(ii)** and **(iii)** the two momentum vectors mv_w and mv_r after two separate collisions.

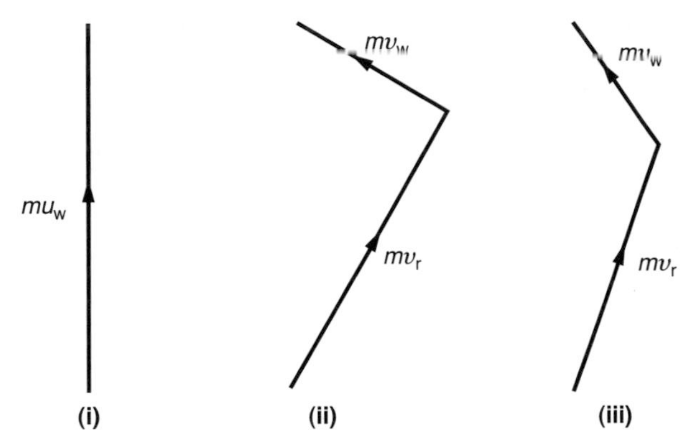

(a) Explain what measurements you need to make on the diagrams to show whether momentum and kinetic energy are conserved.

(b) Make the measurements. There is of course some uncertainty in your measurements: within those limits are momentum and k.e. conserved in these two collisions?

(c) Explain why the angle between mv_w and mv_r after the collision must be 90° if kinetic energy is conserved.

12.23 A particle of mass $2m$, travelling along a straight line with velocity v, explodes into two parts each of mass m. The explosion causes the k.e. of the system to be doubled. Assume that the explosive forces act along the initial line of travel.

(a) Show that velocities of $2v$ and zero for the two parts are consistent with the principles of conservation of momentum and of energy.

(b) *Prove* that the resulting velocities of the two parts are $2v$ and zero. [Difficult.]

12.2 Force and rate of change of momentum

In this section you will need to

- remember that the impulse of a force is calculated as the average force times the time for which the force acts, i.e. impulse = $F\Delta t$
- use the impulse–momentum equation $F\Delta t = \Delta(mv)$
- understand that Newton's third law applies at every instant of an interaction
- use Newton's second law in the form: rate of change of momentum of a body = resultant force acting on the body (i.e. $\Delta p/\Delta t = F_{res}$)

- use the principles of conservation of momentum and mechanical energy to analyse collisions and explosions
- understand that in an elastic collision no kinetic energy is transferred to internal energy
- understand that a totally inelastic collision is one in which the colliding bodies stick together.

12.24 What is the impulse of the forces in the following situations:

(a) a man pulling a garden roller to the east with a horizontal force of 300 N for 10 s

(b) a rock of weight 20 N moving vertically downwards for 5.0 s

(c) a hammer hitting a nail with a vertical force of 800 N for 0.60 ms?

12.25 A tennis ball of mass 58 g is moving horizontally, at right angles to the net, with a speed of 25 m s^{-1} (about 60 m.p.h.). A player hits it straight back so that it leaves his racket with a speed of 30 m s^{-1} (about 67 m.p.h.). What is

(a) the size of the change of momentum of the ball

(b) the impulse of the force that the racket exerts on the ball?

12.26 The graph shows how a varying force acts on a body.

(a) Calculate the impulse of the force for the first 2.0 s.

(b) Calculate the impulse of the force from 2.0 s to 6.0 s.

(c) Estimate the impulse of the force from 6.0 s to 9.0 s.

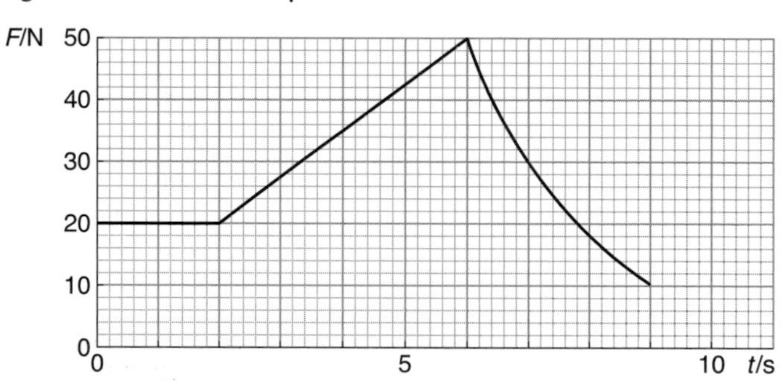

12.27 A stationary snooker ball of mass 0.21 kg is struck by a cue which exerts an average horizontal force of 70 N on it. The cue is in contact with the ball for 8.0 ms. Calculate the speed of the ball after the impact.

12.28 The graph describes the momentum of a tennis ball of mass 58 g as it is struck during a rally. Estimate the average force exerted by the tennis racket on the ball.

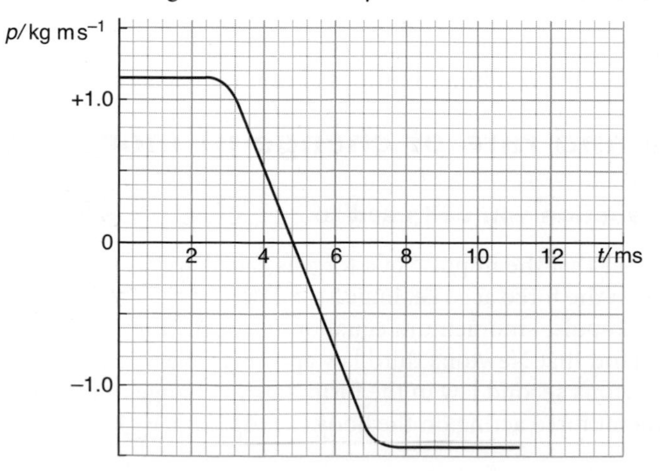

12.29 **(a)** Calculate the average force exerted by a golf club on a golf ball of mass 46 g, if the ball leaves the club at a speed of $80\,\mathrm{m\,s^{-1}}$ and the contact between the club and the ball lasts for 0.50 ms. Suggest an object which would have a weight approximately equal to this force.

(b) Sketch a graph to show how the force on the golf club might vary with time. Add scales to both axes.

12.30 The diagrams show two strides for a sprinter during a 100 m race together with graphs showing how the *horizontal* push of the ground on the sprinter varies with time during each stride.

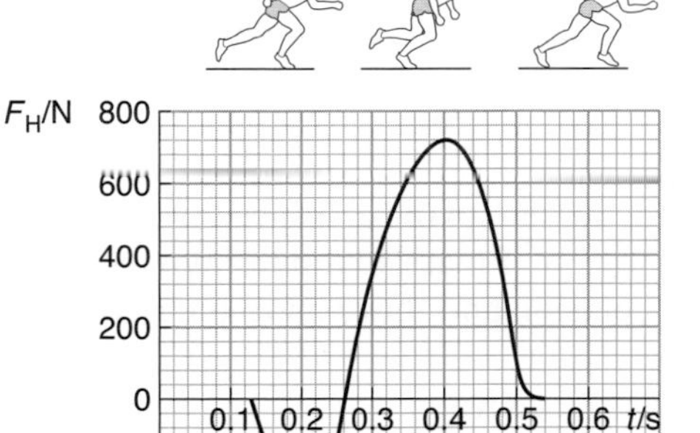

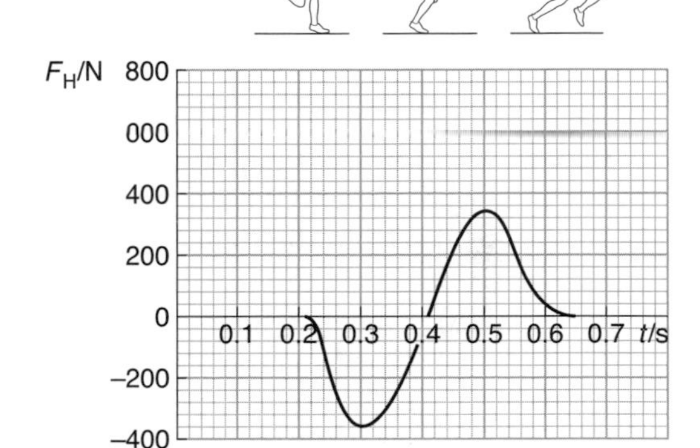

(a) Explain why the area under the graph represents the change of momentum of the sprinter during each stride.

(b) For stride A, estimate this change of momentum and hence deduce the increase in velocity of the sprinter if he has a mass of 80 kg.

(c) Describe what is happening to the sprinter during stride B.

(d) Sketch a graph of the horizontal push of the ground on the sprinter for a stride after the end of the race during which he is slowing down.

12.31 See question 12.12.

(a) What is the physical significance of the slope of these momentum–time graphs?

(b) Use the graph to calculate **(i)** the force which truck A exerts on truck B **(ii)** the force which truck B exerts on truck A.

(c) Comment on your answers to **(b)**.

12.32 The first two diagrams show two free-body force diagrams, one for a man and one for the Earth. The man has just jumped off a wall and landed on the ground.

(a) Describe the forces P and P'.

(b) P and P' vary with time as shown in the other two diagrams. Explain the shapes of the two graphs and state how they are related.

(c) Copy the P–t graph and indicate on it the size of the force W.

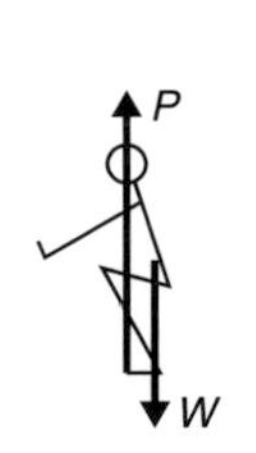

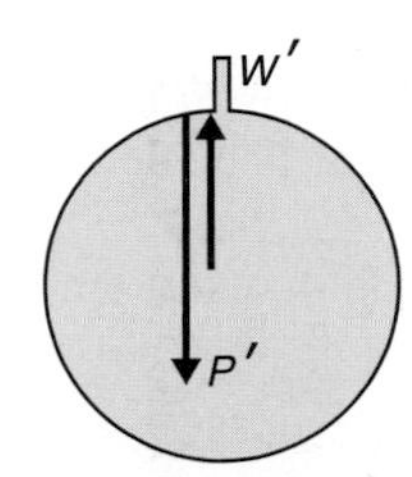

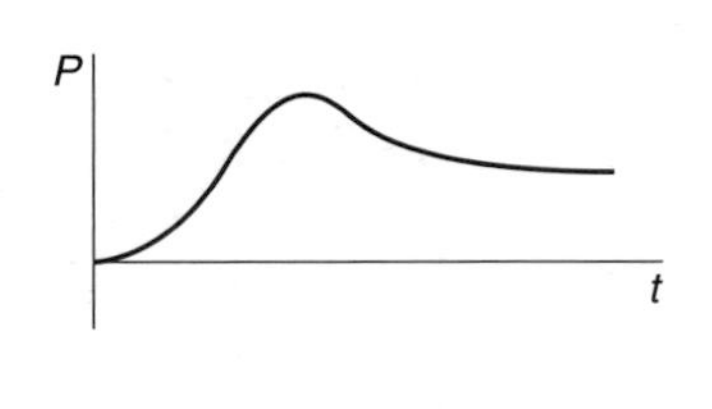

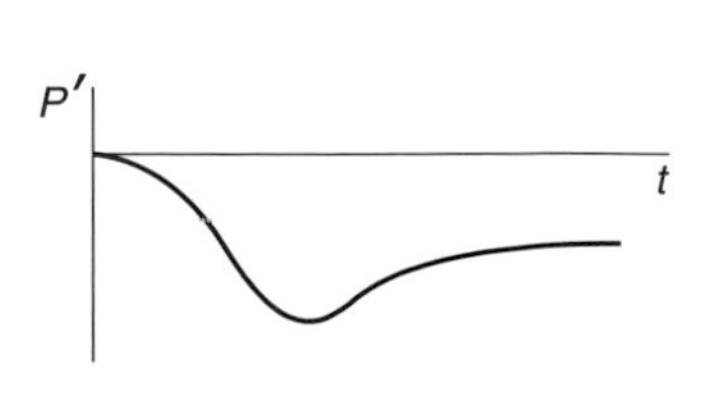

12.33 The graph shows how the momentum of a body varies with time. Estimate the resultant force F_{res} acting on the body **(a)** at 1.0 s **(b)** at 2.0 s.

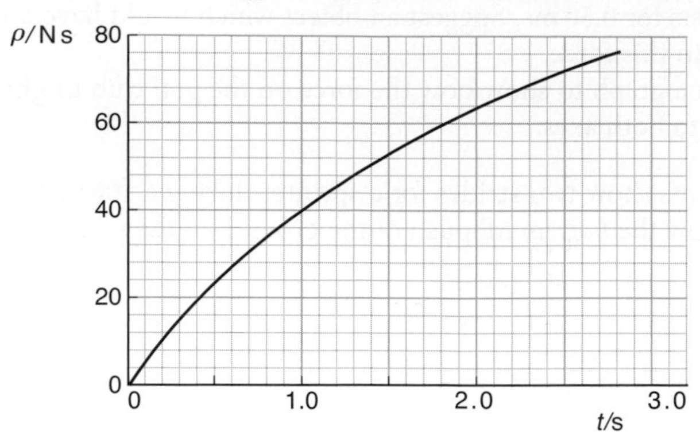

12.34 A railway wagon of mass 10 tonnes is seen to be moving at a speed of 2.0 m s^{-1} to the right. After 3.0 s it collides with another wagon of mass 40 tonnes moving at a speed of 1.0 m s^{-1} to the right. They link after the collision and move on to the right at the same velocity. [1 tonne = 1000 kg.]

(a) Calculate their common velocity after the collision.

(b) Draw a graph of momentum against time for the period $t = 0$ to $t = 7$ s if the collision took 1.0 s, i.e. from 3.0 s to 4.0 s. Assume that during the collision the momentum of each wagon changes uniformly with time.

(c) Calculate the gradient, with units, of the two graphs during the collision.

12.35 The photograph in question 1.34 shows a car being used to test seat-belts when crashing into a solid concrete block at 20 m s^{-1}. For a car of mass 1600 kg the average stopping force is 640 kN. Calculate how long it takes for the car to come to rest in this crash.

12.36 Gas atoms can be slowed down by bombarding them with infrared photons. Photons have momentum: $p = h/\lambda$, where λ is the wavelength associated with the photon and h is the Planck constant (6.63×10^{-34} J s).

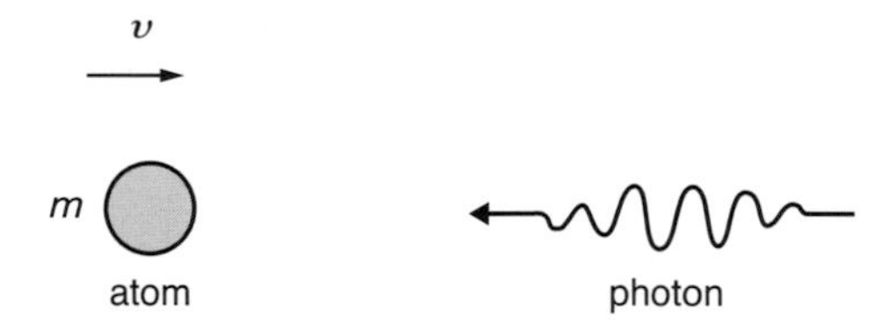

Derive an expression for Δv, the *change* of speed of an atom of mass m, initially moving at a speed v, after it has absorbed the oncoming photon and show that your expression has the unit of speed.

12.37 During a heavy storm, rain falls at the rate of 22 mm per hour; the speed of the individual raindrops is 25 m s^{-1}.

(a) If the rain strikes a flat roof and then flows off the roof with negligible speed, what is the force exerted per square metre on the roof area? [The density of water is 1000 kg m^{-3}.]

(b) What depth of water would have to stand on the roof to exert ten times this force per unit area?

12.38* The table gives some data about three rockets (the V2 was first developed in the 1940s). Copy the table and fill in the spaces.

rocket	thrust/N	mass flow/kg s^{-1}	exhaust speed/m s^{-1}
V2	2.5×10^5	180	
Atlas	1.8×10^6		1800
Saturn V		1.4×10^4	2400

12.39 70 kg of air pass through an aircraft jet engine every second. The exhaust speed of the air is 600 m s^{-1} greater than the intake speed. Calculate

(a) the change of momentum of the air in one second

(b) the force exerted on the air to change its momentum in this way

(c) the thrust produced by four such engines.

12.40 Gas molecules, each of mass 4.8×10^{-26} kg, collide with a flat surface. The average speed of the molecules perpendicular to the surface is 550 m s^{-1} both before and after they collide with it.

(a) Calculate the change of momentum of a molecule as a result of one collision.

(b) If the force on a square millimetre of the surface is 0.10 N, how many molecules collide with that square millimetre every second?

(c) What pressure is produced by this molecular bombardment?

12.41 The push F of a horizontal water jet hitting a wall can be calculated as $F = v^2 A\rho$, where v is the speed of the water, A is the cross-sectional area of the jet and ρ is the density of water. Show that the unit of the right-hand side of the equation is N.

13 Circular motion and gravitation

Data: 1 radian $\equiv \frac{360}{2\pi}$ degrees = 57.3°

gravitational field strength at Earth's surface $g_0 = 9.81\,\mathrm{N\,kg^{-1}}$

universal gravitational constant $G = 6.67 \times 10^{-11}\,\mathrm{N\,m^2\,kg^{-2}}$

mass of Earth $m_E = 5.97 \times 10^{24}\,\mathrm{kg}$

radius of Earth $r_E = 6.37 \times 10^{6}\,\mathrm{m}$

mass of Sun $= 1.99 \times 10^{30}\,\mathrm{kg}$

13.1 Describing circular motion

In this section you will need to

- use the equations $s = \theta/t$, $T = 1/f$, $v = r\omega$ and $\omega = 2\pi f$ where s is the arc length, θ is the angle (in radians), T is the time period, f is the frequency of rotation, and ω is the rate of rotation (in radians per second)
- understand that a body moving in a circle at a constant speed is accelerating towards the centre of the circle
- use the expressions $a = v^2/r$ and $a = r\omega^2$ for centripetal acceleration.

13.1 Calculate the angular speed **(i)** in degrees per second **(ii)** in radians per second, of
(a) a fan blade rotating at 2.5 r.p.m. (revolutions per minute)
(b) the minute hand of a clock.

13.2 All points on the Earth rotate with an angular speed of $7.27 \times 10^{-5}\,\mathrm{rad\,s^{-1}}$.
(a) Show how you could have calculated the angular speed for yourself.
(b) What is the speed of a point on the equator? [Use $r_E = 6.37 \times 10^{6}\,\mathrm{m}$.]
(c) What is the speed of a person in London, which is at latitude 51.5°?

13.3 A helicopter blade is designed to rotate at such an angular velocity that the tip of the blade moves at less than the speed of sound in air, which is $340\,\mathrm{m\,s^{-1}}$. The blades rotate at 260 r.p.m. Calculate the maximum possible length of the blades.

13.4 A compact disc (CD) player varies the rate of rotation of the disc in order to keep the track from which the music is being reproduced moving at a constant linear speed of $1.30\,\mathrm{m\,s^{-1}}$. Calculate the rates of rotation of a disc of diameter 12.0 cm when the music is being read from **(a)** the outer edge of the disc **(b)** a point 2.55 cm from the centre of the disc.
Give your answers in $\mathrm{rad\,s^{-1}}$ and in $\mathrm{rev\,min^{-1}}$.

13.5 The diagram shows a particle moving in a circle ABCDA of radius 8.0 m at a constant speed. It completes one revolution in 5.0 s. What is

(a) its average speed for one revolution
(b) its average speed from A to B
(c) its average velocity for one revolution
(d) its average velocity from A to C
(e) its average velocity from A to B
(f) its change of velocity from A to C
(g) its change of velocity from A to B?

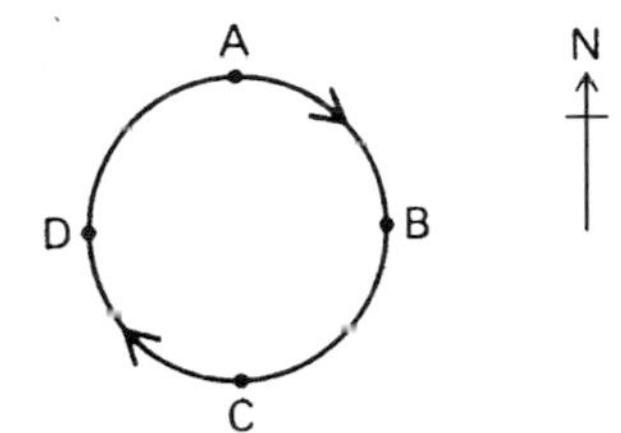

13.6 The London Eye completes two revolutions in one hour. Each capsule follows a circular path at a constant speed of $0.26\,\mathrm{m\,s^{-1}}$.

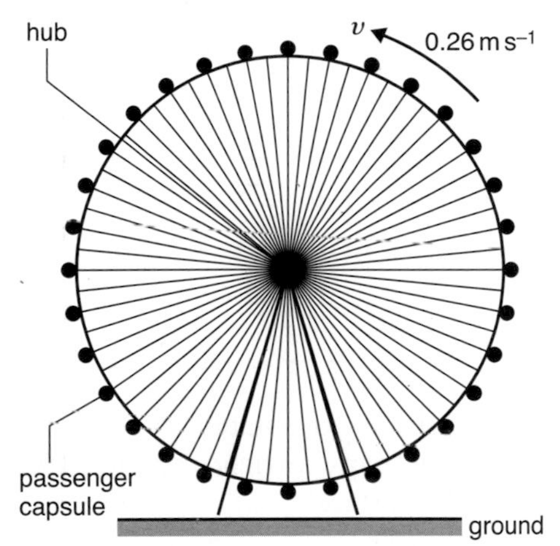

(a) What is the radius of this path?
(b) How big is the centripetal acceleration of a passenger in one of the capsules?

13.7 A geosynchronous satellite makes a complete circular orbit once every 24 hours. The radius of the orbit is 4.2×10^7 m.

(a) What is its angular velocity?
(b) What is its speed?
(c) What is its centripetal acceleration?

13.8* Riders in a bobsleigh are said to be subject to high *g*-forces. Calculate the centripetal acceleration of a bobsleigh that is moving at $33\,\mathrm{m\,s^{-1}}$ and is on a banked corner of radius 38 m.

13.9 When a wet sock is spun in a washing machine, the drum rotates at high speed. Water in the sock escapes through the holes in the drum.
In the diagram the radius of the drum is 0.24 m and the drum rotates at 820 revolutions per minute.

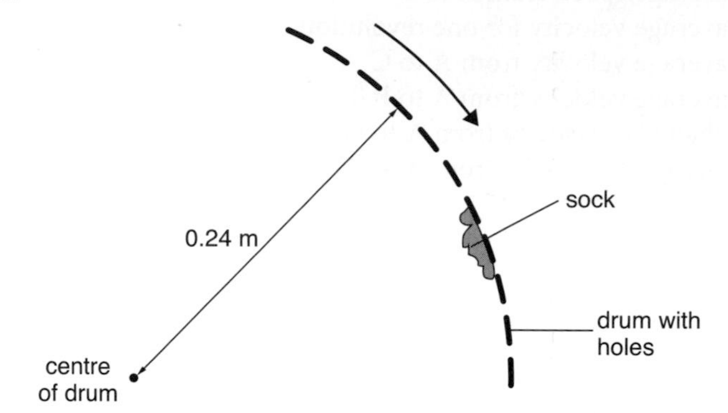

(a) Calculate the speed of the rim of the drum.
(b) Estimate the acceleration of the sock.
(c) Explain why the water escapes through the holes.

13.10 A space station is made roughly in the shape of the inner tube of a bicycle tyre. The diagram shows a sectional view across a diameter of the station.

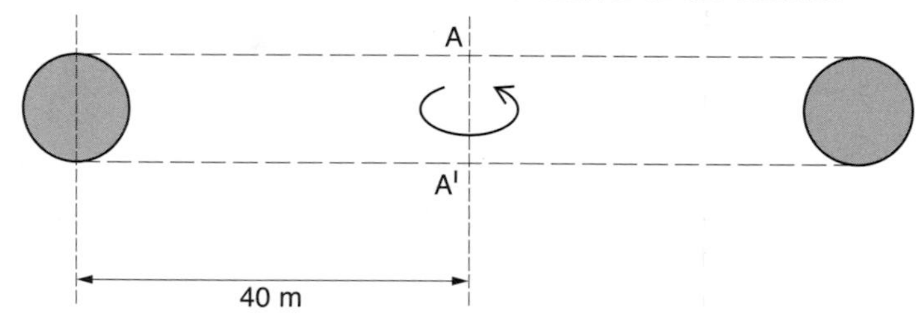

(a) With what period must the space station be rotated about the axis AA′ in order that someone living inside it may experience a centripetal acceleration equal to g_0, the value of g at the Earth's surface?
(b) What are the problems likely to be experienced by an astronaut living and working in such a rotating space station?

13.11* The designer of a loop-the-loop ride in an amusement park wants to create an acceleration of $4g$ at the bottom of a loop of radius 8.0 m.
At what speed should the car be moving to produce this acceleration?

13.2 Centripetal forces

In this section you will need to

- understand that the resultant force acting on a body moving in a circle at a constant speed is directed towards the centre of the circle
- draw free-body force diagrams and use Newton's second law in the form $mv^2/r = F_{res}$ for uniform circular motion.

13.12 A child of mass 30 kg is playing on a swing. Her centre of mass is 3.2 m below the supports when she moves through the bottom of her swing at 6.0 m s^{-1}.
(a) Draw a free-body force diagram for the child at this moment.
(b) Calculate **(i)** her centripetal acceleration **(ii)** the push of the seat of the swing on her.

13.13 In throwing the 'hammer' an athlete whirls a steel ball in a circle of radius 1.9 m at 2.4 rev s^{-1}. The steel ball has a mass of 7.3 kg.
(a) What is the size of the ball's acceleration?
(b) Calculate the centripetal pull of the wire on the ball.
(c) The athlete feels a centrifugal pull when whirling the hammer. Is there always a centrifugal force associated with a centripetal force? Explain.

13.14 In the free-body force diagram for a bicycle and cyclist seen from the front, the push of the road on him has been resolved into two components: N the normal contact push of the ground on him and F the sideways frictional push of the ground on him.
(a) The total mass of cyclist plus bicycle is 95 kg. Explain why $N = 930$ N. [Use data.]
(b) The cyclist is moving at 18 m s^{-1} in a circle of radius 50 m. Calculate F.
(c) What is **(i)** the resultant force acting on the bicycle and cyclist **(ii)** the resultant push of the ground on him?

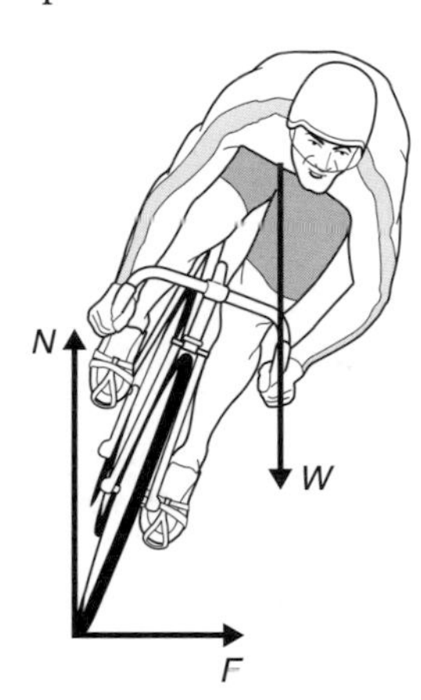

13.15 Draw a sketch of a man of mass 75.0 kg standing on a weighing machine (a pair of scales calibrated in newtons) at the Earth's equator where, because the Earth is rotating, he has a centripetal acceleration of 0.034 m s^{-2}.
(a) What is the pull of the Earth on him? Take g to be 9.780 N kg^{-1}.
(b) Draw a free-body diagram for the man and add the forces acting on him.
(c) Calculate the push of the weighing machine on him.
(d) What will the weighing machine record as his weight? Explain in words why this is not equal to mg.

13.16 The designer of an amusement park ride wants a train to have a centripetal acceleration of 19.6 m s^{-2} at the top of a loop of radius 7.0 m.
(a) Calculate the minimum speed he must ensure the train has at the top of the loop.
(b) The diagram shows two forces acting on a train of four cars at the top of the loop. The resultant centripetal force is $P + W$.
(i) Describe the two forces as the push or pull of something on the train.
(ii) Calculate W and P for a train of mass 1600 kg. [Use data.]

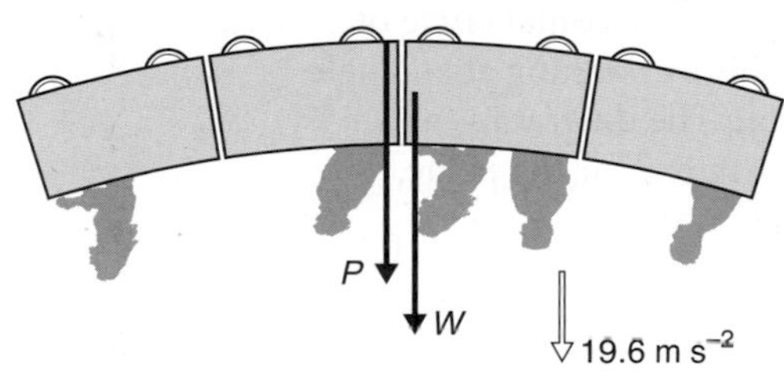

13.17* The diagram shows two views of a piece of apparatus that enables each of F, m, r and ω to be measured for an object (here a steel ball) moving in a horizontal circle, thus providing experimental evidence that $F = mr\omega^2$.
The bulb B lights up when the steel ball touches either the terminals RS or the terminals PQ.

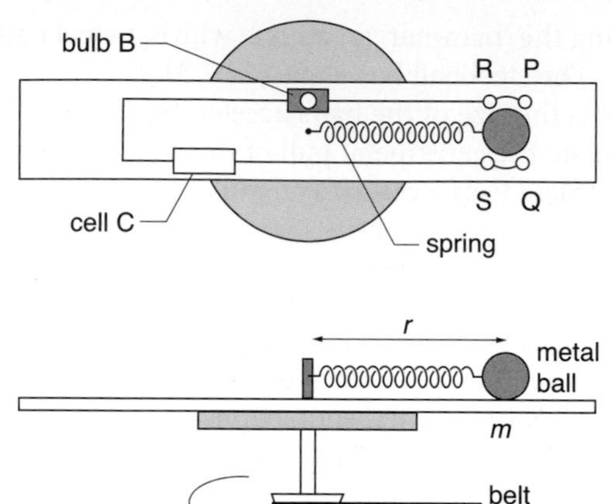

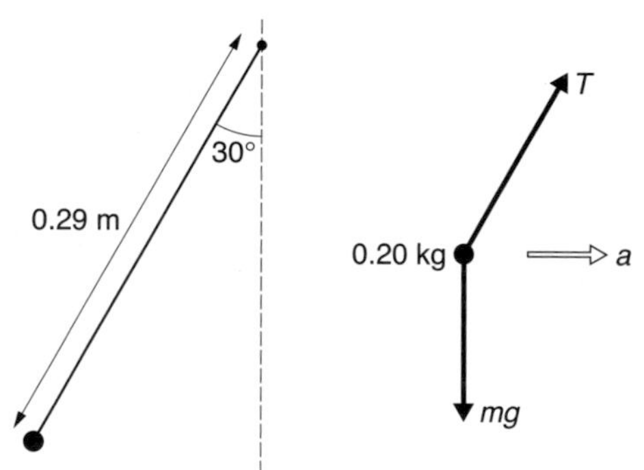

Describe how you would use the apparatus to test the relationship. You may use other routine laboratory apparatus as well as that shown in the diagram.

13.18 A metal bob of mass 0.20 kg is whirled in a horizontal circle on the end on a string once every second as shown in the diagram. A free-body force diagram for the bob is also shown.

(a) Calculate
(i) the speed of the bob and
(ii) its centripetal acceleration.
(b) Hence show that $T = 2.3\,\text{N}$.

13.19 An aeroplane of mass m is moving at a constant speed v in a horizontal circle of radius r. It does this by banking at an angle θ to the horizontal. The diagram is a free-body force diagram for the plane. Show that $v = \sqrt{(rg\tan\theta)}$.

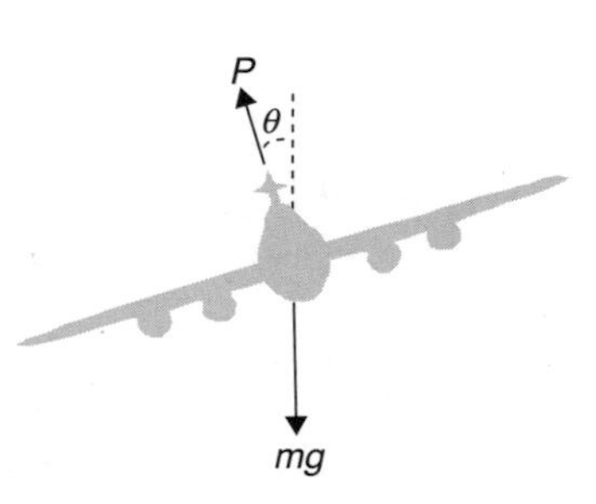

13.20* A massive ball is rotating in a horizontal circle on the end of a steel wire. The radius of the wire is 1.0 mm and the breaking stress of steel is 280 MPa, but the stress in the wire must not rise above a quarter of this for safety reasons.

(a) Calculate the maximum permissible tension in the wire.

(b) The mass of the ball is 12 kg. Considering the vertical forces, with the maximum permissible tension in the wire, find the angle which the wire makes with the vertical.

(c) If the length of the wire is 0.50 m, what is the maximum allowable angular speed, in revolutions per second, at which the mass can be spun?

(d) At what rate of rotation will the wire break?

13.3 Gravitational forces and fields

In this section you will need to

- use Newton's law of gravitation for two masses m_1 and m_2 separated by a distance r: $F = Gm_1m_2/r^2$
- understand that gravitational forces are only noticeable when one of the attracting bodies is at least of planetary size
- remember that the laws for gravitational forces apply to both point and spherical masses
- understand that the Earth's gravitational field is radial and of size $g = Gm_E/r^2$, where m_E is the mass of the Earth
- understand that g is a vector quantity
- use the equation $g = F/m$ for the gravitational field strength at a point where F is the gravitational force on a mass m placed at the point
- draw diagrams to show field lines and equipotential surfaces for gravitational fields
- remember that the difference in gravitational potential energy (g.p.e.) for a body of mass m between two points separated by a vertical distance Δh in a uniform g-field is $mg\Delta h$
- understand that the gravitational potential difference between two points is independent of the mass of the body moving between the points.

13.21 **(a)** Give a numerical example to show that the gravitational force between two bodies is very small unless one of the bodies has a very large mass, e.g. a planet or star. [Use data.]

(b) Why does only *one* of the attracting bodies need to have a very large mass?

13.22 Express the unit for G, $\mathrm{N\,m^2\,kg^{-2}}$, in base SI units.

13.23 An astronaut stands on the surface of the Moon. He is carrying a life-support pack of mass 62 kg. The mass of the Moon is 7.3×10^{22} kg and its radius is 1.6×10^6 m. Calculate the push of the pack on him.

13.24 The weight of a mass of 5.00 kg at the Earth's surface is measured to be 49.1 N.

(a) Use $F = Gm_1m_2/r^2$ with the values of G and the radius of the Earth from the data to calculate a value for the mass of the Earth.

(b) Deduce the mean density of the Earth.

When scientists first measured G they said that they were 'weighing the Earth'.

(c) Explain what they meant.

13.25* Isaac Newton, by guessing that the average density of the Earth was between 5000 and 6000 kg m^{-3}, was able to calculate a value for 'big Gee', the gravitational constant G. (The radius of the Earth was fairly well known at the time.)

(a) Show that

$$G = \frac{3g}{4\pi r_E \rho_E}$$

(b) Use the data to show that the value of G which Newton would have calculated was between 6 and 7×10^{-11} N m^2 kg^{-2}.

13.26 The values of g on the surface of the planets Mars, Saturn and Jupiter are: 3.7 N kg^{-1}, 8.9 N kg^{-1} and 23 N kg^{-1}, respectively. A mass of 20 kg is lifted 1.6 m from the surface of each planet. Calculate for each planet

(a) the force needed to lift the mass

(b) the gain in gravitational potential energy of the mass

(c) the gravitational potential difference, i.e. the change in g.p.e. per unit mass.

13.27 A multi-storey car park has six levels each 3.00 m above the other. Sketch the car park and label the entry level 1.

(a) What is the change in g.p.e. of a 1400 kg car as it moves **(i)** from level 1 to level 4 **(ii)** from level 6 to level 4?

(b) Label the levels on your diagram with values for the gravitational potential, giving level 1 a potential of 0 J kg^{-1}.

13.28 The diagram shows part of the gravitational field near the surface of the Earth. On this scale the field is essentially uniform.

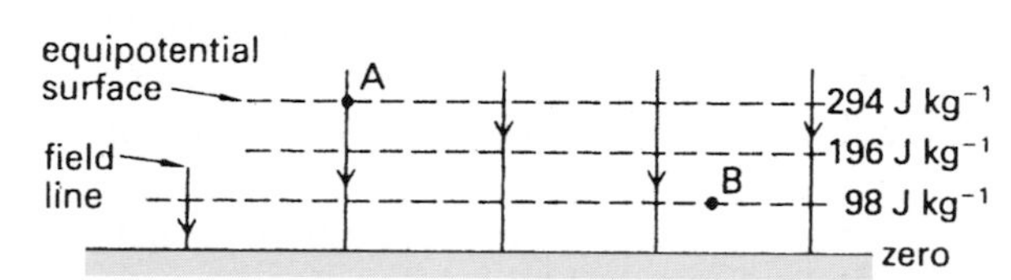

(a) Through what height would you need to raise a mass of 1.0 kg for its g.p.e. to increase by 98 J?

(b) Hence explain why the distance apart of the equipotential surfaces is 10 m.

(c) How much work must be done on a mass of 30 kg to move it from B to A?

13.29 Copy the diagram used in the previous question and draw two possible paths along which you could throw a heavy stone to pass through both A and B.

13.30* The graph shows how g varies from the Earth's surface to 30×10^6 m from the centre of the Earth. (The values of g are negative because gravitational forces act inward towards the Earth.)
By taking at least three pairs of values from the graph, check that gr^2 = constant, i.e. show that it does describe the inverse square relationship between g and r.

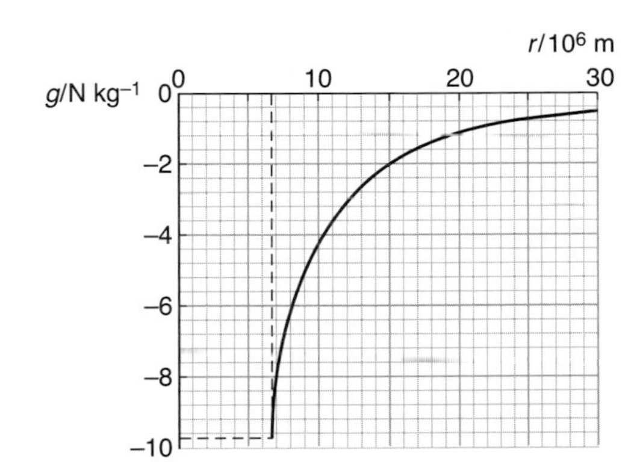

13.31* Using the graph in the previous question, copy and complete the following table:

$r/10^6$ m	7.0	10.0	14.5	21.0
$-g$/N kg^{-1}	8.8	4.0	2.0	1.0
lg ($r/10^6$ m)				
lg ($-g$/N kg^{-1})				

Plot a graph of lg(r/m) on the x-axis against lg($-g$/N kg^{-1}) on the y-axis.
(a) What does the shape of the graph tell you?
(b) Calculate the slope of the graph.
(c) How are g and r related?

13.32 The diagram shows a space vehicle S at a point between the Earth and the Moon at which the pull of the Earth on it is equal in size to the pull of the Moon on it. By equating the forces using Newton's law of gravitation express the mass of the Earth m_E as a multiple of the mass of the Moon m_M.

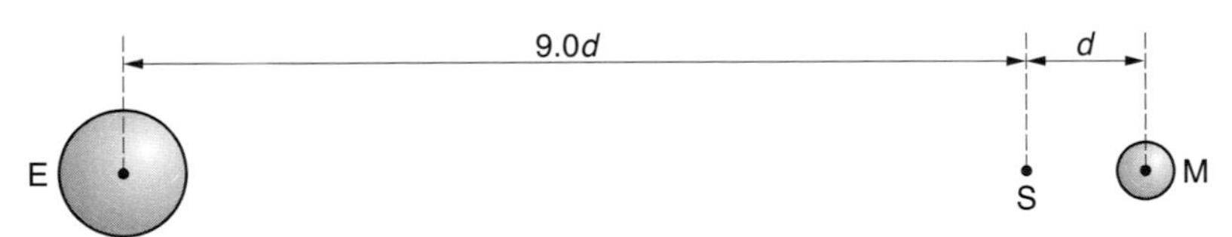

13.33 Suppose that the Sun shrank to become a neutron star of radius 1.5 km but retained its present mass. Calculate the gravitational field strength at its surface. [Use data.]
(In fact, the Sun would undergo changes before it became a neutron star which would involve it losing a considerable proportion of its present mass.)

13.34 **(a)** Calculate, using $g = Gm_E/r^2$, the Earth's gravitational field at its surface and 30 km above its surface.
(b) Show that the change is about 1% and predict at what height it will reduce by approximately a further 1%. Explain how you made your prediction.

13.35* The diagram shows a region near the Earth with inward field lines and circular equipotential dashed lines. At $r = 8.0 \times 10^6\,\text{m}$, $V = -5.0 \times 10^7\,\text{J kg}^{-1}$, and at $r = 10 \times 10^6\,\text{m}$, $V = -4.0 \times 10^7\,\text{J kg}^{-1}$.

A 'space tourist' in a space vehicle of total mass 1500 kg moves away from Earth on a path through A and B. On his return journey he passes in free fall through level B at a speed of $4.0\,\text{km s}^{-1}$.

(a) What is the gain in g.p.e. of the space vehicle as it moves up from A to B?

(b) On the return journey calculate

(i) the kinetic energy of the space vehicle at level B

(ii) the kinetic energy of the space vehicle at level A

(iii) the speed of the space vehicle as it passes through level A.

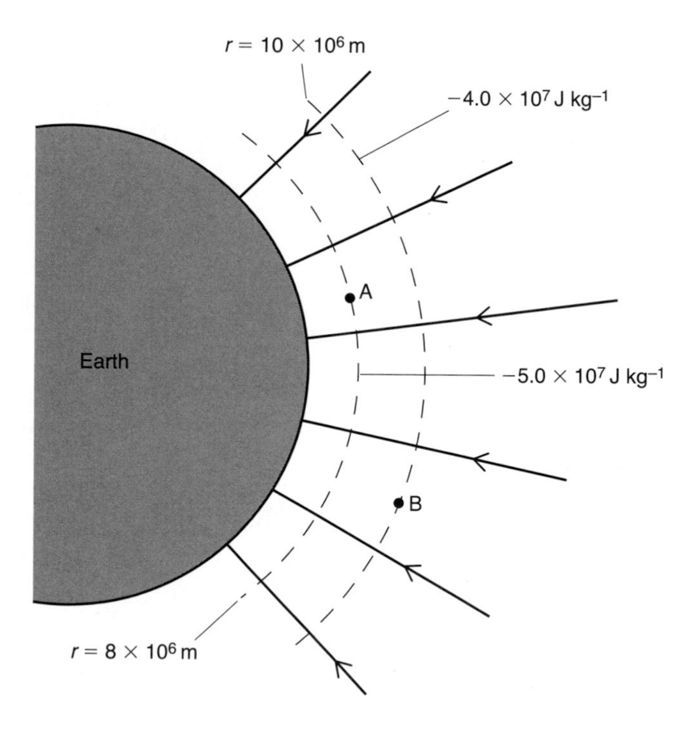

13.36 Show that the kinetic energy which must be given to a body of mass m at the Earth's surface if it is to escape the Earth's gravitational field, i.e. reach an infinite distance from the Earth, is equal to $mm_E G/r_E$.

(a) Hence derive an expression for the speed v_e at which a body must be projected from the Earth's surface (ignoring the atmosphere) if it is to escape from the Earth's gravitational field.

(b) Calculate a value for v_e and comment on the fact that it does not depend on the mass of the body. [Use data.]

13.37 **(a)** Calculate an escape speed for the Moon. Take the mass of the Moon to be $7.34 \times 10^{22}\,\text{kg}$ and its radius to be $1.64 \times 10^6\,\text{m}$.

(b) The temperature of the Moon's surface rises to about 400 K when sunlight falls on it. At this temperature the average speed of oxygen molecules is over $500\,\text{m s}^{-1}$. Suggest why the Moon has no atmosphere.

13.38* **(a)** Show that the gravitational pull of the Sun on the Earth is 180 times that of the Moon on the Earth. [Take the mass of the Sun to be 2.0×10^{30} kg, the mass of the Moon to be 7.3×10^{22} kg, the Sun–Earth separation to be 1.5×10^{8} km and the Earth–Moon separation to be 3.8×10^{5} km.]

(b) Suggest why, despite this, the Moon is more influential than the Sun in producing tides on the Earth.

(c) What are 'spring' and 'neap' tides? Explain why each occurs approximately twice during each month.

13.4 Satellite motion

In this section you will need to

- understand that the centripetal acceleration, v^2/r, of bodies in circular orbits in a gravitational field is equal to the size of the g-field at that orbit
- use Newton's second law combined with his law of gravitation to solve satellite problems
- explain that a person feels weightless when he or she is in a state of free fall.

13.39* A non-scientist asks the question: 'Why doesn't an orbiting satellite fall towards the Earth?'
The physicist's answer is: 'It does!'
This answer doesn't make sense to the non-scientist. How might the physicist explain what he means? (You might like to describe Newton's idea of firing a shell at increasing speeds from the top of a very high mountain.)

13.40 The photograph shows an astronaut who is moving in a circular orbit 350 km above the Earth's surface. (There is a space shuttle close by!)

(a) What is **(i)** the Earth's gravitational field at this height **(ii)** the acceleration of the astronaut towards the centre of the Earth? [Use data.]

(b) Explain **(i)** why the astronaut does not crash to the ground **(ii)** why the astronaut feels 'weightless'.

(c) Calculate **(i)** the speed of the astronaut in this orbit **(ii)** how long it takes the astronaut to circle the Earth.

13.41 A geosynchronous satellite moves in a circle of radius 4.2×10^7 m every 24 hours.

(a) Use this data to calculate **(i)** the speed **(ii)** the centripetal acceleration of the satellite.

(b) Write down the size of the Earth's gravitational field at the satellite's orbit.

13.42 **(a)** Use Newton's law of gravitation and apply his second law to an object of mass m that is orbiting a planet of mass M at a constant speed v in a circular path of radius r (see diagram).

(b) Express its speed as circumference divided by orbital period T, and substitute for v.

(c) Hence show that T^2 is proportional to r^3. (This is Kepler's third law.)

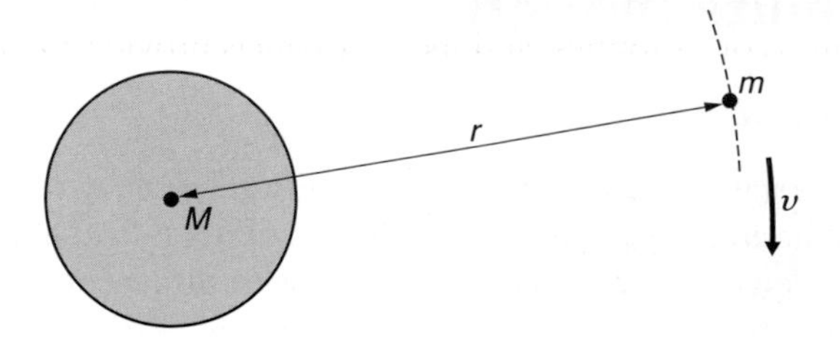

13.43* The relationship between the period of revolution T of a body in a circular orbit and its distance r from the gravitationally attracting body at the centre of the orbit (Kepler's third law) is

$$4\pi r^3 = GMT^2$$

where M is the mass of the central body.

(a) Show that the units of both sides of this expression are the same.

(b) Use it to calculate the mass of the Sun, given that the mean Earth–Sun distance is 1.5×10^8 km. [Use data.]

(c) Jupiter has many moons, four of which can be seen using a good pair of binoculars or a small telescope. Explain, stating what measurements you would have to make, how observations of the motion of the moons of a planet enable the mass of the planet to be calculated.

13.44* The period T of a satellite around its gravitationally attracting central mass (a star or planet) is related to its mean distance r from the centre of this mass by the relationship $T^2 \propto r^3$.

(a) Use the following data about four of the moons of the planet Uranus to plot a suitable graph to test this relationship.

T/hours	60.5	99.5	209	323
$r/10^3$ km	192	266	436	582

(b) Suppose a further moon of Uranus has been discovered with a period of 170 hours. Use your graph to find its mean distance from the centre of Uranus.

(c) Use Kepler's third law for circular orbits from the previous question to show that the gradient of your graph is $4\pi/GM$, where M is the mass of Uranus.

(d) Hence calculate the mass of Uranus. [Use data.]

13.45 **(a)** Calculate the orbital radius of a synchronous communications satellite, i.e. one which has a period of 24 hours so that it appears to remain stationary above one point on the Earth's surface. Express this radius as a multiple of the Earth's radius to the nearest whole number. [Use data.]

(b) Why must the satellite's orbit lie in the plane of the equator?

(c) Draw a scale diagram to estimate the angle above the horizon that a receiving aerial at the same longitude and at latitude 45° must point in order to receive signals from the satellite.

13.46 The Hubble space telescope moves in a circular orbit 570 km above the Earth's surface. It has a mass of 11 tonnes, and its solar panels provide a maximum power of 2.8 kW.

(a) What is the gravitational field at this height? [Use data.]

(b) Calculate **(i)** the speed of the Hubble space telescope and **(ii)** how many times a day it orbits the Earth?

The solar flux from which Hubble generates its electrical power is delivered to a pair of large solar panels.

(c) If the solar flux is 1.4 kW m^{-2}, what minimum total area must these panels have?

(d) Suggest what makes your answer to **(c)** well below the area needed in practice.

13.47* Search the internet for information about the International Space Station and other large man-made Earth satellites.

14 Oscillations

14.1 Simple harmonic oscillators

In this section you will need to

- remember that the time for one complete oscillation is called the period T, the number of oscillations per unit time is called the frequency f and that $f = 1/T$
- remember that the angular velocity ω (in radians per second) is given by $\omega = 2\pi f$
- understand that, for a body moving with simple harmonic motion (s.h.m.):
  - the period is independent of the amplitude
  - the displacement varies sinusoidally with time
  - the maximum velocity occurs when the displacement is zero
- use the equations
  - $x = x_0 \cos 2\pi ft$
  - $v = -v_0 \sin 2\pi ft$ where $v_0 = 2\pi f x_0$
  - $a = -a_0 \cos 2\pi ft$ where $a_0 = 2\pi f v_0$
- use the relationships $a \propto -(\text{constant})x$ and $a = -(2\pi f)^2 x$
- describe how to produce displacement–time graphs for oscillating objects
- understand the relation between sinusoidal x–t, v–t and a–t graphs.

14.1 Estimate the time periods of each of the following motions and hence calculate (to 1 significant figure) their frequencies:

(a) a child on a playground swing

(b) a baby rocked in its mother's arms

(c) the free swing of your leg from the hip.

14.2 The graph shows two periods of a simple harmonic oscillation. On a copy of the graph

(a) **(i)** mark with a P any place where the speed is a maximum

(ii) mark with a Q any place where the speed is zero.

(b) Explain where the acceleration is a maximum and where it is a minimum.

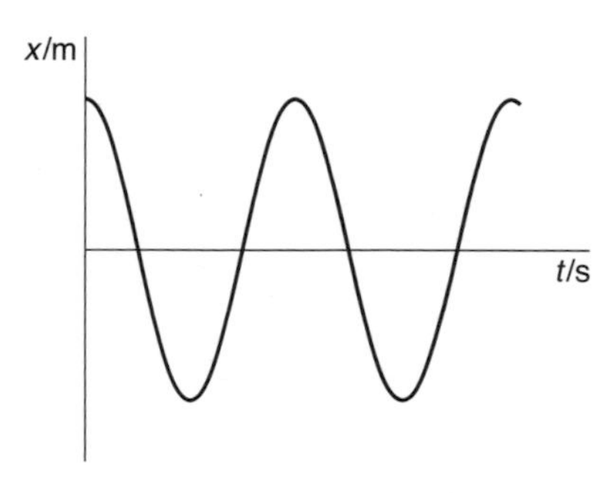

14.3* Write an explanation which you could give to a non-scientist of what is meant by simple harmonic motion. Use a situation with which he or she will be familiar to illustrate your explanation.

14.4 The graph shows half a period of an s.h.m. oscillation.

(a) What is **(i)** the amplitude **(ii)** the frequency of this oscillation?

(b) **(i)** Explain how you would use the graph to determine the maximum speed of the oscillating body and **(ii)** find this speed.

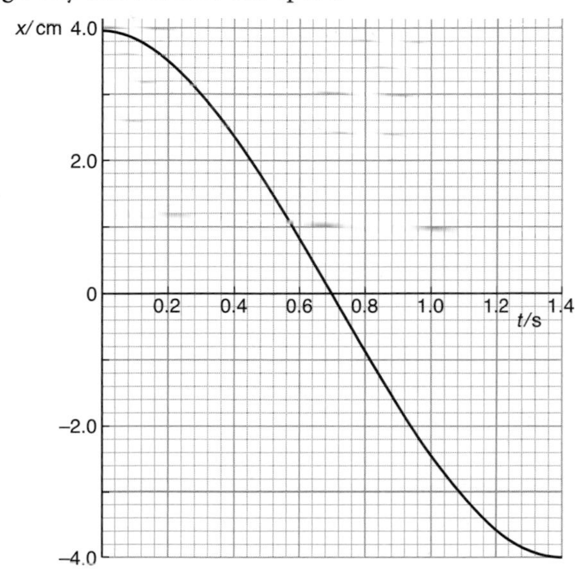

14.5 **(a)** Find out **(i)** the frequency of the note middle C on a piano **(ii)** the frequency of a typical heart beat **(iii)** an approximate value for the period of the tidal rise and fall of the sea.

(b) Hence find **(i)** the period of a middle C note **(ii)** the period of the heart beat **(iii)** the tidal frequency.

14.6 The graph shows how the acceleration a varies with displacement x for a particle undergoing s.h.m.

(a) What is the frequency of this oscillation?

(b) Calculate the maximum speed with which the particle moves.

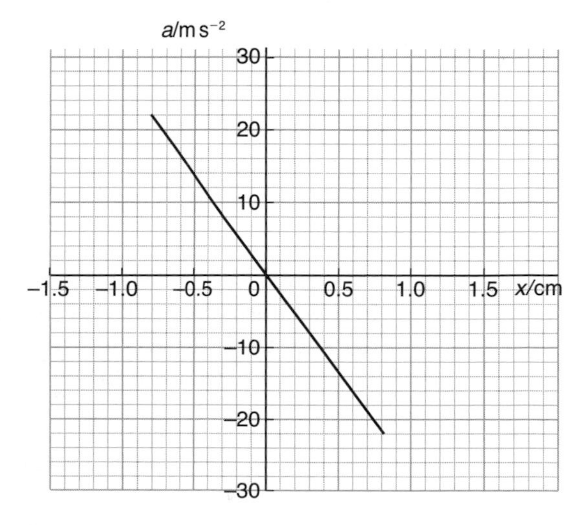

14.7 The equation defining linear s.h.m. is

$a = -(\text{constant})x$

(a) What units must the constant have?

(b) Two s.h.m.s, A and B, are similar except that the constant in A is nine times the constant in B. Describe how these s.h.m.s differ.

14.8 For an s.h.m. (for which $x = x_0 \cos 2\pi ft$) of period 2.00 s and amplitude 16.0 cm, calculate the displacement (remembering that $2\pi ft$ is in radians not degrees) **(a)** at $t = 0$ **(b)** at $t = 1.00$ s **(c)** at $t = 0.50$ s **(d)** at $t = 0.25$ s.

14.9 A bored student holds one end of a flexible plastic ruler against the laboratory bench and flicks the other end, setting the ruler into oscillation. The end of the ruler moves a total distance of 8.0 cm as in the diagram and makes 28 complete oscillations in 10 s.
(a) What are the amplitude x_0 and frequency f of the motion of the end of the ruler?
(b) Calculate the maximum speed of the end of the ruler.

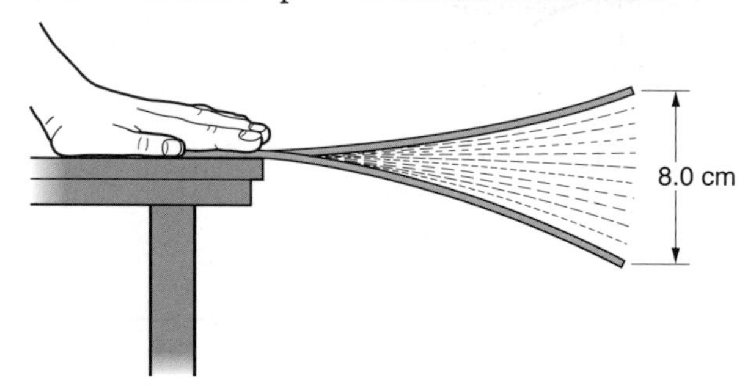

14.10 A body oscillates with s.h.m. described by the equation
$x = (1.6\,\text{m}) \cos(3\pi\,\text{s}^{-1})t$
(a) What are **(i)** the amplitude **(ii)** the period of the motion?
(b) For $t = 1.5$ s, calculate **(i)** the displacement **(ii)** the velocity **(iii)** the acceleration of the body.

14.11 A spring of spring constant k (it obeys Hooke's law when extended *or* contracted) is attached to a wall. At the other end of the spring is a mass m that slides on a horizontal frictionless surface. The mass rests in equilibrium at O.

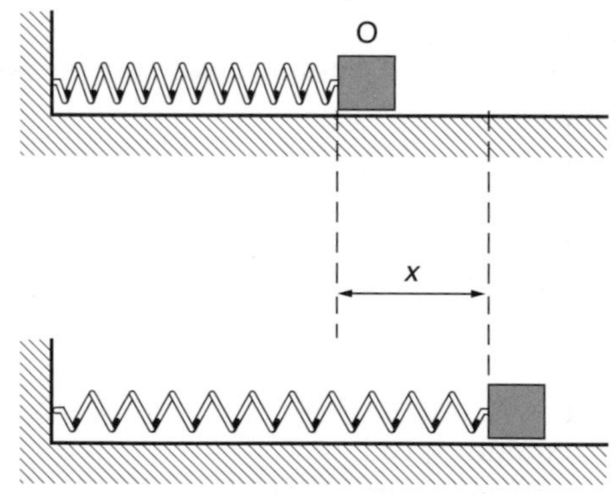

(a) When displaced a distance x to the right, what is
(i) the horizontal pull F of the spring on the mass
(ii) the size of the acceleration a of the mass?
(b) With what period will this mass oscillate?

14.12 A mass hangs in equilibrium from a light vertical spring, and when disturbed oscillates with s.h.m.

(a) Explain whether the period of oscillation T would be unchanged, increased or decreased if, separately,

(i) the mass was doubled

(ii) a second identical spring was placed in parallel with the original spring

(iii) the spring was cut in two, and one half of it used to replace the original spring?

(b) Explain why the period would be the same on the Moon as on Earth.

14.13 The diagram shows three sinusoidal (sine-shaped) graphs for displacement x, velocity v and acceleration a for s.h.m.

(a) Explain the relationship between

(i) the x–t and the v–t graphs

(ii) the v–t and the a–t graphs

(iii) the x–t and the a–t graphs.

(b) Illustrate your answer to **(iii)** by sketching a graph of x against a.

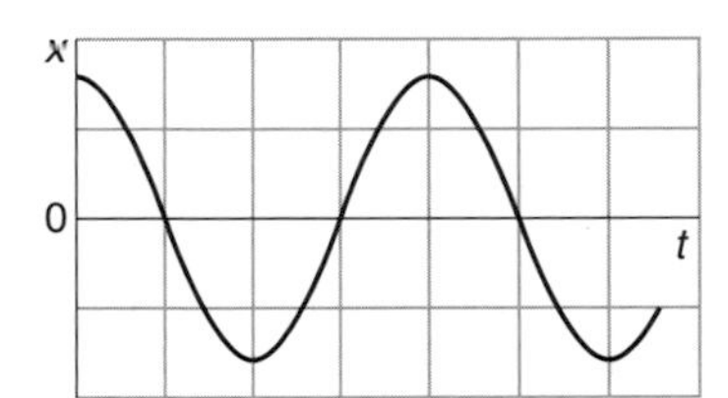

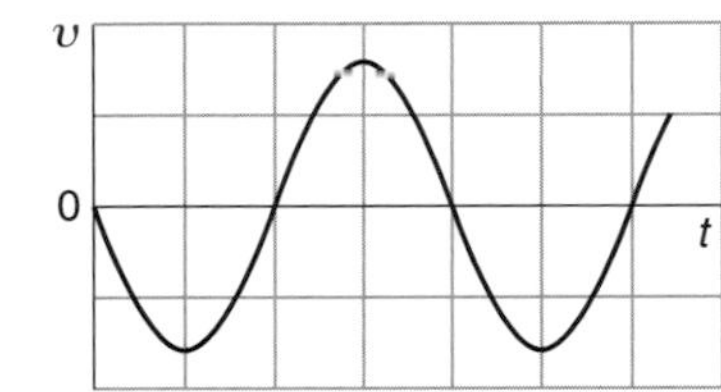

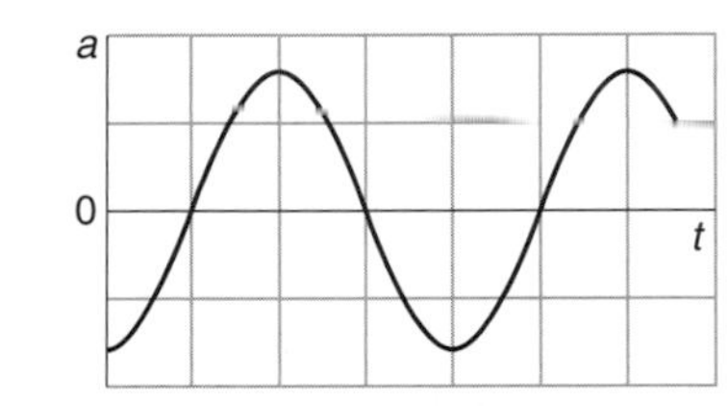

14.14 In order to test how well pilots can recognise objects when seated in a juddering helicopter they are subjected to vibrations of frequency from 0.1 Hz to 50 Hz.
If a pilot is being tested with vibrations of frequency 35 Hz and amplitude 0.60 mm, what is

(a) his maximum velocity

(b) his maximum acceleration during the test?

14.15 A potential difference which alternates sinusoidally is applied to the Y-plates of an oscilloscope. A stationary trace, with an amplitude of 4.0 div and a 'wavelength' of 1.5 div, is obtained with the time base set at $1.0\,\text{ms div}^{-1}$. When the time base is switched off the trace becomes a vertical line. Calculate the maximum speed of the spot of light on the screen when producing the vertical line. Take 1.0 div to equal a length of 10 mm.

14.16 A loudspeaker produces musical sounds by oscillating a light diaphragm. When the amplitude of oscillation is limited to 1.2×10^{-3} mm, what is the maximum frequency if the acceleration of the diaphragm should not exceed $10\,\text{m s}^{-2}$, i.e. g?

14.2 Pendulums and mass–spring oscillators

In this section you will need to

- use the equation $T = 2\pi\sqrt{(\ell/g)}$ for the period of a simple pendulum
- use the equation $T = 2\pi\sqrt{(m/k)}$ for the period of a mass–spring oscillator.

14.17 What is the length of a 'seconds pendulum' (i.e. a simple pendulum with a period of 2.000 s) at a place where $g = 9.816\,\text{m s}^{-2}$?

14.18 The pendulum bob of a grandfather clock swings through an arc of length 196 mm from end to end. The period of the swing is 2.00 s.
(a) What is **(i)** the amplitude **(ii)** the frequency of the bob?
(b) With what speed does the bob pass through the centre of the swing?

14.19* Describe in detail, giving the dimensions of the apparatus and any precautions you would take, how you would use a small steel sphere and a length of thread to determine the acceleration of free fall in the laboratory.

14.20 What is the period of a pendulum on Mars, where the free fall acceleration is about 0.37 times that on Earth, if the pendulum has a period of 0.48 s on Earth?

14.21 The clock in the Big Ben tower in London is regulated by a pendulum consisting of a massive bob attached to the bottom of a metal rod. To ensure that it keeps time with Greenwich Mean Time (GMT), engineers place coins on top of the bob (or remove them) to adjust the frequency of the pendulum.
Explain what effect adding a coin to the bob will have on the frequency, and whether this will make the clock run faster or slower.

14.22 A fisherman's scale stretches 2.6 cm when a 1.9 kg fish is hung from it.
(a) What is the spring stiffness?
(b) Calculate the frequency with which the fish will vibrate if it is pulled down a little and released.

14.23* Astronauts need to measure their mass while in 'free fall' in space. To do this they strap themselves firmly into a 'chair' that can be set oscillating.

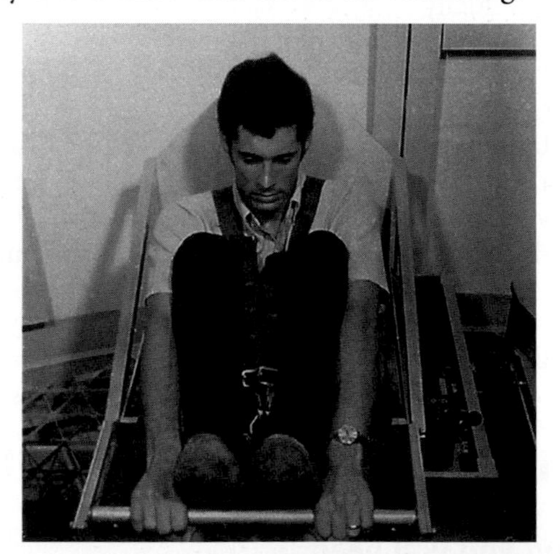

In a situation where the mass of the 'chair' is 6.0 kg and the stiffness of the springs controlling the oscillations is $1.2\,\text{kN}\,\text{m}^{-1}$, what is the mass of the astronaut who oscillates with a time period of 1.6 s?

14.24 A man of mass 80 kg bounces on the seat of a motorcycle. He finds that he is at the bottom of his bounces every 0.50 s. Calculate the spring constant of the suspension. State any assumptions you make.

14.25 **(a)** Two identical springs of stiffness k are connected first in series and then in parallel supporting a mass m and are set into vertical oscillation. Determine the ratio

$$\frac{\text{period when in series}}{\text{period when in parallel}}$$

(b) What mass should be added to the parallel connection in order to make the periods equal?

14.26 The diagram shows one of the cylinders of an internal combustion engine. The length of the 'stroke' is 86 mm and the mass of the piston, which oscillates approximately with s.h.m., is 580 g.

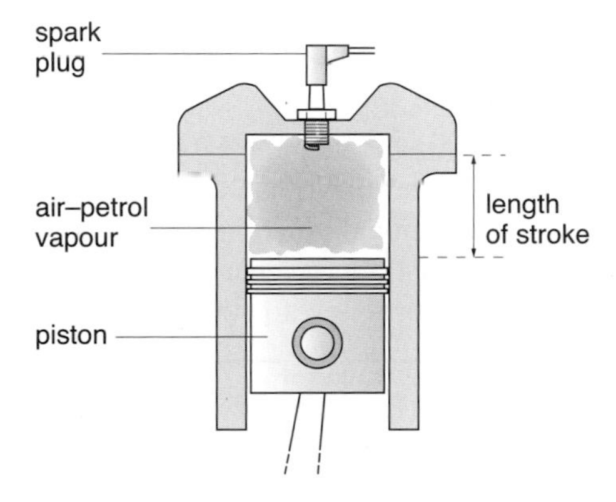

(a) When the engine rotates at 3600 rev min^{-1}, the piston makes 3600 oscillations each minute. Show that the maximum acceleration of the piston is then 6100 m s^{-2}.

(b) Hence calculate the maximum force pushing or pulling the piston.

14.27 A 'baby bouncer' is a device for amusing babies before they can walk. It consists of a harness suspended from the lintel of a doorway.
Such a device has ropes 1.20 m long which stretch to 1.42 m when a baby of mass 8.5 kg is placed in the harness. The baby is then pulled down 8.0 cm and released.

Calculate

(a) the spring constant for the baby bouncer

(b) the period of the baby's motion

(c) the baby's maximum speed.

14.3 Energy transfer and resonance in oscillators

In this section you will need to

- draw energy flow diagrams (Sankey diagrams) to illustrate energy transfer in simple harmonic oscillators
- describe the nature of damped harmonic oscillators
- understand that the energy stored in a Hooke's law spring is proportional to the square of the extension of the spring (see also section 7.3)
- use the equation $W = \frac{1}{2}kx^2$ for the energy stored in a spring
- understand the condition for resonance in mechanical oscillators
- describe an experiment to demonstrate the resonant condition of a mechanical system.

14.28 The diagram shows the energy transfer in an undamped simple harmonic oscillator. The p.e. might be elastic or gravitational potential energy.
Draw a similar diagram to illustrate the energy transfers in a damped harmonic oscillator and explain any features new to your diagram.

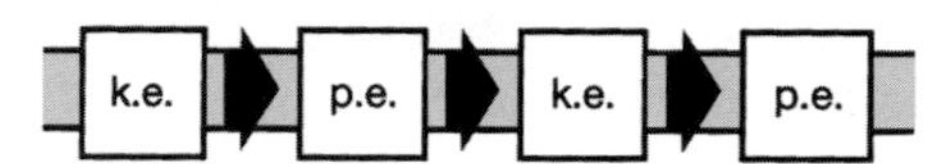

14.29* The graph shows the energy stored in a spring of stiffness $30\,N\,m^{-1}$. (This is like the expendable springs used in school and college laboratories.)

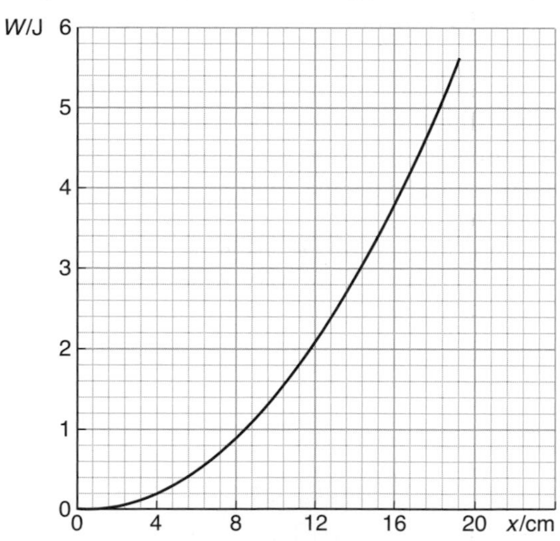

(a) Check that the graph has been correctly drawn.
(b) Explain why the graph cannot be extrapolated to predict the energy stored for extensions up to, for example, 50 cm.

14.30 The graph shows how the kinetic energy of a heavy pendulum varies with time. During these swings the total mechanical energy can be taken to be constant.

(a) Make a rough copy of the graph and add a line showing the total mechanical energy of the pendulum.

(b) Add a dashed line showing how the gravitational potential energy of the pendulum varies with time.

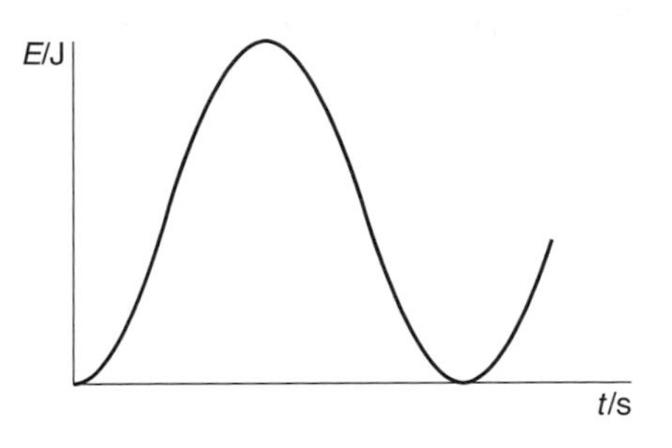

14.31 A certain spring has the same stiffness, 200 N m^{-1}, when extended and compressed. At extensions or compressions x of 0.1 m and 0.2 m it stores energy $W = F_{av}x$ of 1.0 J and 4.0 J.

(a) Sketch (*do not use graph paper*) a graph showing W (x-axis) against x (y-axis) for values of x from −0.2 m to +0.2 m.

(b) A block of wood of mass 200 g is attached to one end of this spring. The block is placed on a horizontal frictionless surface (as in the diagram for question 14.11) and is pulled so that the spring has an extension of 20 cm and then released.

(i) Add to your graph a second curve to show how the kinetic energy of the block varies with the displacement x.

(ii) Calculate the speeds of the block when x = 0, ±0.1 m and ±0.2 m.

14.32* Modern 'bungee' jumping began in Bristol in the UK in 1979, when a man of mass 80 kg using an elastic rope of unstretched length 45 m jumped from the Clifton suspension bridge 75 m (about 250 ft) above the water and came within a few centimetres of the river Avon on his first fall.

(a) Use the principle of conservation of mechanical energy to calculate the stiffness of the rope to which he was attached. [Use g = 10 N kg^{-1}.]

(b) What was the period of his subsequent oscillations?

14.33 The diagram shows a system of light pendulums which oscillate perpendicular to the page as if they were suspended from points along the line AC.

(a) Describe the motion of the pendulums when the heavy metal sphere sets the cord ABC oscillating with s.h.m.

(b) Suggest how you could alter the damping of the small spheres and describe the result of making them more heavily damped.

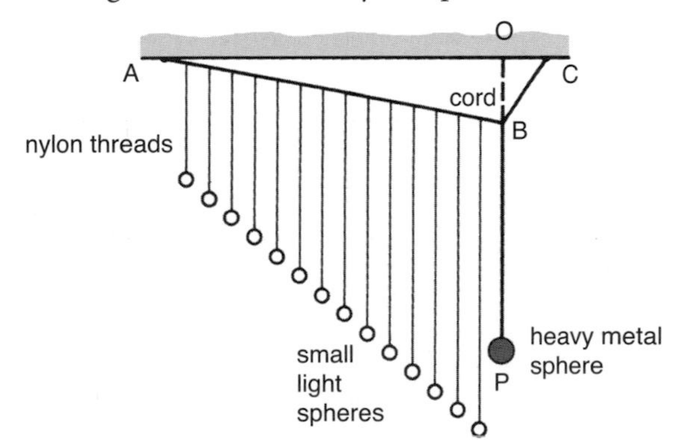

14.34 At certain rates of rotation, parts of a washing machine, e.g. a side panel, may vibrate strongly. Explain the physical reason for this strong vibration and suggest how it might be reduced.

14.35 Our legs and the front legs of many four-legged animals swing like pendulums, the period of which depends on the length of the leg. Suppose that a giraffe has a leg which is similar in shape to that of a horse, but that it is twice as long. Find the ratio of the following quantities for the giraffe and the horse:

(a) the period of the leg's free swing
(b) the frequency of the leg's free swing
(c) the length of a stride
(d) the speed of walking.

14.36* Resonance and the resonant transfer of energy are key features of many objects and devices. Use reference books, encyclopaedias and search engines to find out more about:

- microwave cooking
- MRI scans
- TV and radio tuning
- musical instruments
- bridge designs.

15 Radioactivity

Data: electronic charge $e = 1.60 \times 10^{-19}$ C

rest mass of electron $m_e = 9.11 \times 10^{-31}$ kg

Planck constant $h = 6.63 \times 10^{-34}$ J s

Avogadro constant $N_A = 6.02 \times 10^{23}$ mol^{-1}

electron volt 1 eV = 1.60×10^{-19} J

rest mass of a nucleon (p or n) = 1.67×10^{-27} kg

speed of electromagnetic waves in a vacuum $c = 3.00 \times 10^{8}$ m s^{-1}

unified atomic mass constant u = 1.66×10^{-27} kg

activity of radioactive source: 1 becquerel (Bq) = 1 disintegration per second

15.1 The nuclear atom

In this section you will need to

- understand that an atom consists of a nucleus surrounded by one or more electrons
- understand that α-particle scattering provides evidence that the atom has a nucleus
- understand that the nucleus contains protons and neutrons
- use the notation $^{A}_{Z}X$ when referring to nuclides, where Z is the atomic (proton) number and A is the mass (nucleon) number of the nuclide of element X
- use the term neutron number N when describing nuclides and realise that $A = N + Z$
- use the atomic mass unit u and remember that it is equal to one twelfth of the mass of the nuclide $^{12}_{6}C$.

15.1 The element copper comes 29th in a list of the elements. What information does this fact alone give you about atoms of copper?

15.2 The diameters of most atoms are about 3×10^{-10} m.
Roughly how many atoms thick is
(a) a piece of gold foil, of thickness 6×10^{-7} m
(b) a strand of copper wire of diameter 0.10 mm?

15.3 The proton number Z of carbon is 6; the neutron number of its three common isotopes are 6, 7 and 8.
(a) What are the mass numbers A of these isotopes
(b) What is the mass, in kg, of its most massive isotope? [Use data.]

15.4 A solid gold ring has a mass of 25 g. Gold has only one stable isotope, of nucleon number 197. How many atoms are there in the gold ring? [Use data.]

15.5* In a series of experiments Geiger and Marsden fired α-particles at a thin sheet of gold foil. Most α-particles were effectively undeflected, but a few were deflected through large angles.

(a) What did Rutherford deduce about the nature of atoms?

(b) The diagram shows two paths which α-particles would take when they approach a nucleus. They are labelled 'a' and 'd'. Copy the diagram and add the paths which α-particles would take if they passed through the points labelled 'b', 'g' and 'k'.

(c) A diagram like this gives a false impression of what happens. Why?

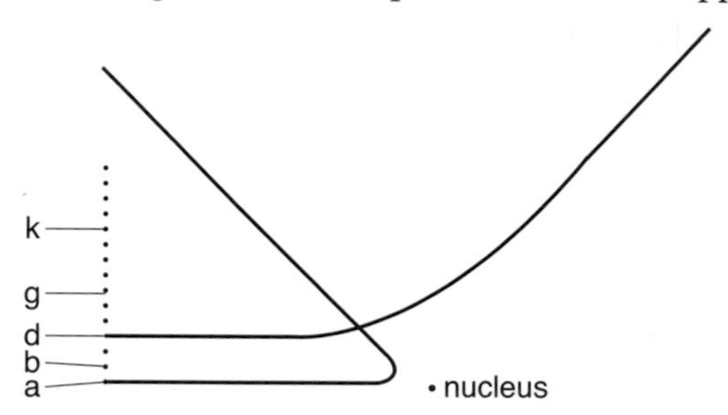

15.6 The diagram shows an α-particle being deflected by the nucleus of an atom. Copy the diagram and mark on it

(a) the point in the α-particle's path where its speed is least (label it X)

(b) the direction of the push of the nucleus on the α-particle at points A, B and C.

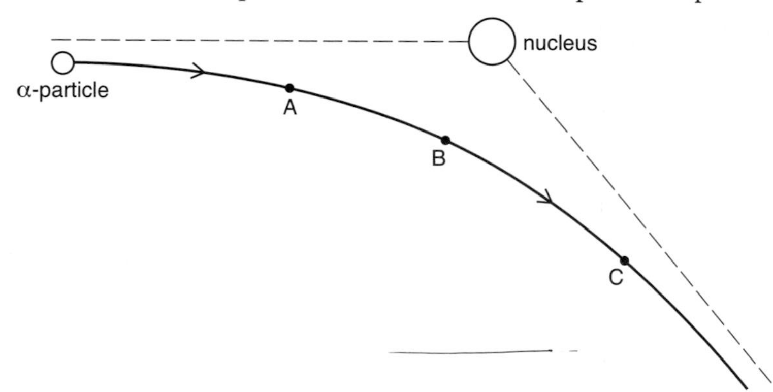

15.7 An atom of uranium-238 has 92 protons, 146 neutrons and 92 electrons.

(a) What fraction of the mass is not in the nucleus? [Use data.]

(b) If the nucleus of this atom may be thought of as a sphere of radius 7.4×10^{-15} m, what is the average density of the material of the nucleus?

15.8 Electron scattering at 'low' energies provides information about the radii of nuclei. Draw a sketch (similar to the one in question 15.5) to show the paths of two electrons which are travelling towards a proton and are deflected by it.

15.9 **(a)** Measure the diameter of a penny.

(b) If a penny represents the nucleus of a gold atom, how large, on this scale, would an atom be? [Take the diameter of a nucleus to be 2.0×10^{-14} m and that of an atom to be 3.0×10^{-10} m.]

15.10 The average density of the matter in a nucleus is given by its mass divided by its volume.

(a) In the nuclide $^{A}_{Z}X$, what is the number of nucleons?

(b) If each nucleon has a mass m, what is the total mass of the nucleus?

(c) Suppose the radius of a nucleon is r_0. What is the sum of the volumes of the separate nuclei?

(d) The volume of the nucleus will be similar to the sum of the volumes of the separate nuclei. Use your answers to **(b)** and **(c)** to write down an approximate expression for the density of the nuclear matter.

(e) Why does your answer to **(d)** show that the density of nuclear matter is the same for all nuclei, whatever their mass and charge?

(f) Use the answer to **(d)** to calculate the value of the density of nuclear matter. Take the value of r_0 to be 1.0×10^{-15} m.

15.11 A neutron star is a dead star consisting entirely of neutrons. Its mass may be about the same as that of the Sun (2×10^{30} kg). A typical diameter is 18 km.
Calculate the density of the material.

15.12* The Stanford Linear Accelerator can accelerate electrons to an energy of up to 20 GeV. When fired at protons the electrons *emerge with different energies and at different angles.*

(a) This is called deep inelastic scattering. Which part of the phrase in italics shows that the scattering is inelastic?

(b) The experiment shows that the proton has a structure. How is this experiment similar to Rutherford's α-particle scattering experiment?

(c) What did physicists mean when they said that the experiment showed that the proton was not a fundamental particle?

15.13 Tin (symbol Sn and proton number 50) has more stable isotopes than any other element. The number of neutrons can be 62, 64, 65, 66, 67, 68, 69, 70, 72 or 74. Give a list of the symbols for these nuclides.

15.14 The mass of an atom of sodium, Na, is 3.82×10^{-26} kg.
What is the mass of one mole of sodium atoms? [Use data.]

15.15 This is a list of the commonest nuclides of several different elements with increasing proton number Z:

$$^{12}_{6}C \quad ^{45}_{21}Sc \quad ^{93}_{41}Nb \quad ^{147}_{62}Sm \quad ^{208}_{82}Pb$$

For each nuclide calculate the ratio N/Z, where N is the number of neutrons (the neutron number) and comment on your answer.

15.2 Unstable nuclei

In this section you will need to

- remember the nature of the alpha (α), beta (β^- and β^+) and gamma (γ) radiations together with their relative charges and masses
- understand the mechanism of electron capture as another method by which one nuclide changes to another
- write nuclear equations of the form $^{4}_{2}He + ^{14}_{7}N \rightarrow ^{17}_{8}O + ^{1}_{1}H$ in which both charge and mass are conserved.
- remember what you studied in Chapter 9 relating to energy levels in atoms, the quantisation of energy and the use of the equations $c = f\lambda$ and $E = hf$.

15.16 Copy this table and fill in the gaps to show the properties of these four types of radiation from unstable nuclei:

	α	β^+	β^-	γ
electric charge / C			-1.6×10^{-19}	
mass / kg		9.1×10^{-31}		
ionising power	very high			
what it is				a photon

15.17 Samarium-147 (proton number 62, symbol Sm) decays by α-emission. The following is a list of neighbouring elements with their proton numbers (in brackets): cerium (58), praseodymium (59), neodymium (60), promethium (61), europium (63), gadolinium (64). Explain what isotope samarium must decay into.

15.18 Cobalt-56 (symbol $^{56}_{27}$Co) is an isotope that decays by β^+-emission.

(a) What name is given to a β^+-particle?

(b) How many protons, neutrons and electrons are there in each neutral atom of this isotope?

(c) The neighbouring elements in the Periodic Table are $_{26}$Fe and $_{28}$Ni. What is the nuclide into which $^{56}_{27}$Co decays?

15.19 Copy the grid which gives the number of neutrons, *N*, in a nucleus (*y*-axis) against the number of protons, *Z* (*x*-axis), i.e. it is a grid of neutron number against atomic number in the region $N = 81$, $Z = 57$.

(a) Explain why the arrow shows an α-decay and write the nuclear equation for this decay.

(b) Add two labelled arrows to your grid to show

(i) a possible β^--decay

(ii) a possible β^+-decay.

Give the appropriate nuclear equations.

(c) Discuss how the arrows you have drawn for β-decays support the fact that nuclides with an excess of neutrons tend to undergo β^--decay whereas those with too few neutrons tend to undergo β^+-decay.

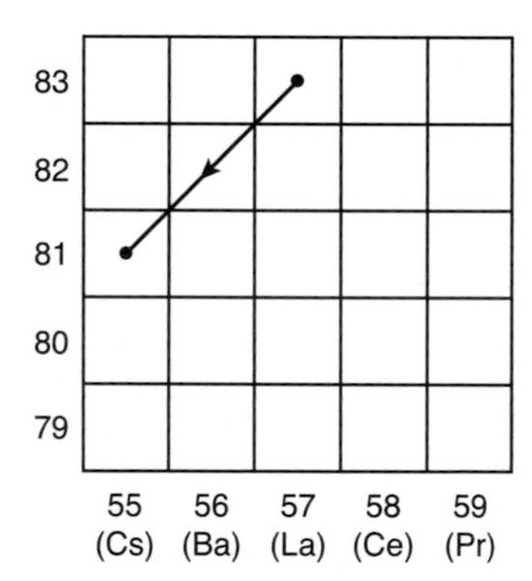

15.20 Draw a grid similar to that in the previous question with the same range of proton number *Z*, but with nucleon numbers *A* ranging from 136 to 140 (*y*-axis).

(a) Draw and label lines on this grid to represent each of α, β^- and β^+ decay.

(b) Give the appropriate nuclear equations.

15.21 $^{238}_{92}$U decays by emitting a succession of α-particles and β^--particles to form $^{206}_{82}$Pb. Deduce how many of each particle are emitted, explaining your working.

15.22 Aluminium-26 ($^{26}_{13}$Al) decays by electron capture.

(a) Write down the equation which describes this decay. The neighbouring elements, with their atomic numbers, are magnesium ($_{12}$Mg) and silicon ($_{14}$Si).

(b) Write down a second equation which shows that the emission of a β^+-particle (a positron) would produce the same nuclide.
(c) Suggest how electron capture and β^+-decay can be distinguished in a laboratory.

15.23 **(a)** When a β^--particle is produced in radioactive decay, is there any change in the number of orbital electrons in the atom **(i)** immediately **(ii)** eventually? Explain.
(b) Repeat this question for the case of α-decay.

15.24 The diagram shows a plot of neutron number N against proton number Z for known nuclides. Each small black square shows that there is a stable nuclide with particular values of N and Z. The stable nuclides form a 'fuzzy' curve of increasing slope.

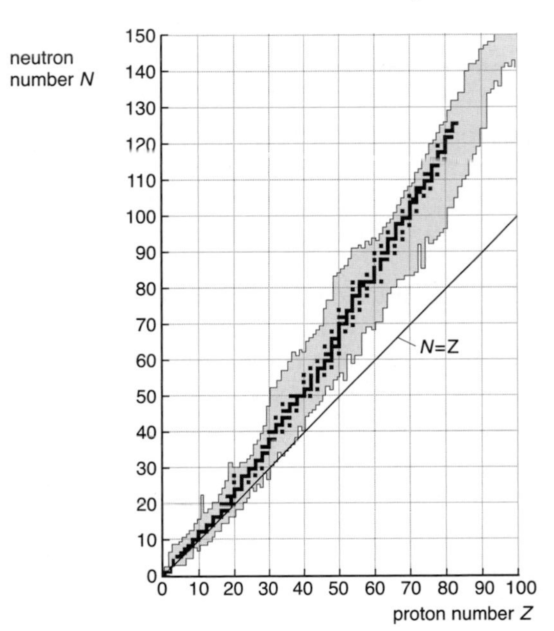

(a) For $Z = 15$, estimate a typical value for N, and calculate the value of N/Z.
(b) Repeat this procedure for $Z = 30, 45, 60, 75$ and sketch a graph to show how N/Z varies with Z. Write a sentence to describe what your graph shows.

15.25 The diagram shows the different energy levels when a nucleus of cobalt-60, decays to an excited state of the nucleus of nickel-60 by emitting a β^--particle.

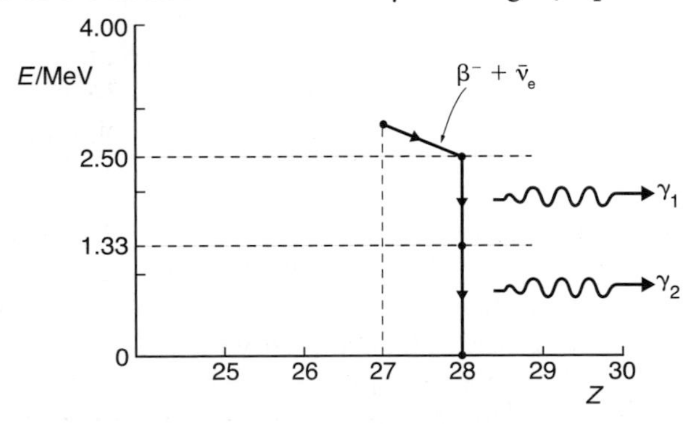

(a) Describe what then happens. (It happens almost instantaneously.)
(b) Calculate the wavelengths of the two γ-photons

15.26 The nuclei of $^{226}_{88}$Ra (radium) undergo α-decay to nuclides of radon (Rn). There are two possible modes of decay: (i) producing α-particles of energy 4.78 MeV when decaying to the ground state and (ii) producing α-particles of energy 4.59 MeV when decaying to an excited state.

(a) Sketch a graph (similar to the one in the previous question) of energy (y-axis) against proton number (x-axis) to illustrate these decays.

(b) Label the decay paths and the result of the subsequent return to the ground state of the excited nucleus.

(c) Calculate the energy in joules of the emitted photon.

15.3 Properties of α, β and γ radiations

In this section you will need to

- remember the properties of the three types of radiation
- remember the relative ionising properties of the radiations
- remember the typical penetrating abilities of the radiations
- describe how to use a GM (Geiger–Müller) tube and scaler in the study of ionising radiations
- describe how to use a diffusion cloud chamber to study the properties of α-particles
- describe how to use an ionisation chamber to study the properties of α-particles
- understand that magnetic fields may be used to deflect β-particles
- understand how to use the equation $I = I_0 e^{-\mu x}$ for the attenuation of radiation where μ is the absorption coefficient for a material absorbing γ-rays.

15.27 The photograph shows tracks left by α-particles in a cloud chamber.

(a) Explain what these white lines consist of.

(b) Each line fades within a few seconds. What has happened to it?

(c) If the source had emitted β-particles or γ-rays there would have been no tracks, or a few very faint tracks. What property of α-particles causes them to leave such obvious tracks?

(d) How can you deduce that the source is emitting α-particles of two different energies?

(e) Suggest why a few of the tracks seem to have small 'kinks' in them at the very ends of their paths.

15.28 The 'radium' sources available in school and college laboratories consist of a mixture of nuclides as a result of the decay of radium. They therefore emit α-particles, β^--particles and γ-rays. A student places sheets of different materials of different thicknesses between the radium source and a GM tube placed close to the source and measures the number of events per minute registered by the GM tube. Explain each of the following observations:

(a) When the GM tube is moved a few centimetres further away from the source the count-rate falls considerably.
(b) When the GM tube is moved back to its original position, and a sheet of paper is placed between the GM tube and the source, the count-rate again falls considerably.
(c) The paper is removed, and the GM tube is moved to a position 10 cm from the source. When a sheet of aluminium of thickness 1.0 mm is placed between the GM tube and the source, the count-rate falls by about 30%.
(d) Each of three more sheets of aluminium of the same thickness cause the count-rate to fall even further.
(e) Two further sheets of aluminium produce little effect on the count-rate, but a sheet of lead of thickness 10 mm has a considerable effect.

15.29* A hundred years ago Rutherford and Royds performed an experiment in Manchester that demonstrated conclusively that α-particles are identical to helium nuclei. Write an account of their experiment. (Diagrams would add a great deal to your account.) Quote the URL of any website or the title and author(s) of any textbook you use to help you.

15.30 Radon is a radioactive element which is a gas at room temperature. It emits α-particles with an energy of 6.3 MeV. Each particle loses on average 30 eV of energy for each pair of ions it creates in collisions with air molecules.
(a) How many ion pairs would you expect an α-particle from radon to create?
(b) What eventually happens to each α-particle?

15.31* **(a)** Bubble chamber experiments show that when a radioactive nuclide decays by β-emission the direction of recoil of the nucleus, and the direction of the emitted β-particle, are not directly opposite (as shown in the diagram). Why does this suggest that a third particle is emitted?

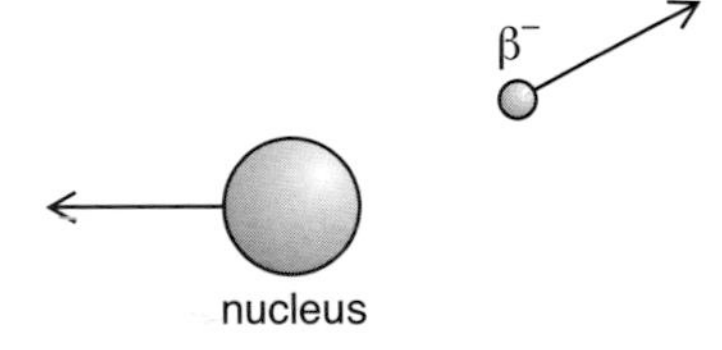

(b) This particle does not leave a track in the bubble chamber. What can you deduce from this?
(c) Why might you expect the β-particles emitted by a particular source to have a whole range of energies?
(d) What names are given to the 'invisible' emitted particles accompanying β^+-emission and β^--emission?
(e) $^{35}_{16}S$ emits a β^--particle. Write a nuclear equation for this decay. (The neighbouring elements are $_{15}P$ and $_{17}Cl$.)

15.32* A student used the apparatus shown in the diagram to investigate, using sheets of aluminium, the absorption of β^--particles from a particular source.

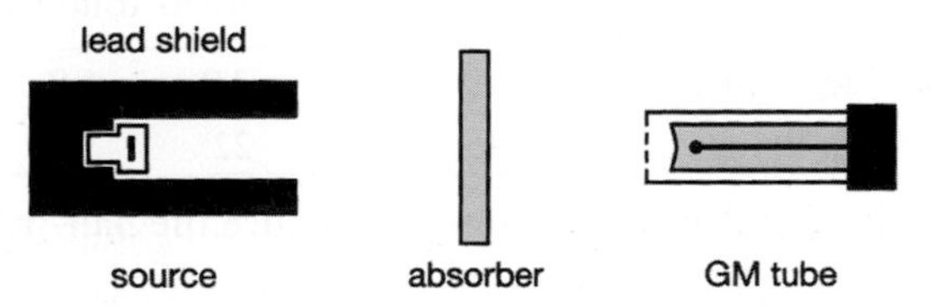

The sheets were stamped with the thicknesses x in mg cm^{-2} as shown in the table on the following page. Each count was taken for a time t.

x/mg cm^{-2}	0	200	387	554	702
counts	19 960	4436	2243	500	224
t/minutes	0.25	1.0	2.0	3.3	10

(a) Create a table to show, for each thickness x, the number of counts per minute, c, and $\ln(c/\text{min}^{-1})$.

(b) Plot a graph of $\ln(c/\text{min}^{-1})$ against x/mg cm^{-2} and deduce a value for the absorption coefficient (μ in the equation $c = c_0 e^{-\mu x}$) of aluminium.

(c) Why did the student have the counter running for a longer time for the later readings?

15.33 The graphs show the percentage of **(i)** α-particles from one source that penetrate a given distance x in air (upper scale on x-axis), and **(ii)** β^--particles from one source that penetrate a given thickness of aluminium (lower scale on x-axis). What can you deduce from the graphs about the particles emitted by the two sources?

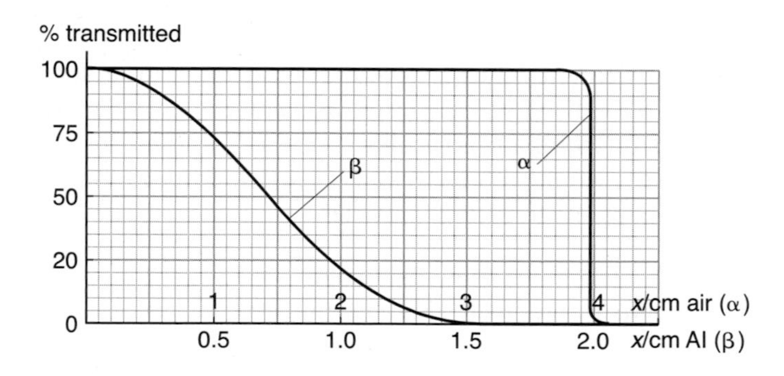

15.34 The half-value thickness for cobalt-60 γ-rays passing through lead is 10 mm.

(a) Draw a graph to show how the intensity of a parallel beam of these γ-rays would vary with thicknesses of lead from zero to 30 mm.

(b) The cobalt-60 γ-ray sources used in school or college laboratories are often contained in lead cylinders whose thickness is about 4 mm. Show that the intensity of the γ-radiation is reduced about 25% by this thickness of lead.

15.35 A laboratory source of γ-rays is often a tiny quantity of cobalt-60 emitting 1.8×10^5 γ-photons per second. Calculate how many γ-photons pass through each of your eyes (taken to have a circular cross-section of radius 10 mm) every second when you look at a laboratory source of γ-rays from a distance of **(a)** 0.50 m **(b)** 2.50 m.

15.36* Sheets of lead of thickness x are placed between a γ-source and a γ-detector, the output of which is a current in pA (10^{-12} A). The results are as follows:

lead thickness x/mm	0	2.0	4.0	8.0
detector current I/pA	31	22	14	7

(a) Plot a graph of I/pA against x/mm and deduce the half-thickness of lead for these γ-rays.

(b) Make tables of $\ln(I/\text{pA})$ and $\ln(x/\text{mm})$ and plot a second graph to deduce the half-thickness.

(c) State which is the better method, and explain your choice.

15.4 Radioactive decay

In this section you will need to

- understand that radioactive decay is a random process but that laws do govern radioactive decay if the activity is large enough
- remember that the activity A of a radioactive source is equal to the number of disintegrations per second and is measured in becquerel
- use the equation $A = -\lambda N$ which relates activity A to the number N of undecayed nuclei, and where λ is the radioactive decay constant
- remember the relationship $\lambda t_{½} = \ln 2$
- understand the mathematics in section 10.7
- use the equation describing radioactive decay $N = N_0 e^{-\lambda t}$
- use radioactive decay curves to find the half-life $t_{½}$ and the decay constant λ of radioactive nuclides
- 'take logarithms' of both sides of the equation $N = N_0 e^{-\lambda t}$ to give $\ln N = -\lambda t + \ln N_0$
- explain how to measure the half-life of long-lived radioactive nuclides
- apply the laws of radioactive decay to situations where radioactive nuclides are used in practice, for example as tracers or in carbon dating.

15.37 The following list shows nine of zinc's known isotopes. Five of these are stable and occur naturally and four are radioactive with short half-lives, given below the list.

$^{63}_{30}Zn$	$^{64}_{30}Zn$	$^{65}_{30}Zn$	$^{66}_{30}Zn$	$^{67}_{30}Zn$	$^{68}_{30}Zn$	$^{69}_{30}Zn$	$^{70}_{30}Zn$	$^{71}_{30}Zn$
38 min	stable	244 d	stable	stable	stable	56 min	stable	2.4 min

(a) Explain what is meant by **(i)** 'isotope' and **(ii)** 'half-life'.
(b) Calculate the decay constant, λ, for **(i)** Zn-65 and **(ii)** Zn-69.
(c) Suggest why Zn-63 decays by β^+-emission but Zn-71 decays by β^--emission.

15.38 $^{24}_{11}Na$, an isotope of sodium which emits β^- particles, has a decay constant of $1.28 \times 10^{-5}\,s^{-1}$. Suppose a sample initially contains 6.00×10^{10} nuclei.
(a) What is its initial activity?
(b) What is its half-life, in hours?
(c) After 30 hours how many undecayed nuclei will there be?
(d) After 30 hours what will be its activity?

15.39* Each of you reading this question has something like 200 g of potassium in your body, which amounts to about 3×10^{24} potassium atoms. This potassium contains 0.012% of a naturally occurring isotope potassium-40 with a half-life of 1.3×10^9 years (1.3 billion years).
(a) What is the decay constant of potassium-40 in s^{-1}?
(b) Calculate the number of unstable nuclei there are in 200 g of potassium.
(c) What is the activity of potassium in your body?
(d) Are you concerned when you sit close to friends?

15.40 The graph shows how the activity of a sample of a radioactive gas $^{220}_{86}$Rn (derived from thorium hydroxide $^{232}_{90}$Th) varies with time as the thorium decays.

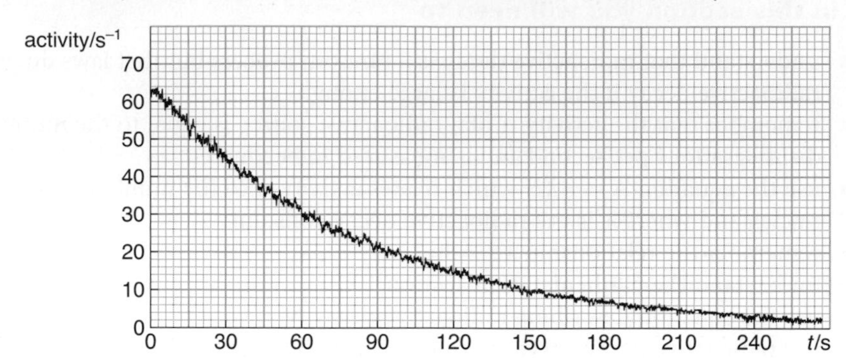

(a) Explain how the graph shows both the random and the systematic nature of radioactive decay.

(b) How many α- and β^--particles are emitted when $^{232}_{90}$Th decays to $^{220}_{86}$Rn? Explain your answer.

(c) Make measurements (note the plural) from the graph to find the half-life of $^{220}_{86}$Rn.

15.41 An exponentially decaying quantity has a value of 654 at 11 a.m. and a value of 587 at 12 noon.

(a) What is the ratio 587/654?

(b) What will be the value of the quantity at **(i)** 1 p.m. **(ii)** 2 p.m.?

(c) What was its value at 10 a.m.?

15.42 Caesium-137 has a half-life of 30 years. How many caesium-137 nuclei will be needed to give an activity of 2.0×10^5 Bq?

15.43* The graph in question 15.40 shows how the activity A of a radioactive nuclide $^{220}_{86}$Rn varies with time t.

(a) **(i)** From the graph set up a table of A and t and add values for $\ln(A/\mathrm{s}^{-1})$.

(ii) Plot a graph of $\ln(A/\mathrm{s}^{-1})$ against t.

(iii) Deduce the value of the decay constant λ from your graph and calculate the half-life $t_{½}$.

(b) Explain whether this value of $t_{½}$ is more reliable than that read directly from the graph of A against t in question 15.40.

15.44 If radioactivity is a random process, how can there be mathematical laws which can be used to calculate what happens?

15.45* In a piece of living timber a fraction 1.25×10^{-12} of the carbon is in the form of the radioactive isotope carbon-14 (^{14}C, half-life 5730 years). When the tree (or an animal) dies the fraction of radioactive carbon decreases.

(a) Calculate the number of atoms of carbon-14 in a 5.00 g sample of carbon taken from a living tree. [Use data.]

(b) What is the decay constant of carbon-14? Hence calculate the activity A of the sample, and find the (average) number of disintegrations to be expected from the sample in 10 minutes.

(c) The average number of disintegrations obtained from a 5.00 g sample of ancient timber in 10 minutes is found to be 520. How old is the timber?

(d) These data do not explain that the background count, an average of 100 counts in 10 minutes, has been allowed for in the calculations. Write briefly how this, and any other assumption made in the calculations, makes this form of 'carbon dating' an inexact science.

15.46 Modern methods of radiocarbon dating take only a very small sample (about a milligram) of carbon. This is ionised and mass spectrometer techniques measure the ratio $^{14}C/^{12}C$ precisely. In modern timber this ratio is 1.25×10^{-12}. How old is a piece of timber for which the ratio is 0.36×10^{-12}?
[$\lambda = 1.21 \times 10^{-4}\,y^{-1}$ for ^{14}C.]

15.47 In order to find the volume of water in a central heating system a small quantity of a solution containing the radioactive isotope sodium-24 (half-life 15 hours) is mixed with the water in the system. The solution has an activity of $1.6 \times 10^4\,s^{-1}$.
When 30 hours have elapsed it is assumed that the sodium-24 has mixed thoroughly with the water throughout the system, and a 100 ml sample of the water is drawn off and tested for radioactivity. It is estimated that the activity of the sample is $2.0\,s^{-1}$.

(a) What is the total volume of water in the central heating system?

(b) Why is a short-lived isotope used in this experiment?

15.5 Nuclear medicine

In this section you will need to

- explain that the cumulative effect of background radiation varies from place to place
- understand the precautions which need to be taken to minimise our exposure to radiation
- understand the use of radioactive nuclides in medicine as tracers for diagnosis and for volume measurement
- understand the use of radioactive nuclides in medicine for therapy.

15.48* $^{99m}_{43}Tc$ is formed from the β^--decay of molybdenum-99 (Mo-99).

(a) Write a nuclear equation for this process.

(b) Molybdenum-99 has a half-life of 67 hours and is produced in nuclear reactors by the process of neutron capture. After five days what fraction remains?

15.49* The half-lives of $^{99}_{42}Mo$ and $^{99m}_{43}Tc$ are 67 hours and 6.0 hours. After the delivery of a sample of molybdenum-99 to a hospital the technetium-99m is chemically separated from the molybdenum for use in tracer studies every day for five or six days. Draw up a weekly plan for tracer studies, such as might be designed by a hospital physicist, starting with the delivery of molybdenum on Monday morning.

15.50 The most widely used radionuclide used as a tracer in diagnostic studies when attached to a suitable chemical compound is a metastable radioisotope of technetium $^{99m}_{43}Tc$. Research the properties of $^{99m}_{43}Tc$ and explain why it is so widely used.

15.51 When $^{99m}_{43}Tc$ returns to its ground state, $^{99}_{43}Tc$, γ-rays are emitted.

(a) Calculate the decay constant of technetium-99m. [For data see question 15.49.]

(b) Each γ-ray has an energy of 140 keV. Show that if the activity of a sample of technetium-99m is about $9 \times 10^{11}\ s^{-1}$, it has an emitted γ-power of 20 mW.

15.52 Suppose a worker in the nuclear industry is working near a radioactive source which has an activity *A*.

(a) Explain why, if *t* is the time for which she is exposed to the radiation and *d* is her distance from the source, the dose *D* of radiation received is proportional to At/d^2.

(b) The damage done also depends on the nature of the radiation. The dose equivalent *H* is given by $H = QD$ where *Q* is the quality factor for the radiation. Explain why the quality factor for α-particles (20) is much higher than the quality factor for γ-rays (1).

15.53* The diagram shows how the average annual radiation dose equivalent of 2 mSv (millisievert) is made up for people living in the UK.

(a) What percentage of the average dose comes from artificial sources?

(b) The dose received varies considerably over the British Isles. Suggest why some regions are regions of high dose.

(c) In the year following the Chernobyl accident the average dose over the whole of Britain increased by about 0.1 mSv. Express this as a percentage of the average dose from **(i)** natural sources **(ii)** artificial sources.

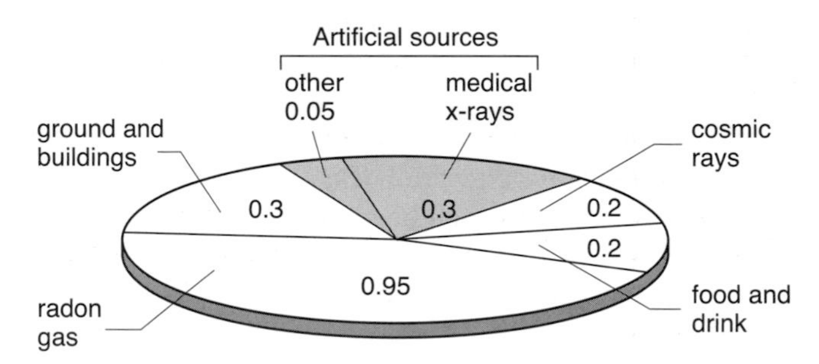

15.54 A radioactive source is to be mixed with some mud in preparation for a study of the movement of sediment in an estuary. The team involved wear protective clothing, handle the source with very long pincers and complete the task as quickly as possible after removing the γ-source from its lead-lined box. Explain how they are reducing the dose they receive to as small a value as practicable.

15.55* People living near Chernobyl when the explosion took place absorbed ^{131}I from the fall-out into their thyroid glands. ^{131}I has a radioactive half-life of 8 days, but as well as decaying it is also removed from the body by excretion. In 15 days half is excreted.

(a) What is the decay constant, in d^{-1} (per day), for **(i)** radiation **(ii)** excretion?

(b) What is the total decay constant?

(c) What is the effective half-life for ^{131}I in the thyroid gland?

15.56 Iodine-123 has a half-life of 13 hours. In the diagnostic treatment of the thyroid gland, a solution of sodium iodide of activity 500 kBq is prepared and given as a drink to the patient whose thyroid gland takes up the iodine. Calculate

(a) the decay constant λ for iodine-123

(b) the number of atoms of iodine-123 needed to produce an activity of 500 kBq

(c) the mass of iodine-123 that contains this number of atoms. [Use data.]

15.57* Chromium-51 has a half-life of 28 days. Some chromium-51 in the form of sodium chromate is used to measure the volume of red cells in a patient's blood.
A sample of 10 ml of the patient's blood is 'labelled' with this tracer and injected into the blood stream. The activity of the injected sample is 210 kBq.
After 10 minutes or so a 10 ml blood sample is removed. This sample is found to have an activity of 1100 Bq.

(a) What is the volume of red blood cells in the patient?
(b) Give two assumptions you made in your calculation.

15.58 In investigations of the blood flow in parts of the brain, a patient may be injected with a β^+-particle emitter such as $^{18}_{9}F$, $^{15}_{8}O$, $^{12}_{7}N$ or $^{11}_{6}C$. These nuclides are created in a cyclotron on the hospital site because they all have short half-lives.

(a) Considering the values of N and Z for these nuclides, why might you expect them all to be β^+-particle emitters?
(b) Give two reasons why short half-lives are desirable.

The emission of the positrons enables doctors to monitor brain activity. This is called positron emission tomography (PET).

(c) Why is it necessary to base PET on the use of positrons, rather than electrons?

(See also questions 16.12 and 16.13.)

15.59 The nuclide $^{238}_{94}Pu$ decays by emitting α-particles of energy 5.5 MeV with a half-life of 88 years. It is proposed to use this nuclide to power a heart pacemaker.

(a) What is the initial activity of the source if it is to be inserted with a power of 70 mW?
(b) Calculate how many Pu-238 atoms would provide this initial power.
(c) What would be the mass of such a source?
(d) For how long will the output of the source be more than 50 mW?
(e) Into what nuclide does Pu-238 decay? (Neighbouring elements are $_{93}Np$, $_{92}U$ and $_{91}Pa$.)
(f) For safety reasons, what properties would this nuclide and its decay products need to have?
(g) Use a search engine to find out the half-lives of the nuclide and the next two of its decay products.

15.60 In the treatment of some cancers needles filled with radioactive material are inserted into the malignant tissue. A typical needle is shown in the diagram. Caesium-137, whose half-life is 28 years, is often used in the needle. What will be its activity if the mass of the caesium is 2.0 mg? [Use data.]

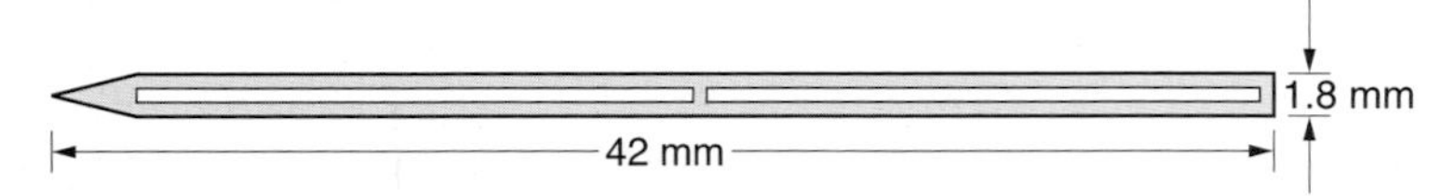

15.61 $^{131}_{53}I$ has a half-life of 8.0 days and decays by β^--emission to an isotope of xenon. It can be used in thyroid therapy, i.e. to kill or damage cancerous cells in the thyroid. For this purpose high doses $\approx$ 2000 MBq are used.

(a) Write a nuclear equation for the β^--decay of iodine-131.
(b) Calculate the number of atoms in a 2000 MBq source of iodine-131 and hence the mass of the isotope in this source.
(c) Explain why iodine-131 can be used for thyroid therapy but the pure γ-emitting isotope iodine-123 cannot be used.

16 Nuclear physics

Data: *You will need to use this data in many questions in this chapter. [Use data.] is not stated on each occasion.*

Avogadro constant $N_A = 6.02 \times 10^{23}\,mol^{-1}$

unified atomic mass constant $u = 1.660\,43 \times 10^{-27}\,kg$

mass of proton $m_p = 1.672\,52 \times 10^{-27}\,kg = 1.007\,28\,u$

mass of neutron $m_n = 1.674\,82 \times 10^{-27}\,kg = 1.008\,67\,u$

mass of electron $m_e = 9.109\,08 \times 10^{-31}\,kg = 0.000\,51\,u$

electron volt: $1\,eV = 1.60 \times 10^{-19}\,J$

rest mass of a nucleon (p or n) $= 1.67 \times 10^{-27}\,kg$

speed of electromagnetic waves in a vacuum $c = 3.00 \times 10^{8}\,m\,s^{-1}$

electronic charge $e = 1.60 \times 10^{-19}\,C$

$1\,u \equiv 931.5\,MeV$

Planck constant $h = 6.63 \times 10^{-34}\,J\,s$

up quark u: charge $\frac{2}{3}e$; down quark d: charge $-\frac{1}{3}e$

16.1 Mass and energy

In this section you will need to

- use the mass–energy equivalence equation $\Delta E = c^2 \Delta m$
- understand that all changes in energy result in changes of mass, and vice-versa
- remember how to convert masses and energies between the units: u, kg, J, eV
- explain the terms binding energy and mass defect of a nucleus
- explain β^- and β^+ decay in terms of a surplus or deficit of neutrons
- understand what is meant by antimatter, that antiparticles may annihilate each other, and that pairs of antiparticles can be formed.

16.1 Express the following masses in unified atomic mass units, u:

(a) $1.674\,82 \times 10^{-27}\,kg$

(b) $9.109 \times 10^{-31}\,kg$

(c) $1.992\,51 \times 10^{-26}\,kg$.

16.2 The mass of a proton is 1.007 276 u.

(a) Express this mass in kilograms.

(b) What is the energy equivalent of this mass, in joules?

(c) Express this energy in MeV.

16.3 Using the values of c and e given in the data, show that $1\,\text{u} \equiv 934\,\text{MeV}$.
(More precise values of c and e give $1\,\text{u} \equiv 931.5\,\text{MeV}$ – the value given in the data.)

16.4 An α-particle with a kinetic energy of 5.48 MeV is emitted from a nucleus.

(a) What additional mass does the moving α-particle have as a result of this energy? Give your answer in u.

(b) The rest mass of the α-particle is 4.0015 u. What is its additional mass as a percentage of its rest mass?

16.5 A β-particle with the same kinetic energy as the α-particle in the previous question (5.48 MeV) is emitted from another nucleus.

(a) What additional mass does the moving β-particle have as a result of this energy? Give your answer in u.

(b) The rest mass of the β-particle is 0.000 55 u. What is its additional mass as a multiple of its rest mass?

16.6* It requires 1.75 GJ to raise the temperature of a steel bar of mass 5.0 tonnes from 300 K to 1100 K.

(a) What is the mass equivalent of this energy?

(b) What is its additional mass as a fraction of its rest mass?

(c) Compare this answer with the answer to **(b)** of the previous question. What would you say to someone who said 'This mass–energy equivalence is nonsense – whoever felt the mass of something increase just because it had more energy? We all know the mass of something is constant.'

16.7 The diagram shows $^{15}_{6}\text{C}$, a nuclide whose atomic mass is 15.010 60 u.

(a) Add up the masses of its protons, neutrons and electrons. [Use data.]

(b) Your answer to **(a)** is larger than the atomic mass of the atom. How much larger?

(c) There is more mass because the particles have more energy when they are separated (it takes energy to pull them apart). What is the binding energy for this nuclide, i.e. how much more energy do the particles have when they are separate? Give your answer in MeV.

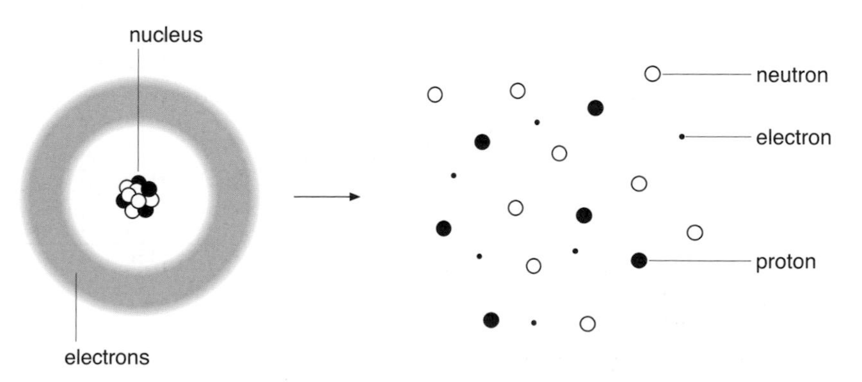

16.8* The argument in parts **(a)** and **(b)** of the previous question can be summarised by the relationship:

$$Zm_p + (A - Z)m_n + Zm_e > \text{atomic mass of nuclide}$$

(a) Describe what this inequality is stating, and show that the difference for the nitrogen-15 atom (mass 15.000 10 u), for which $Z = 7$, is 116 MeV.

(b) Calculate the equivalent difference for the oxygen-15 atom (mass 15.003 07 u).

(c) Suggest which atom (nitrogen-15, carbon-15 or oxygen-15) is likely to be the most stable and explain how you made your choice.

16.9 In 1932 Cockroft and Walton produced the first artificial nuclear disintegration. They directed a beam of low-energy protons at a lithium target in a cloud chamber. The diagram illustrates the outcome of their experiment.

(a) What evidence is there in the diagram to suggest that the two helium nuclei (α-particles) have the same initial energy?

(b) Write a nuclear equation for this event.

(c) Cockroft and Walton measured the energy released in the reaction to be about 17 MeV, a value that agreed with that given by the masses of the particles. Why were these values of mass important at the time?

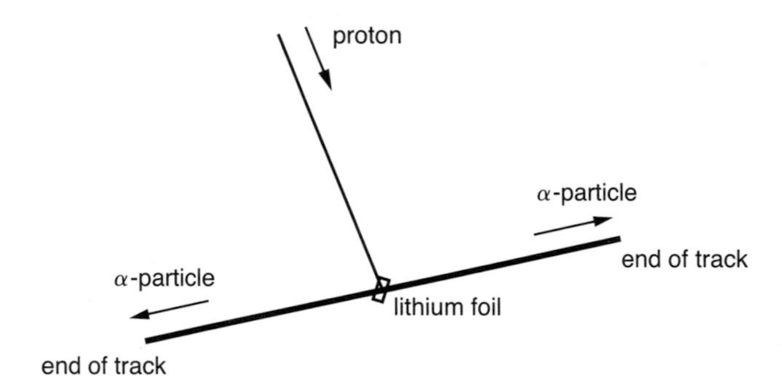

16.10 The graph shows the binding energy per nucleon B/A against the nucleon number A from nitrogen-14 to uranium-238. (Not all nuclides lie exactly on the line and at values of A below 10 there is no smooth relationship.)

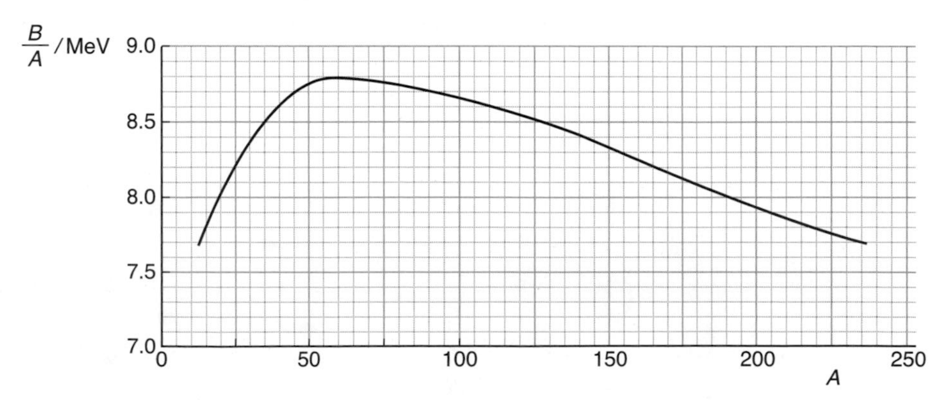

(a) At what value A_{max} does the curve peak?

(b) Estimate the binding energy of this nucleus.

(c) What is the mass difference in kg between this nucleus and its component parts?

16.11 When a $^{224}_{88}$Ra nucleus emits an α-particle and decays to a $^{220}_{86}$Rn nucleus, there is a decrease of 0.006 22 u in the rest masses of the particles.

(a) How much kinetic energy, in MeV, do the particles have after the decay?

(b) The kinetic energy of the α-particle has been measured to be 5.68 MeV. What other particle has kinetic energy?

16.12 The positron is the antiparticle of the electron. It has the same rest mass as an electron. When a positron is emitted in β^+-decay it immediately annihilates with an electron to produce a pair of γ-rays. Mass–energy is conserved.

(a) Why must *two* γ-rays be produced?

(b) How much energy is available in this annihilation for the creation of the two γ-rays? Express your answer in MeV and in J.

(c) What are the directions of the two γ-rays relative to one another?

(d) Calculate the wavelength of the two γ-rays.

16.13* The physics developed in the previous question forms the basis of positron emission tomography (PET) scans.

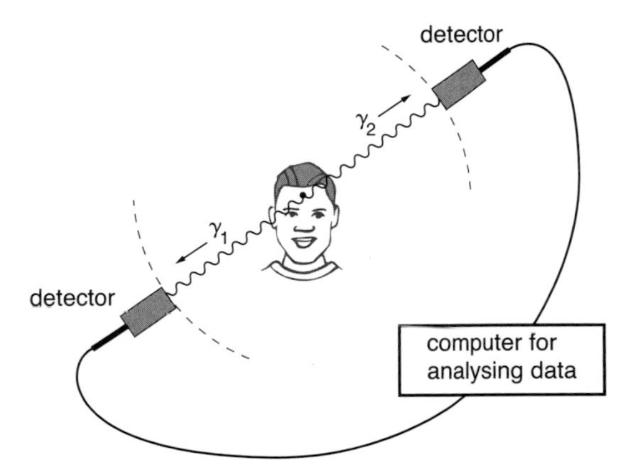

(a) The isotope oxygen-15 decays by β^+-emission. Write a nuclear equation for this decay.

(b) This isotope can be taken up by the blood and has a half-life of about 2 minutes. In the brain the β^+-particles annihilate with electrons. If the difference of arrival times of the two resulting γ-photons at the detectors is 0.30 ns, calculate the difference in distances from the point of annihilation in the brain to the detectors.

(c) Use a search engine to find out more about PET scans.

16.14 The following two equations could be written down for the β^+ and β^- decay of the scandium nucleus:

$$^{47}_{21}\text{Sc} \rightarrow {}^{47}_{20}\text{Ca} + {}^{0}_{1}\beta \quad \text{and} \quad {}^{47}_{2}\text{Sc} \rightarrow {}^{47}_{22}\text{Ti} + {}^{0}_{-1}\beta.$$

The masses of the scandium, calcium and titanium atoms are 46.952 40 u, 46.954 50 u and 46.951 76 u, respectively.

Will the scandium decay by β^+ or by β^- emission? Explain your answer.

16.15 The bubble chamber photograph on the next page shows the spontaneous production of an electron–positron pair from an incoming high-energy photon.

(a) The total rest mass of the created particles is 2 × 0.000 549 u. Calculate **(i)** the minimum energy **(ii)** the maximum wavelength of the incoming γ-ray.

(b) What will be the speed of the e^+ and e^- particles (assumed equal) produced in this decay if the incoming γ-photon had 1.00% more than this minimum energy?

(c) The magnetic field used in taking this photograph was directed into the page. Deduce which particle is the electron and which the positron.

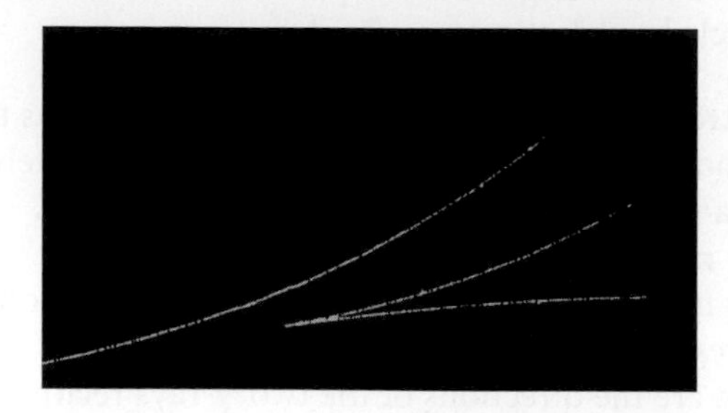

16.16 Uranium-233 decays by α-particle emission to give thorium-229 according to this equation:

$$^{232}_{92}\text{U} \rightarrow {}^{229}_{90}\text{Th} + {}^{4}_{2}\text{He}$$

The masses of the atoms involved are given in the table.

atom	^{233}U	^{229}Th	^{4}He
mass/u	233.039 50	229.031 63	4.002 604

(a) What is the change in mass of the particles involved in this reaction?

(b) How much energy is available as kinetic energy?

(c) The energy of the α-particles is measured to be 4.82 MeV. Why is this less than your answer to **(b)**?

16.17 An isotope of thorium $^{228}_{90}$Th decays by α-emission to an excited state of radium. The masses of the nuclei are: Th 227.9793 u; Ra 223.9719 u; α 4.0015 u.

(a) Write a nuclear equation for this decay.

(b) Deduce the speed of the emitted α-particle, stating any assumption you make.

16.18 The nuclear equation below describes how argon-37 decays by electron capture, i.e. its nucleus captures an inner-shell atomic electron, to form a stable isotope of chlorine:

$$^{37}_{18}\text{Ar} + {}^{0}_{-1}\text{e} \rightarrow {}^{37}_{17}\text{Cl} + \nu_e$$

(a) The masses of the atoms of argon and chlorine are 36.966 77 u and 36.965 90 u. Calculate the energy released in this reaction.

(b) The electron neutrino ν_e carries away nearly all of this energy. What happens to the rest?

16.19 An atom of phosphorus-32 decays by β^--emission into an atom of sulphur-32, and in the process 1.71 MeV of nuclear energy is transferred to the kinetic energy of the particles. The atomic mass of sulphur-32 is 31.972 07 u. Calculate the atomic mass of phosphorus-32.

16.20 The dashed line on the graph shows how the kinetic energy of a proton would vary with the square of its speed if k.e. = $\frac{1}{2}mv^2$ for all speeds v. The curved line shows how the kinetic energy actually varies with the square of the speed because of relativistic effects.

(a) Use the graph to calculate the rest mass m_p of the proton.

(b) For $v^2 = 6.0 \times 10^{16}\,m^2\,s^{-2}$ (i.e. a speed of $2.45 \times 10^8\,m\,s^{-1}$) show that the *increase* in mass Δm of the proton above its rest mass m_p is about 7×10^{-28} kg.

(c) What is $\Delta m/m_p$ for this speed?

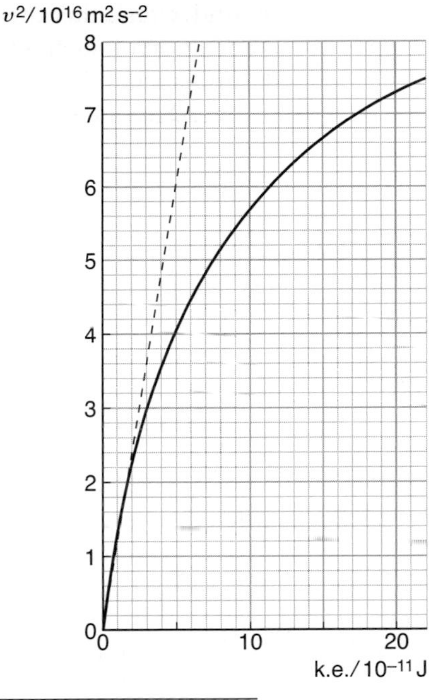

16.21 A nucleus of radon-220 ($^{220}_{86}Rn$) decays by α-particle emission.

(a) Write down the equation for this reaction given that the element with atomic number 84 is polonium (Po).

The masses of the atoms are given in the table:

nuclide	radon-220	helium-4	polonium-216
mass/u	220.011 40	4.002 604	216.001 92

(b) What is the decrease in mass in this reaction?

(c) How much kinetic energy (in MeV) will the two products share?

(d) The mass of the α-particle is 54 times less than the mass of the polonium nucleus. Explain why the speed of the α-particle is 54 times greater than the speed of the polonium nucleus.

(e) Hence explain why the kinetic energy of the α-particle is 54 times greater than the kinetic energy of the polonium nucleus.

(f) What is the kinetic energy (in MeV) of the α-particle?

16.2 Nuclear power

In this section you will need to

- understand that when both nuclear fusion and nuclear fission take place there is an increase in the binding energy per nucleon of the nuclei involved in the reaction
- understand that the absorption of slow neutrons by uranium-235 results in the fission of the resulting nuclide
- understand what is meant by a chain reaction and that it is possible if a reaction induced by one neutron results in the emission of several neutrons
- remember that only 0.7% of naturally-occurring uranium is uranium-235 and that the rest is uranium-238
- understand the use of moderators and control rods in a nuclear pile and which materials might be used for each
- calculate the power available from a nuclear reactor, given the masses of the nuclides involved

- understand that the bombardment of nuclei by fast-moving particles can result in the creation of nuclides which do not occur naturally
- understand that accelerators are needed to give energy to particles and that greater amounts of energy provide more information.

16.22 In a fission process a uranium-235 nuclide splits into two less massive fragments.

(a) Suppose the fragments have mass numbers 91 and 141 (there are also some free neutrons). Use the graph in question 16.10 to find the binding energies per nucleon for the uranium-235 nucleus and the fragments.

(b) Use these figures to calculate **(i)** the binding energy of the uranium-235 nucleus **(ii)** the sum of the binding energies of the two fragments.

(c) How much energy would this fission make available?

16.23 One other possible pair of fragments from the fission of uranium-235 has mass numbers 95 and 137. Does this fission make available more or less energy than the fragments in the previous question? (Use the graph in question 16.10.)

16.24* When a uranium-235 nucleus absorbs a neutron the resulting nuclide splits into two massive fragments and also emits at least one neutron. (See the diagram.)
One possible neutron-induced fission reaction for uranium-235 is

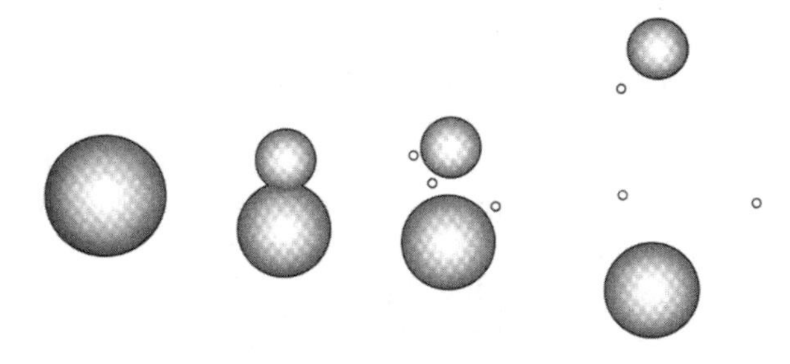

$$^{235}_{92}\text{U} + ^{1}_{0}\text{n} \rightarrow ^{141}_{56}\text{Ba} + ^{92}_{36}\text{Kr} + 3^{1}_{0}\text{n}$$

The masses of the atoms involved are given in the table:

atom	$^{235}_{92}\text{U}$	$^{141}_{56}\text{Ba}$	$^{92}_{36}\text{Kr}$
mass/u	235.0439	140.9143	91.9263

(a) Calculate the change of mass when one of these reactions occurs.

(b) Hence calculate the energy released in MeV.

(c) In 1.00 kg of uranium-235 how many moles are there?

(d) In 1.00 kg of uranium-235 how many atoms are there?

(e) How much energy, in MeV, is available when 1.00 kg of uranium-235 undergoes fission?

(f) Express this energy in **(i)** J **(ii)** kW h.

(g) Suppose a power station is transferring energy at a rate of 1000 MW. How long could it do this using 1 kg of ^{235}U?

16.25 **(a)** Calculate the ratio N/Z (where N is the neutron number and Z the proton number) for the two fission fragments $^{141}_{56}\text{Ba}$ and $^{92}_{36}\text{Kr}$ resulting from the uranium-235 in the previous question.

(b) Refer to question 16.24 and explain whether these fragments are stable and, if not, whether they decay by emitting β^--particles or β^+-particles.

(c) $^{141}_{56}Ba$ and $^{92}_{36}Kr$ eventually decay to $^{141}_{59}Pr$ and $^{92}_{40}Zr$, respectively. How many β^--particles are emitted in each of these decay processes?

16.26 **(a)** Referring to the nuclear equation in question 16.24, explain what is meant by a *chain reaction*.

(b) The possibility of a chain reaction depends on whether the neutrons emitted in a fission event cause more than one fission in another uranium nucleus. For a given mass of uranium what shape is most likely to allow a chain reaction to occur?

(c) Explain what is meant by *critical mass*.

16.27* **(a)** Naturally occurring uranium contains only 0.7% of uranium-235. The rest is uranium-238, which does not undergo fission. To produce 'enriched' uranium, the proportion of uranium-235 must be increased. Use a search engine to discover how this enrichment is achieved.

(b) The nuclide $^{239}_{94}Pu$ also undergoes fission, and is therefore a suitable fuel for a nuclear reactor. (Plutonium-239 is produced in fission reactors when fast neutrons interact with uranium-238 nuclei.)

(i) Why is it easier to separate ^{239}Pu from ^{238}U than to separate ^{235}U from ^{238}U?

(ii) Write two nuclear equations, each involving β^--decay, that lead to the production of $^{239}_{94}Pu$ when neutrons interact with $^{238}_{92}U$. (Element 93 is neptunium, Np.)

16.28* The diagram shows a section of a nuclear pile. The fuel, the moderator and the control rods are labelled.

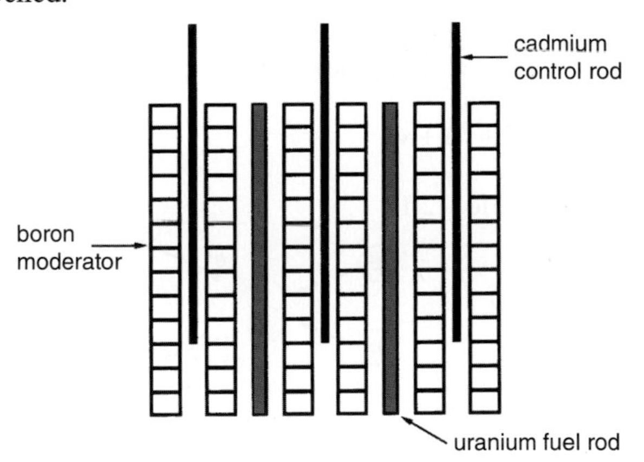

(a) Describe the function of **(i)** the moderator **(ii)** the control rods.

(b) Some fission fragments in such a nuclear pile can be extracted and used for industrial applications such as the γ-irradiation of food or the powering of RTGs – radioactive thermo-electric generators. Other nuclides with long half-lives must be stored. Describe the problems associated with the long-term storage of radioactive nuclear waste.

16.29 The fission of one atom of uranium-235 releases about 200 MeV of energy. Suppose a nuclear power station that uses uranium-235 has an output of 1000 MW and is 40% efficient. Calculate

(a) the rate at which energy is converted in the reactor

(b) the number of atoms of uranium-235 that it uses per hour

(c) the quantity of uranium-235 atoms (in mol) that it uses per hour
(d) the mass of uranium-235 that it uses per hour.

16.30* Naturally occurring uranium contains only 0.72% of uranium-235 with essentially all the rest being uranium-238. The half-lives of ^{235}U and ^{238}U are about 7×10^8 years and 4.5×10^9 years. Three billion years ago, a natural fission reaction occurred in what is now a uranium mine in Gambia, West Africa.

(a) Show that there was twenty times more ^{235}U three billion years ago than now.
(b) Estimate how many times more ^{238}U there was then than now.
(c) What percentage of natural uranium was ^{235}U at that time?

16.31 Refer to question 16.22 and the graph that accompanies question 16.10. The calculations explain how nuclear energy from fission results from the spontaneous fission of nuclides with high nucleon number such as uranium-235. Nuclear fusion involves the fusion of nuclides with low nucleon numbers, e.g. two deuterium nuclei yielding an isotope of helium:

$$^{2}_{1}\text{H} + ^{2}_{1}\text{H} \rightarrow ^{3}_{2}\text{He} + ^{1}_{0}\text{n}$$

(a) Explain how the low-nucleon-number part of the graph suggests that energy will be released in fusions of this kind.
(b) The masses of the nuclei of ^{2}H and ^{3}He are respectively 2.013 553 u and 3.014 932 u. Calculate **(i)** the loss of mass in this reaction using the data **(ii)** the energy available from this reaction.
(c) How many fusions per second would be needed to produce a power output of 2000 MW?
(d) What mass of deuterium would be needed each day?

16.32* In the interior of the Sun, nuclear reactions take place whose net result is the conversion of hydrogen into helium with the release of energy. The three stages in this proton–proton cycle are as follows:

Stage 1: $^{1}_{1}\text{H} + ^{1}_{1}\text{H} \rightarrow ^{2}_{1}\text{H} + ^{0}_{1}\beta + \nu$
Stage 2: $^{1}_{1}\text{H} + ^{2}_{1}\text{H} \rightarrow ^{3}_{2}\text{He} + \gamma$
Stage 3: $^{3}_{2}\text{He} + ^{3}_{2}\text{He} \rightarrow ^{4}_{2}\text{He} + 2^{1}_{1}\text{H}$

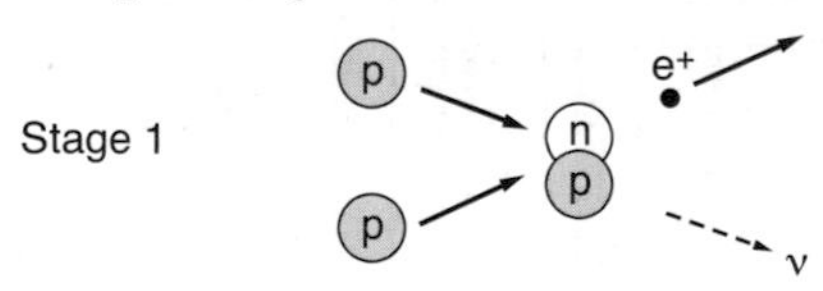

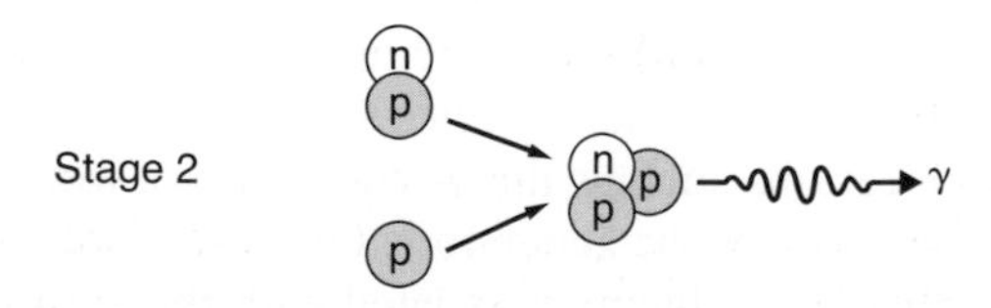

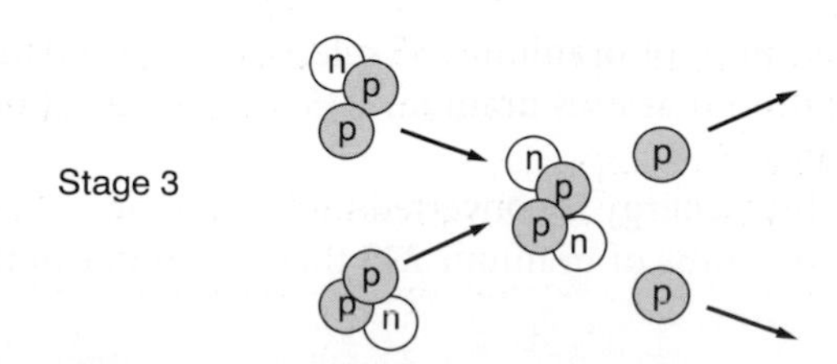

The net effect of this process is to convert four protons into one helium-4 nucleus and release a large amount of energy. The overall reaction is:

$$4\,{}^{1}_{1}\text{H} \rightarrow {}^{4}_{2}\text{He} + 2\,{}^{0}_{1}\beta$$

together with the emission of two neutrinos and two gamma photons.

(a) Confirm that charge is conserved in this overall nuclear reaction.

(b) What is **(i)** the total mass loss **(ii)** the total energy released in the overall reaction? [Use data for the masses of the proton and positron; the mass of the helium nucleus is 4.001 51 u.]

16.33 There is some extra energy available in the conversion of hydrogen into helium described in the previous question, as each ${}^{0}_{1}\beta$-particle (a positron) annihilates with an electron to produce a pair of γ-photons. What is the energy of each photon?

16.34 The first stage in the fusion process described in the diagram for question 16.32 is the fusion of two ionised hydrogen atoms (protons). The energy needed to do this is 430 keV.

(a) This energy comes from the k.e. of the protons colliding at very high temperatures. What value of T in $\frac{3}{2}kT$ will provide this energy? [k is the Boltzmann constant $= 1.4 \times 10^{-23}\,\text{J K}^{-1}$.]

(b) Why are these fusion reactions referred to as thermonuclear reactions?

16.35 The output power of the Sun is about 4×10^{26} W. Referring to questions 16.32 and 16.33, at what rate must the Sun be producing helium nuclei?

Questions 17.44 to 17.57 in the next chapter deal with the effects of fusion reactions in stars.

16.3 Particle physics

In this section you will need to

- remember that quarks are fundamental particles and have an electric charge which is a multiple of $\pm\frac{1}{3}e$
- understand that the four fundamental forces are gravitational, electromagnetic, weak nuclear and strong nuclear and that these have different ranges and have different effects on different groups of particles
- understand that exchange particles are a mechanism by which the four fundamental forces operate, and remember which particles mediate the different kinds of force
- draw Feynman diagrams to illustrate the operation of the four fundamental forces
- remember that electrons, muons and neutrinos (and their antiparticles) form a group called leptons, and that these are fundamental particles
- remember that protons and neutrons (and their antiparticles) are baryons, that pions and kaons (and their antiparticles) are mesons, and that these are all hadrons, and are not fundamental particles
- understand that the up and down quarks have charge and baryon number and that the strange quark has charge, baryon number and strangeness
- understand the use of cloud chambers and bubble chambers to analyse collisions between particles
- remember that charge Q, baryon number B and strangeness S are conserved with the exception that strangeness may vary by 1 in weak interactions
- be able to analyse particle equations to deduce missing information about charge, baryon number and strangeness
- use information about Q, B and S to test whether a reaction is possible.

16.36 **(a)** What are the charges on the up and down quarks (u and d)?
(b) What charge is carried by a baryon with quark structure **(i)** uud **(ii)** udd?
(c) Suggest a quark structure for a baryon with charge $-e$.

16.37 **(a)** List the four interactions (fundamental forces) met with in physics.
(b) Give the exchange particle(s) for each.
(c) Are the exchange particles fundamental particles?

16.38 **(a)** Express the rest mass of a proton as a rest energy. [$m_p = 1.6725 \times 10^{-27}$ kg, $e = 1.6022 \times 10^{-19}$ C and $c = 2.9979 \times 10^8$ m s^{-1}.]
(b) Explain why this mass can be expressed as 938 MeV/c^2.

16.39 The table shows six fundamental particles all of which are leptons. (There are another six leptons, namely the antiparticles of those in the table.)

e^-	μ^-	τ^-
ν_e	ν_μ	ν_τ

(a) List these six particle symbols and give names to the particles they represent.
(b) What are the lepton numbers for **(i)** the six listed in the table **(ii)** their antiparticles?
(c) Which fundamental force does *not* affect leptons?

16.40 Is electric charge conserved in these reactions?
(a) $\nu_e + n \rightarrow p + e^-$ **(b)** $n \rightarrow p + e^- + \bar{\nu}_e$
(c) $\nu_\mu + n \rightarrow p + \mu^-$ **(d)** $\bar{\mu} \rightarrow e^- + \bar{\nu}_e + \nu_\mu$

16.41 Show that lepton number is conserved in each reaction listed in the previous question.

16.42* **(a)** Explain why neutrinos are very difficult to detect.
(b) Use a search engine to write notes about an experiment that does detect neutrinos.

16.43 The table lists the charge Q, the baryon number B and strangeness S for some quarks and antiquarks.

type	Q	B	S
u	$+\frac{2}{3}e$	$+\frac{1}{3}$	0
d	$-\frac{1}{3}e$	$+\frac{1}{3}$	0
s	$-\frac{1}{3}e$	$+\frac{1}{3}$	−1
$\bar{u}$	$-\frac{2}{3}e$	$-\frac{1}{3}$	0
$\bar{d}$	$+\frac{1}{3}e$	$-\frac{1}{3}$	0
$\bar{s}$	$+\frac{1}{3}e$	$-\frac{1}{3}$	+1

All baryons have baryon number +1 or −1 and contain three quarks.
(a) What is the quark structure for the proton and the neutron and their anti-particles?
(b) Use the table to predict the possible structure of four other baryons.
(c) Explain why all baryons must be qqq or $\bar{q}\bar{q}\bar{q}$, i.e. there cannot be a mixture of quarks and antiquarks.

16.44 The table lists the charge Q, the baryon number B and strangeness S for nine particles.

symbol	Q	B	S
p	$+e$	+1	0
n	0	+1	0
π^+	$+e$	0	0
π^-	$-e$	0	0
π^0	0	0	0
K^+	$+e$	0	+1
K^-	$-e$	0	−1
K^0	0	0	+1
$\overline{K}^0$	0	0	−1

What name is given to
(a) the first two particles in the table
(b) the last seven particles in the table
(c) all the particles in the table?

(The tables in questions 16.43 and 16.44 will be useful for some of the questions which follow.)

16.45 All mesons contain a quark and an antiquark and have charge $+e$, $-e$ or zero. Use the table in question 16.44 to predict the quark structure of the seven mesons and give the charge, baryon number and name of each of them. Note that one of the mesons has two possible quark structures. (Only one answer is given.)

16.46 **(a)** Write down three impossible two-quark mesons and explain why they cannot exist.
Recent reports suggest the existence of tetraquarks – particles made up of four quarks, a quark–antiquark up meson plus a quark–antiquark bottom meson, as shown in the diagram. (The bottom quark, b, has a charge $-\frac{1}{3}e$.)

(b) **(i)** Write down the quark structure of this tetraquark.
(ii) What is the charge on such a tetraquark?
(iii) What would be its baryon number?

16.47 The mass of the strange quark is thought to be about 90 MeV/c^2.
(a) Why is this not a more definite value?
(b) What is this mass in kg?
(c) The top quark t (charge $+\frac{2}{3}e$) has a mass of about 3.0×10^{-25} kg. Express this mass in GeV/c^2.

16.48 **(a)** By considering the conservation of charge and baryon number, explain whether the following reactions are possible:
(i) $p + p \rightarrow p + p + \pi^0$
(ii) $p + p \rightarrow p + p + \pi^0 + \pi^0$
(iii) $p + n \rightarrow \pi^+ + \pi^0$
(iv) $\pi^- + p \rightarrow n + \pi^0$
(v) $p + p \rightarrow p + \pi^+ + \pi^- + \pi^0$.
(b) Compare **(i)** and **(ii)**. In **(ii)** an additional π^0 was produced. What might have caused this to happen?

16.49 **(a)** What are the antiparticles of the following particles: **(i)** the electron **(ii)** the proton?
(b) Describe as fully as possible what happens when each particle meets its antiparticle.

16.50* The event that led to the discovery of the omega-minus particle, Ω^-, which has quark structure sss, can be summarised by the equation

$$K^- + p \rightarrow K^+ + K^0 + \Omega^-$$

(a) Write the equation and beneath each particle give its quark content and its nature, i.e. whether it is a baryon or a meson.
(b) Check your quark equation for conservation of charge, baryon number and strangeness.
[Use the information in the tables for questions 16.43 and 16.44.]

16.51 Outside the nucleus a neutron decays to form a proton.
(a) Which other two particles are emitted as a result of this decay?
(b) What happens to the quarks when a proton is formed from a neutron?
(c) Is the proton a stable particle?

16.52 The diagram (a Feynman diagram) represents beta-minus decay within the nucleus.

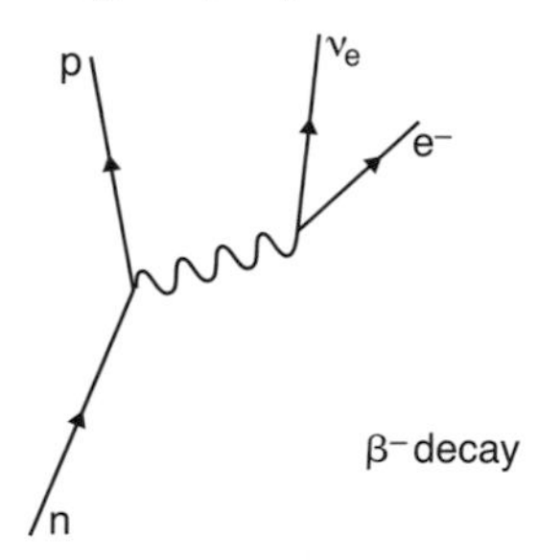

(a) Copy the diagram, adding quarks to the two baryons and labelling the squiggle.
(b) Describe the process of β^--decay **(i)** with an equation using the usual symbols **(ii)** in words.

16.53 Sketch a diagram, similar to that in the previous question, describing β^+-decay.

16.54 The standard model refers to four fundamental forces: gravitational, weak nuclear, electromagnetic and strong nuclear. List these forces and add their exchange particles and an approximate value of their ranges.

16.55 Here are three groups of particles: hadrons, leptons, quarks.
(a) Which two groups of particles are affected by the weak nuclear force?
(b) Which group of particles is not affected by the strong nuclear force?
(c) What kind of particles are affected by the electromagnetic force?

16.56* The result of a collision between an antiproton $\overline{p}$ and a proton p in a bubble chamber is summarised by the equation:

$$\overline{p} + p \rightarrow K^0 + K^- + \pi^+ + \pi^0$$

(a) Use the information in the tables for questions 16.43 and 16.44 to check that in this reaction charge, baryon number and strangeness are conserved. (The antiproton has baryon number −1 and strangeness 0.)
(b) The K^- subsequently interacts with another proton to produce a π^0 plus another particle, X. Predict the quark structure of X.

16.57 Which of the following decays or reactions proceed via the strong force? The Λ^0 has quark structure uds, i.e. strangeness −1.

(a) $K^0 \rightarrow \pi^+ + \pi^-$
(b) $\Lambda^0 \rightarrow p + \pi^-$
(c) $K^- + p \rightarrow \Lambda^0 + \pi^0$
(d) $\Lambda^0 + p \rightarrow \Sigma^+ + n$ (Σ^+ has quark structure uus).

16.58 List the ten baryons which can be formed from any combination of the up, down and charmed (c has charge $+\frac{2}{3}e$) quarks and their antiquarks, along with the charge on each. (Only one answer is given.)

16.59 A particle with a single positive charge enters a bubble chamber containing a large mass of liquid hydrogen (see diagram).

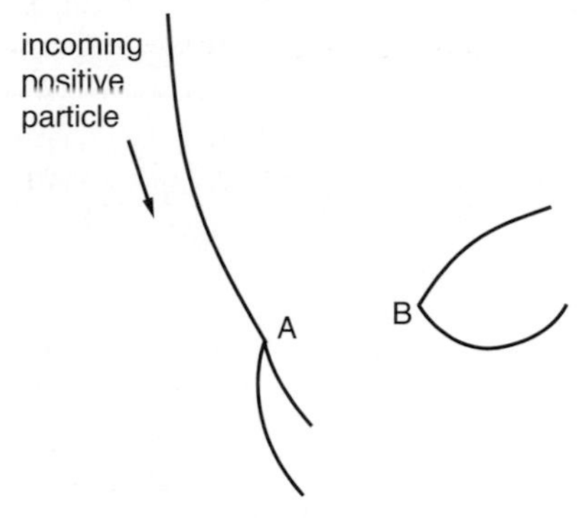

There is a magnetic field into the plane of the diagram which causes the curves of the visible tracks in the bubble chamber.
Account for the tracks produced at A and B.

16.60 The diagram represents the result of a high-energy particle entering a bubble chamber and decaying at A into a neutral particle n and the particle m which causes the track curving up to the right. There is a magnetic field directed into the page.

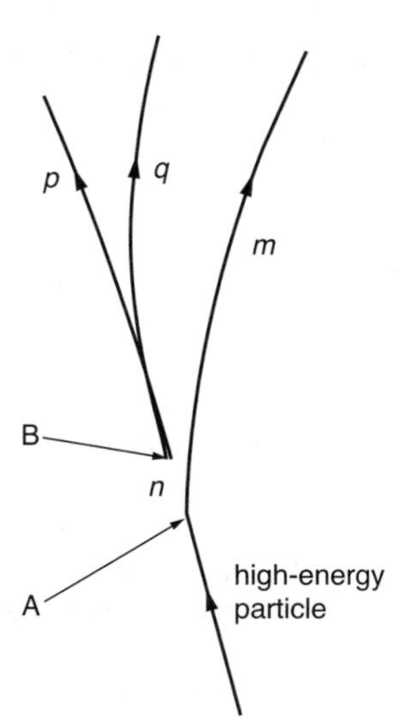

(a) (i) Is the charge on particle *m* positive or negative?
(ii) Why does the particle *n* leave no track?
(iii) The incoming particle must have the same charge as *m*, yet its track appears to be straight. How can this be?

The neutral particle subsequently decays at B into two particles *p* and *q*.

(b) Explain which of *p* and *q* has the greater momentum.

(c) The sum of the masses of *p* and *q* is less than the mass of the neutral particle *n*. How can this be?

16.61 The diagram shows an event (at A and B) recorded in a bubble chamber, in which streams of K^- particles are entering from the left. A K^- particle collides with a proton at A to produce a pion and a Σ^+ (sigma-plus) particle. Shortly afterwards the Σ^+ particle decays at B to produce a second pion and another particle, which leaves no track.

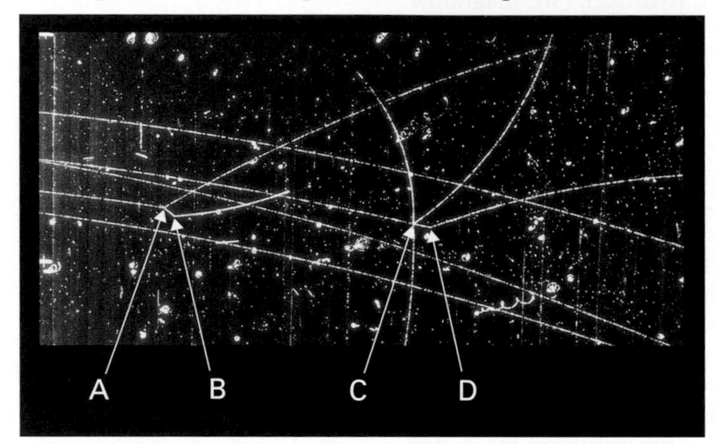

(a) The K^- particle is curving clockwise in the magnetic field in the bubble chamber. Which way is the pion (produced at A) curving? What is therefore the sign of its charge?

(b) Write down an equation for the event which occurs at A.

(c) Use the information in the tables in question 16.43 and 16.44, and the idea of conservation of *Q*, *B* and *S*, to find *Q*, *B* and *S* for the Σ^+ particle.

(d) **(i)** How can you tell that at B a second particle must result from the decay of the Σ^+ particle? **(ii)** What can you deduce from the fact that the second particle leaves no track? **(iii)** Give two reasons why you can deduce that the pion produced at B must have a positive charge.

(e) The decay of the Σ^+ particle is a weak interaction in which strangeness is not conserved. Use the conservation of *Q* and *B* to deduce which of the particles in the second table this particle is.

16.62 The photograph for question 16.61 also shows a second event (at C and D). Another K^- particle collides with a proton at C to produce three pions and a Σ^- particle. Shortly afterwards this particle decays at D to produce another pion and another particle, which leaves no track.

(a) Write down an equation for the event which occurs at C.

(b) What are the quark structures of the Σ^+ and Σ^- particles?

16.63 Strangeness is conserved in strong reactions but may change by 1 in some weak reactions.

(a) By considering the conservation of charge Q, baryon number B and strangeness S in the following strong reactions, explain whether they are possible.

(i) $p + p \rightarrow p + n + K^+ + \overline{K}^0$

(ii) $p + K^- \rightarrow \pi^+ + \pi^+ + \pi^- + \Sigma^-$ (Σ^- contains dds quarks)

(b) In the following weak reactions Q and B are conserved. Is S?

(i) $K^+ \rightarrow \pi^+ + \pi^+ + \pi^-$

(ii) $\Sigma^+ \rightarrow \pi^+ + n$ (Σ^+ contains uus quarks)

(iii) $\Sigma^- \rightarrow \pi^- + n$ (Σ^- contains dds quarks)

There are more questions about charged particles moving in magnetic fields in Chapter 20.

17 Astrophysics

Data: 1° = 3600 arc seconds (or 3600″) = π/180 radians

1 year = 3.15×10^7 s

speed of light in a vacuum $c = 3.00 \times 10^8\,\mathrm{m\,s^{-1}}$

astronomical unit AU = 1.50×10^{11} m (distance from Sun to Earth)

light year (ly) = 9.46×10^{15} m

parsec (pc) = 3.26 ly = 3.08×10^{16} m

luminosity of Sun $L_\odot = 3.90 \times 10^{26}$ W

absolute magnitude of Sun = +4.77

mass of Sun $M_\odot = 1.99 \times 10^{30}$ kg

Stefan–Boltzmann constant $\sigma = 5.67 \times 10^{-8}\,\mathrm{W\,m^{-2}\,K^{-4}}$

Wien's law constant = 2.90×10^{-3} m K

Hubble constant $H_0 = 2.4 \times 10^{-18}\,\mathrm{s^{-1}}$ or $74\,\mathrm{km\,s^{-1}\,Mpc^{-1}}$ (to 2 s.f. only)

gravitational constant $G = 6.67 \times 10^{-11}\,\mathrm{N\,m^2\,kg^{-2}}$

17.1 Astronomical numbers and units

In this section you will need to

- remember that an angle of 1° = 60′ (= 60 arc minutes) = 3600″ (= 3600 arc seconds)
- remember that 1° = π/180 radians
- understand that astronomical distances may be measured in light years (ly), parsecs (pc) and astronomical units (AU), although these are not SI units
- understand the definition of these units, and how each of them may be converted to metres
- remember that the suffix ⊙ indicates that the quantity relates to the Sun, e.g. $M_\odot$ is the mass of the Sun.

17.1 Show that light travels 9.46×10^{15} m in one year.

17.2 If the Sun were to suddenly disappear, how long would it be before we became aware of this on Earth? [Use data.]

17.3* The diagram shows Venus at the point in its orbit round the Sun where, viewed from the Earth, the angle θ is a maximum. The angle EVS is then a right angle.

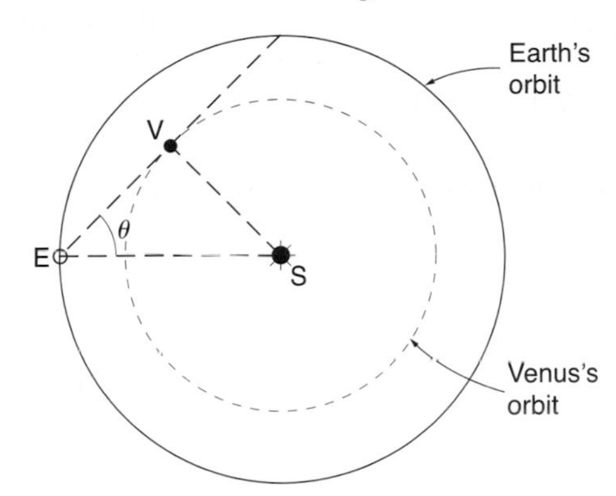

The use of radar enables us to measure distances such as EV. Explain how measuring EV enables us to learn that the distance from the Earth to the Sun, 1 AU, is 1.50×10^{11} m.

17.4 Confirm that 3.26 ly = 3.08×10^{16} m.

17.5 Use a calculator to find how many arc seconds (to seven significant figures) there are in 1 radian.

17.6 Show that the unit of the Stefan–Boltzmann constant σ in $L = \sigma A T^4$ is W m^{-2} K^{-4}. (L is a measure of the power or luminosity of a star of surface area A and surface temperature T.)

17.7 The unit of the Hubble constant, km s^{-1} Mpc^{-1}, may more meaningfully be written as (km/s)/(Mpc). An equivalent unit for H_0 is the s^{-1}.
(a) Show that 1 (km/s)/(Mpc) = 3.24×10^{-20} s^{-1}.
(b) Confirm that the two values for the Hubble constant given in the data are equivalent.

In the rest of this chapter the unit km s^{-1} Mpc^{-1} will be used, with the alternative quantity expressed in s^{-1} in square brackets. You should use whichever unit you are familiar with.

17.8 Show that the expression for the critical mass of the Universe, $\rho = 3H_0^2/8\pi G$, has the unit kg m^{-3}.

17.9 The diagram shows our galaxy, the Milky Way, seen from the side. The distance from the centre of the galaxy to the Sun is about 8500 pc. The Sun lies on one arm of this spiral galaxy and moves round the galactic centre, completing one circle about every 200 My (million years).

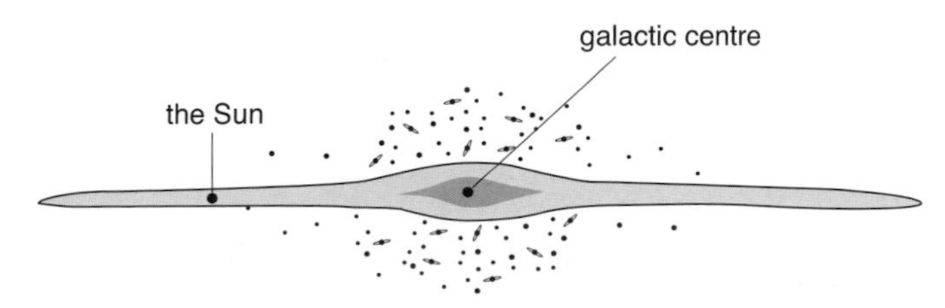

(a) With what speed is the Sun circling the galaxy?
(b) Calculate the centripetal acceleration of the Sun as it rotates around the galactic circle.

17.10 A parsec, pc, is the distance at which the Earth's orbital radius, 1 AU, would subtend an angle of 1 second of arc. Sketch a diagram to illustrate this definition, and prove that 1 pc = 3.08×10^{16} m. (For small angles, $\sin\theta = \tan\theta = \theta$ radians.)

17.2 Observing the Universe

In this section you may (depending on your exam specification) need to

- understand what is meant by M and m, the absolute and apparent magnitudes of a star
- remember that an increase in magnitude is represented by an increasingly *negative* number
- use the distance–modulus equation $m - M = 5\log(d/10\text{ pc})$
- understand what is meant by a standard candle and the importance of Cepheid variables in determining absolute magnitudes
- understand that the luminosity L of a star is a measure of its power in watts
- use Stefan's law $L = \sigma AT^4$ which gives the luminosity L of a star with surface area A and surface temperature (in kelvin) T
- remember that the surface area A of a sphere of radius r is given by $A = 4\pi r^2$
- use the inverse-square law equation $I = L/4\pi d^2$ to find the intensity I of radiation at different distances d from a star
- understand that Wien's law $\lambda_{max}T = 2.90 \times 10^{-3}$ m K relates the wavelength λ_{max} at which the greatest energy is emitted by a star to its surface temperature T
- use the Doppler shift equation $z = \Delta\lambda/\lambda \approx v/c$
- remember that the light from stellar objects moving away from us is red-shifted.

17.11* The diagram shows the positions, E_1 and E_2, of the Earth at an interval of six months, e.g. at 1 January and 1 July. The angle $\theta_2 - \theta_1$ is twice the annual parallax for the star S. (In practice the angles are all extremely small.)
Write a full explanation of what the diagram is describing.

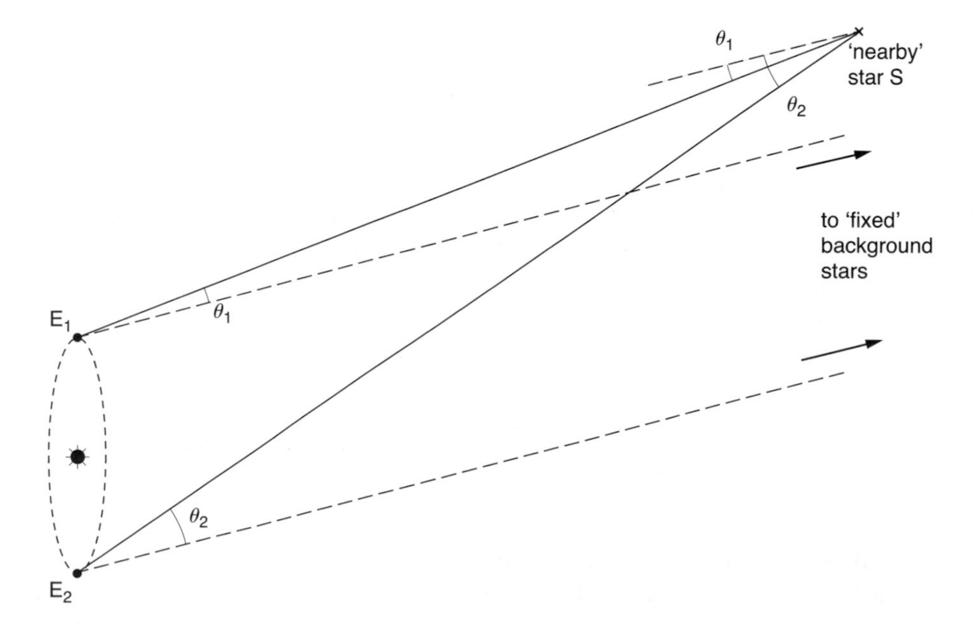

17.12 The distance d (in parsecs) from Earth to a star can be found by calculating the *inverse* of the star's parallax angle p (in seconds of arc or ″). Numerically:

$$d = \frac{1}{p}$$

(a) At what distances, in light years, are stars that exhibit parallax angles of **(i)** 0.772″ (Proxima Centauri, our nearest star) **(ii)** 0.379″ (Sirius, the brightest star)?

(b) **(i)** What parallax 'wobble' does the star Betelgeuse, about 400 light years from Earth, exhibit?

(ii) Use a search engine to discover what are the smallest parallax angles that can be measured.

17.13 A star that lies on an axis perpendicular to the Earth's orbit round the Sun can have a parallax angle p, as shown.

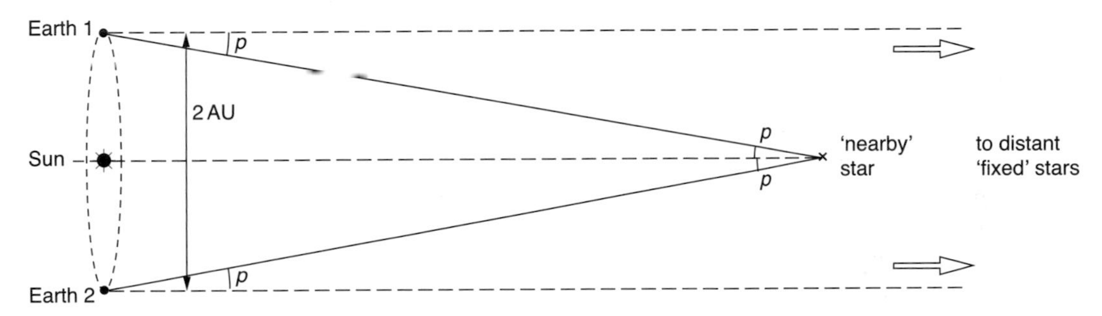

(a) Confirm that, for small angles, p in radians is equal to $1\ \text{AU}/d$ where d is the distance from the Sun to the nearby star.

(b) The current limit for measuring parallax angles is about 0.002 seconds of arc. Calculate **(i)** this angle in radians **(ii)** the furthest distance of stars whose distance from Earth can be directly measured. Give your answer to **(ii)** in both metres and parsecs. [Use data.]

17.14 **(a)** What are meant by **(i)** the absolute magnitude M **(ii)** the apparent magnitude m of a star?

(b) **(i)** Use the distance–modulus equation given at the start of this section to show that the distance d to a star can be expressed as $d = 10\,\text{pc} \times 10^{(m-M)/5}$

(ii) Procyon A has an apparent magnitude of +0.40 and an absolute magnitude of +2.68. How far away, in light years, is Procyon A from Earth?

17.15 The diagram shows the link, to a good approximation, between two different ways of describing the luminosity or brightness of stars. Estimate the luminosity of Procyon A in watts. (See the previous question and the data.)

luminosity

$10^{12}L_\odot$	$10^{10}L_\odot$	$10^{8}L_\odot$	$10^{6}L_\odot$	$10^{4}L_\odot$	$10^{2}L_\odot$	$L_\odot$
−25	−20	−15	−10	−5	0	5

absolute magnitude

17.16 Bellatrix, a bright blue star, has a distance–modulus of 2.7. How far from the Earth is it?

17.17 Alpha Centauri has an apparent magnitude of −0.01 and is known to have an absolute magnitude of +4.34. Show that Alpha Centauri lies 4.4 ly away from Earth.

17.18 Sirius lies 8.6 ly from Earth and has an apparent magnitude of −1.44. Procyon A lies about 11 ly from Earth and has an apparent magnitude of +0.38.
Deduce which of these stars is the brighter, i.e. has the greater luminosity.

17.19* What, in astrophysics, is meant by a 'standard candle'? Use a search engine to find out about **(a)** Cepheid stars **(b)** type Ia supernovae.

17.20 The graph shows how the peak absolute magnitude of one type of variable star, Cepheid variables, is related to the period of oscillation T of the brightness of these stars.

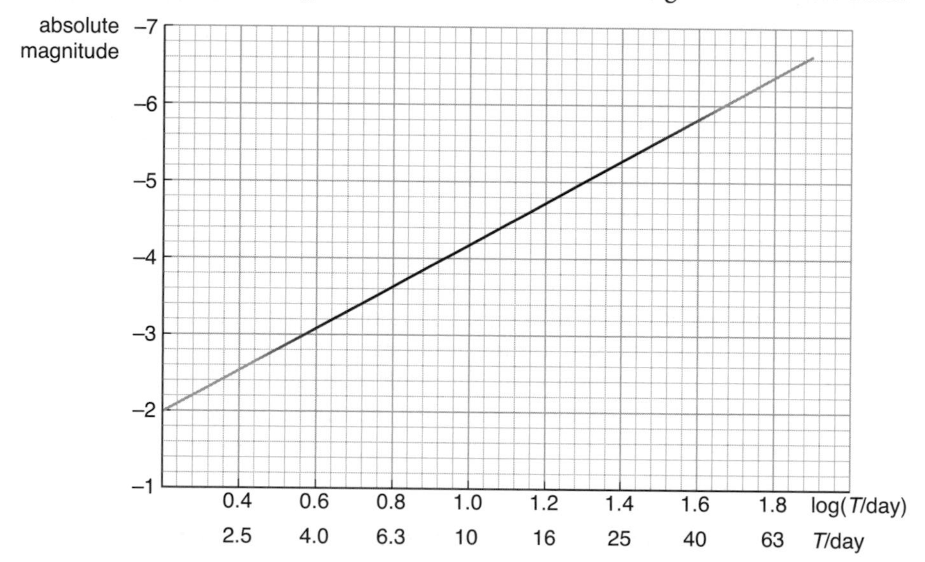

The Pole Star is a Cepheid variable star with a period of 4.0 days.
(a) Use the graph to determine the absolute magnitude of the Pole Star.
(b) The apparent magnitude of the Pole Star is +2.0. How far away from Earth is it?

17.21 The graph shows the variation in the intensity of radiation received on Earth from a Cepheid star in the Andromeda galaxy.
The average luminosity of the star is found to be about $2 \times 10^5 L_{\odot}$.
Calculate the distance from Earth to the Andromeda galaxy.

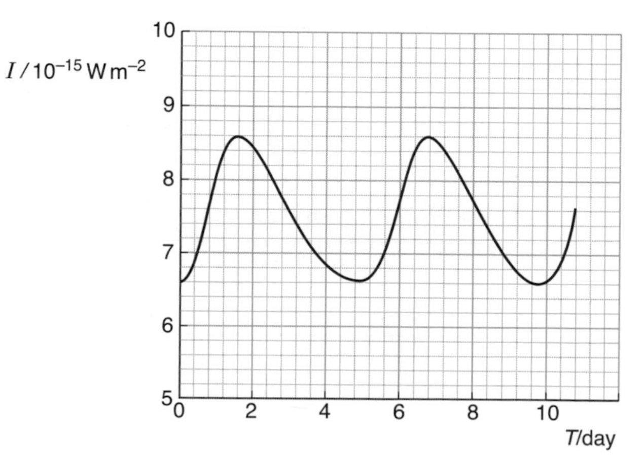

17.22 Refer to the graph in question 17.20.

(a) Estimate the absolute magnitude of a Cepheid star with a period of **(i)** 25 days **(ii)** 5.4 days.

(b) Explain what other measurements are needed in order to calculate the distances to these Cepheid stars.

17.23 Astrophysicists believe that type Ia supernova explosions reach a maximum brightness (absolute magnitude or luminosity) that is the same wherever in the Universe the explosion takes place.
Explain what other information is needed in order to deduce the distance from Earth to such a supernova explosion.

17.24 The graph tells the story of a type Ia supernova explosion. Note that the vertical scale is logarithmic. The maximum brightness is reached within about a day and thereafter the luminosity falls away over a few months.

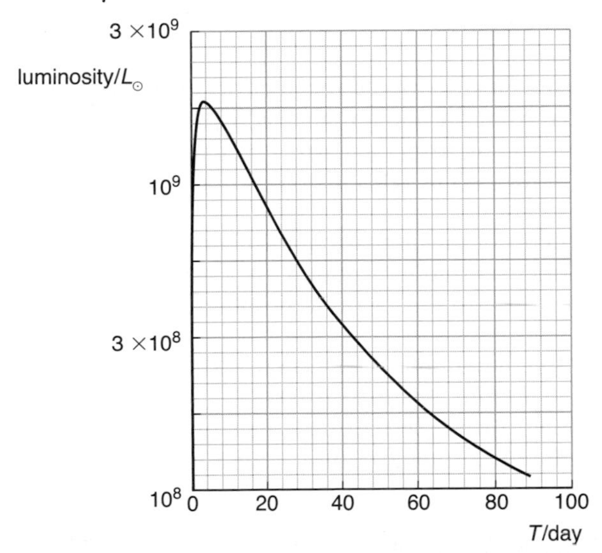

(a) Use the graph and the diagram in question 17.15 to estimate **(i)** the peak luminosity in watts of a type Ia supernova **(ii)** the absolute magnitude of a type Ia supernova.

(b) Read about other types of supernovae. Why do we not use them to estimate how far distant galaxies are from Earth?

17.25* In 1937 a type Ia supernova was observed in a galaxy labelled IC4182. This supernova had a maximum apparent magnitude $m = +8.6$. In 1992, using the Hubble space telescope, other observers recorded a Cepheid variable star in the same galaxy with an apparent magnitude of +22.0 and a period of 42.0 days. Explain why these observations were important in establishing type Ia supernovae as the key to calculating the distances to very distant galaxies.

17.26 The graph shows how the intensity of electromagnetic radiation I from the Sun varies with the distance from its centre d.

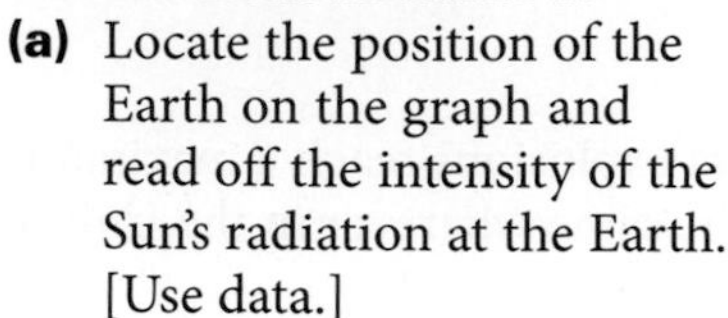

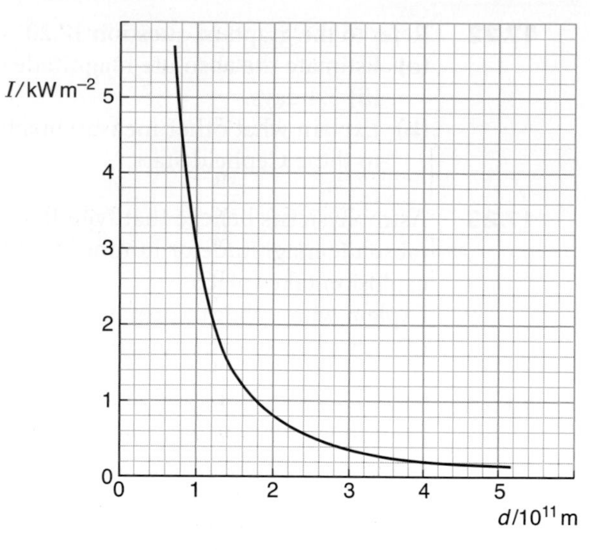

(a) Locate the position of the Earth on the graph and read off the intensity of the Sun's radiation at the Earth. [Use data.]

(b) Confirm that the graph shows an inverse square relationship between I and d.

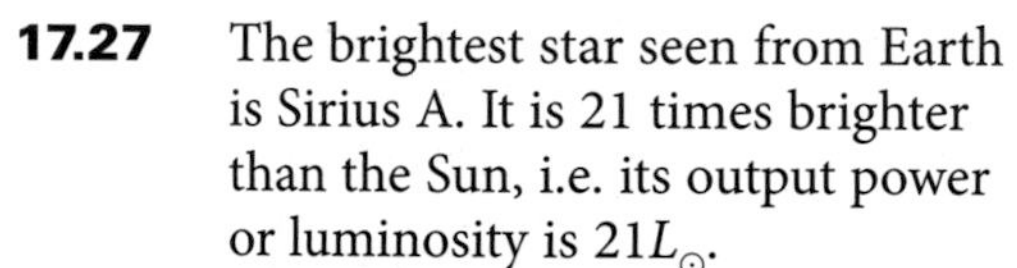

17.27 The brightest star seen from Earth is Sirius A. It is 21 times brighter than the Sun, i.e. its output power or luminosity is $21L_{\odot}$.

(a) How far from Sirius A would the intensity or flux of radiation I be $1.4\,\mathrm{kW\,m^{-2}}$, the same as that from the Sun on the Earth? [Use data.]

(b) Sirius is 8.6 ly away from the Earth. Calculate the intensity of its radiation on Earth in **(i)** $\mathrm{W\,m^{-2}}$ **(ii)** $\mathrm{mW\,m^{-2}}$ **(iii)** $\mathrm{mW\,cm^{-2}}$.

17.28* Explain the benefits of making astronomical observations from above the Earth's atmosphere. (You may find it helpful to use a search engine to learn about the two transparent 'windows' in the Earth's atmosphere.)

17.29 **(a)** Draw a graph illustrating Wien's law $\lambda_{max}T = 2.90 \times 10^{-3}\,\mathrm{m\,K}$ for values of T from 1000 K to 20 000 K.

(b) The wavelength at which the greatest energy is emitted by the Sun is 5.0×10^{-7} m. Calculate the surface temperature of the Sun and mark the position of the Sun on your graph.

17.30 The graph shows the wavelength distribution of the radiation detected by the *COBE* satellite.

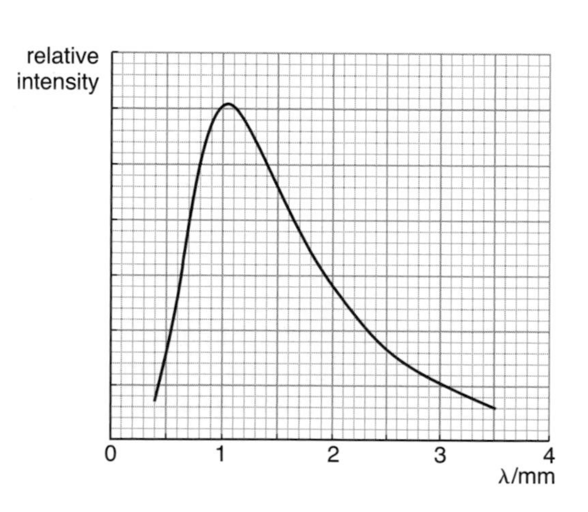

(a) To what part of the electromagnetic spectrum do these radiations belong?

(b) Deduce the temperature of the source of these emissions.

17.31 The intensity of radiation, measured on Earth, from the star Rigel is $5.2 \times 10^{-7}\,\mathrm{W\,m^{-2}}$. Rigel is known to be 240 ly from Earth. Calculate a value for the luminosity of Rigel **(i)** in watts **(ii)** as a multiple of $L_{\odot}$.

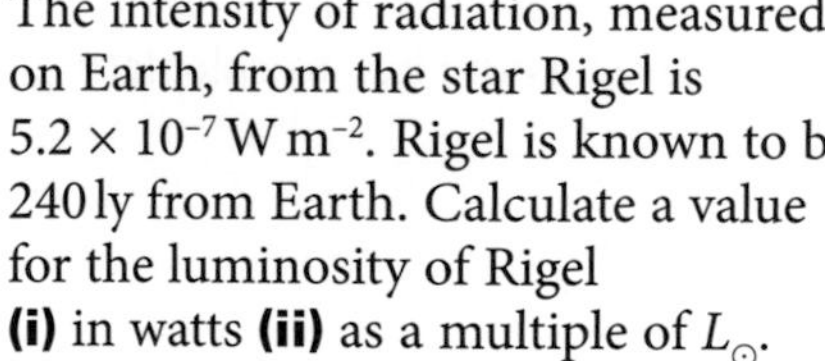

17.32 Barnard's star has a luminosity of 4.0×10^{23} W and a surface temperature of 2700 K. Calculate the surface area of the star and hence find its radius.

17.33 The surface temperature of the Sun is 5800 K. Calculate its radius. [Use data.]

17.34 Betelgeuse has a radius $360R_{\odot}$.
(a) What is the radius of Betelgeuse in AU? Comment on your answer.
(b) Betelgeuse is believed to have a luminosity of about $4 \times 10^4 L_{\odot}$. Calculate its surface temperature and suggest what colour it would have when photographed. [Use data and take $R_{\odot} = 7.0 \times 10^8$ m.]

17.35* The graphs show how the radiation from two stars, with surface temperatures 6000 K and 4000 K, is distributed across the spectrum.
According to Stefan's law, the area under the graphs should be in the same ratio as the fourth power of their temperatures.
(a) Calculate the ratio $(6000\,\text{K})^4 \div (4000\,\text{K})^4$.
(b) Making suitable approximations, count the number of large squares **(i)** beneath the 6000 K curve **(ii)** beneath the 4000 K curve. Does the ratio support Stefan's law?

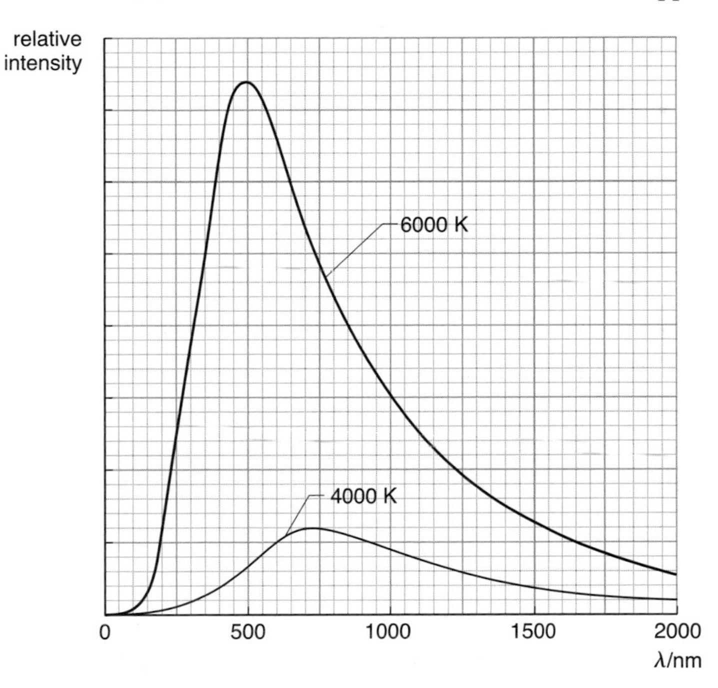

17.36 At what wavelengths are the peaks in the curves in the previous question? Explain whether Wien's law is obeyed.

17.37* The graphs show approximately how the radiation from two stars, with temperatures 6000 K and 12 000 K, is distributed across the spectrum.
Discuss, in terms of Wien's and Stefan's laws, whether the curves have been correctly drawn. (Do not try to make any detailed measurements.)

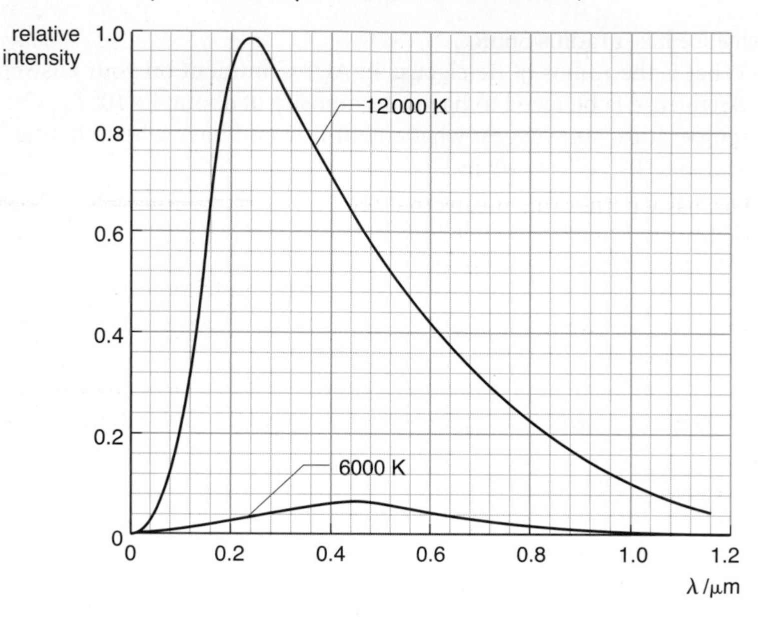

17.38 The stars Betelgeuse and Bellatrix have surface temperatures of 4300 K and 28 000 K respectively.

(a) Calculate the power emitted per unit surface area, L/A, by each star. [Use data.]

(b) The visible spectrum extends from approximately 400 nm to 700 nm. What will be the colour of these two stars?

17.39 The diagram illustrates the Doppler effect for electromagnetic radiation from a source moving at a speed v. The circles labelled 1 to 5 are wavefronts from the source when it was at points 1 to 5 on its path.

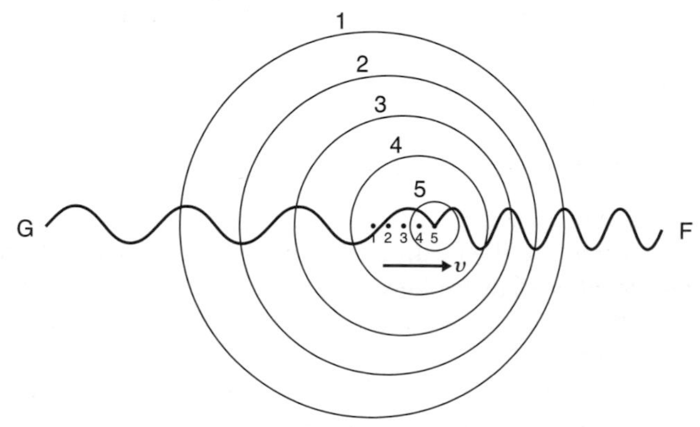

(a) Comment on the wavelength of the radiation observed by an observer **(i)** at F **(ii)** at G.

(b) For $\lambda = 520$ nm and $v = 2.4 \times 10^6$ m s^{-1}, calculate the change in wavelength $\Delta\lambda$ measured by the observer at F.

17.40 The quasar, 3C 273, has a prominent emission line of apparent wavelength 475.0 nm which corresponds to a laboratory line at a wavelength of 410.2 nm. Calculate this quasar's velocity relative to Earth, and express it as a percentage of the speed of light.

17.41 The light from a star in the nearby Andromeda spiral galaxy M31, 2.2 million light years away, is found to be blue-shifted by a factor of 1/1000 at all wavelengths.

(a) Calculate the speed with which this galaxy is approaching Earth.

(b) Discuss whether this means that the radiation from every star in Andromeda will be blue-shifted by the same amount.

17.42 The spectrum of Arcturus, a star in the Milky Way, was photographed twice at an interval of 6 months. An absorption line at 430 nm in the first spectrum showed a red shift of 0.016 nm and the same line showed a blue shift of 0.027 nm in the second spectrum.

(a) Calculate the speeds with which Arcturus and Earth appear to be **(i)** receding from each other **(ii)** approaching each other.

(b) Hence deduce the speed at which the Earth is circling the Sun and calculate a value for the astronomical unit, 1 AU.

17.43 The table gives some information about two stars that have similar surface temperatures of about 9000 K.

star	luminosity/W	distance from Earth/m
Deneb	2.1×10^{31}	1.5×10^{19}
Vega	1.4×10^{28}	2.3×10^{17}

(a) What is the ratio of the intensity of radiation reaching the Earth from these two stars?

(b) Show that Deneb has 1500 times the surface area of Vega.

17.3 Cosmology

In this section you will need to

- understand the processes of fusion which you studied in section 16.2
- familiarise yourself with the Hertzsprung–Russell (H–R) diagram
- understand what is meant by the Main Sequence on an H–R diagram
- understand that an H–R diagram implies a relationship between the luminosity of a star and its lifetime
- understand that Hubble's law, $v = H_0 d$, implies that the speed v of recession of a galaxy is proportional to its distance d from the Earth
- use, as a unit for the Hubble constant H_0, either km s^{-1} Mpc^{-1} or s^{-1}
- understand that a constantly expanding Universe is the implication of H_0 being truly constant

- use the Doppler shift equation $z = \Delta\lambda/\lambda \approx v/c$
- use graphs of the average distance between galaxies against time to discuss the possible outcomes of the Universe.

The process of nuclear fusion (hydrogen 'burning') is investigated in the previous chapter – see questions 16.31 to 16.35.

17.44* The diagram shows the luminosity L and surface temperature T of stars that, like the Sun, are 'burning' hydrogen to form helium. (The vertical scale showing luminosity could be replaced by one showing absolute magnitude – see question 17.15.) These stars lie on what is called the *Main Sequence* of the Hertzsprung–Russell diagram.

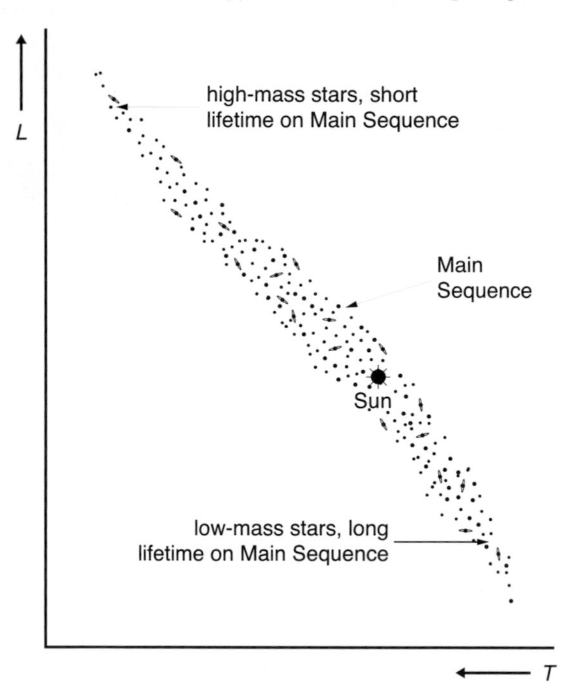

Use a search engine and download a coloured version from the internet. In particular note that:

- both scales are logarithmic, the luminosity scale usually shown as multiples of $L_\odot$ and the temperature scale rising from right to left, i.e. 'backwards'
- a full H–R diagram shows stars other than those on the Main Sequence, e.g. dwarf and giant stars
- much of the information used in questions in the previous section of this chapter is contained in such a coloured H–R diagram. (Some H–R diagrams show the vertical scale in terms of absolute magnitude and the horizontal scale in terms of a star's spectral type.)

17.45 Our star, the Sun, has at present a luminosity of 3.9×10^{26} W. This results from the nuclear fusion of hydrogen to helium.

(a) Calculate **(i)** the mass loss Δm needed in one second to maintain this power output **(ii)** how long before the Sun will have used up 10% of its present mass. [Use data.]

In fact fusion reactions are expected to die away when only about 0.03% of its present mass has been used up.

(b) Using this figure, estimate the lifetime of the Sun on the Main Sequence.

17.46 The following temperatures, doubling each time, illustrate a rising logarithmic sequence:

T/K 2500 5000 10 000 20 000 40 000

On an H–R diagram these temperatures will appear at equal intervals.

(a) Use a calculator to show that the logarithm (to base ten) of these numbers rises linearly from 3.4 to 4.6.

(b) Check that a scale that rises by ×10 each time (e.g. 1000, 10000, etc.) is also logarithmic.

17.47 The diagram is a version of the H–R diagram. (The dashed line shows the position of Main Sequence stars.)

What type of star, and what colour, are each of these stars?

(a) Sirius B **(b)** Rigel **(c)** Aldebaran **(d)** Altair **(e)** Betelgeuse

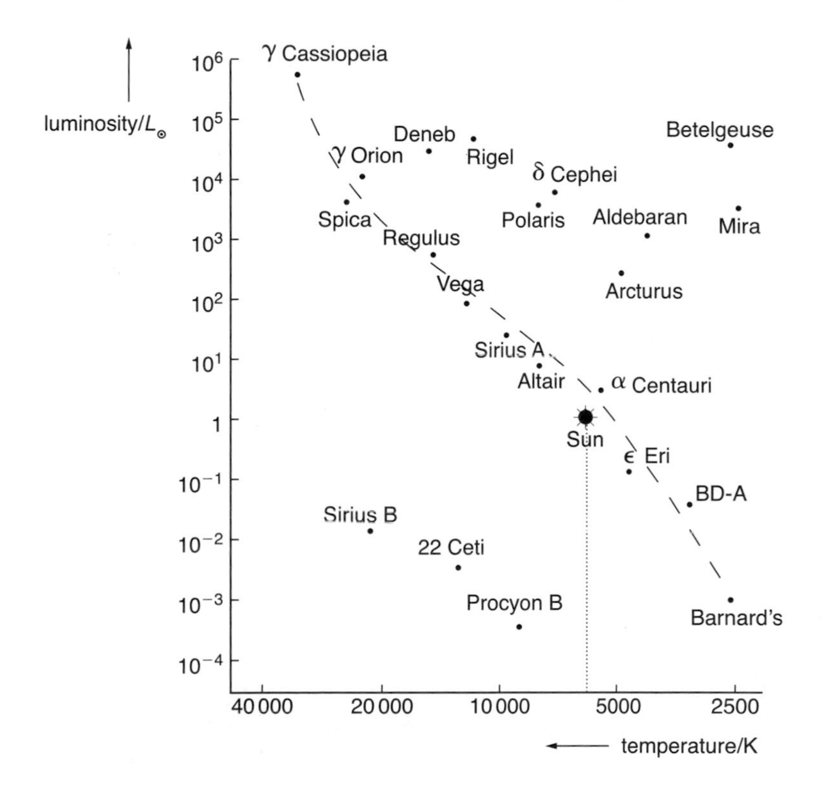

17.48 The luminosity L of stars on the Main Sequence depends on their mass m. Roughly speaking L is proportional to $m^{3.5}$, i.e. m to the power of 3.5.

(a) How many times greater will the luminosity be for a star whose mass is **(i)** twice the mass of the Sun **(ii)** five times the mass of the Sun?

(b) For a star with high luminosity would you expect its rate of 'burning' to be large or small?

(c) Would you expect the time t_{MS} spent on the Main Sequence to be long or short for a star with large mass and therefore high luminosity?

(d) Does your answer agree with the shape of the plot for question 17.44?

17.49* The diagram shows the path that astrophysicists believe that the Sun will follow after it leaves the Main Sequence.

(a) State what sort of star the Sun will evolve into at **(i)** P and **(ii)** at Q.

(b) What will be the future of the Sun after it 'arrives' at Q?

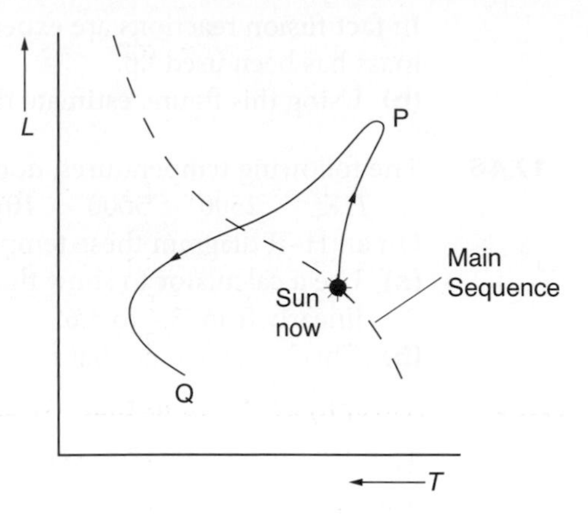

17.50 A star like the Sun will evolve into a red giant of radius 200 times its present radius when it leaves the Main Sequence. Its surface temperature will then be about 4000 K.
Show that the luminosity of such a red giant star is about 9000 times its luminosity while on the Main Sequence. (The present surface temperature of the Sun is 5800 K.)

17.51 The Sun will end up as a cooling dwarf star, about the size of the Earth, consisting of 65% of its Main Sequence mass. Estimate the mean density of such a dwarf star.
[Radius of Earth = 6.4×10^{6} km. Use data.]

17.52 Neutron stars are predicted to have a density of about $4 \times 10^{17}\,\text{kg}\,\text{m}^{-3}$. Calculate the radius of a neutron star with a mass equal to the present mass of the Sun.

17.53 Neutron stars rotate and are observed as pulsars. The pulsar at the heart of the Crab nebula has a period of rotation of 0.033 s and is slowing down at the rate of $12\,\mu\text{s}\,\text{year}^{-1}$. Calculate by how much this pulsar slows down in 100 years and express your answer as a percentage of the pulsar's period.

17.54 The diagram illustrates the lives of stars that have a mass greater than $20\,M_{\odot}$ while on the Main Sequence, and those that have a mass greater than $8\,M_{\odot}$ but less than $20\,M_{\odot}$ while on the Main Sequence. The more massive end up as black holes and the less massive end up as neutron stars.

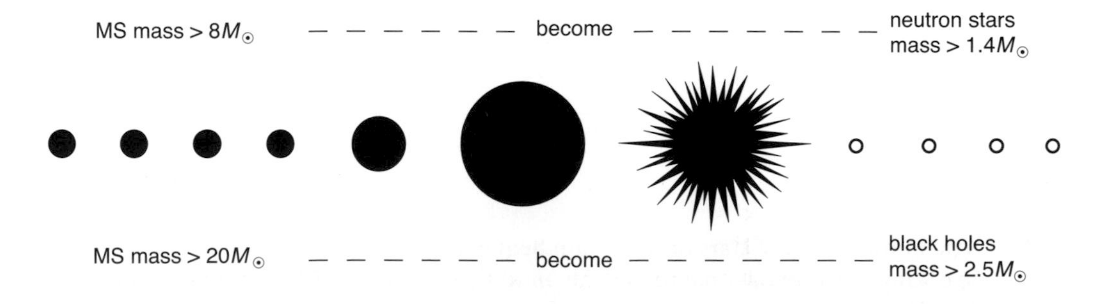

(a) Describe, using reference material, the 'story' of the lives of stars of mass > $20\,M_{\odot}$.

(b) What is meant by a black hole?

(c) Describe how both types of stars are initially formed at the extreme left of the diagram, and state how their lifetimes on the Main Sequence differ.

17.55 The escape speed v_e from the surface of a planet or star of radius r_P and mass m_P is given by

$$v_e = \sqrt{\frac{2Gm_P}{r_P}}$$

Calculate the radius of a black hole of mass 6.0×10^{24} kg (the mass of the Earth). [Use data.]

17.56 An X-ray source of luminosity 1.0×10^{30} W is found to have a peak wavelength of 0.30 nm. Use Wien's and Stefan's laws to calculate the radius of this source and suggest what type of star the source might be.

17.57* Astrophysicists tell us that 'We are all made of stardust.' Justify this statement to a friend who has no knowledge of physics or astronomy.

17.58* What is Olbers' paradox and how do we resolve it?

17.59* Edwin Hubble, an American astronomer working in the 1920s, was able to show that, for selected galaxies

$$v \propto d$$

i.e. the speed with which a galaxy recedes from Earth is proportional to its distance from Earth.

(a) How did Hubble measure **(i)** v and **(ii)** d?

(b) The graph shows some of Hubble's early data. Use a ruler to estimate the best straight line showing the proportionality between v and d and deduce Hubble's value for H_0.

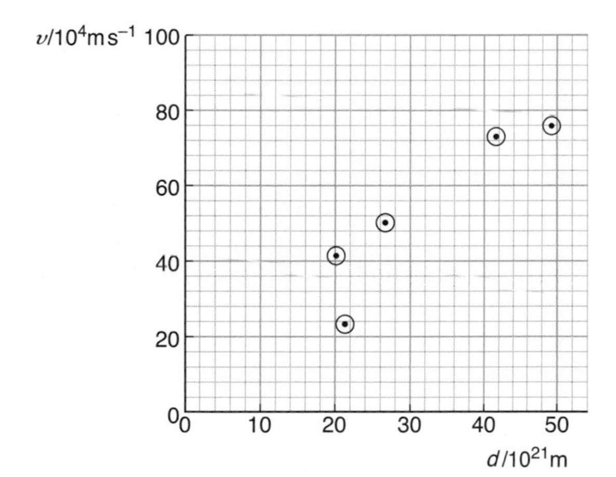

17.60* The modern value for H_0, the Hubble constant, is thought to be within ±5% of 74 km s^{-1} Mpc^{-1} [2.4×10^{-18} s^{-1}].

(a) What is 5% of either of these two quantities?

(b) Comment on Edwin Hubble's early value – the answer to the previous question.

17.61 The galaxy NGC 4889 shows a red shift from a laboratory value of 393.3 nm to 401.8 nm. Calculate

(a) the value of z for this galaxy

(b) the recession speed of NGC 4889

(c) the distance to this galaxy. [Use data.]

17.62* As indicated in question 17.60, the value of H_0 is uncertain. It lies between $78\,\text{km}\,\text{s}^{-1}\,\text{Mpc}^{-1}$ [$2.5 \times 10^{-18}\,\text{s}^{-1}$] and $70\,\text{km}\,\text{s}^{-1}\,\text{Mpc}^{-1}$ [$2.3 \times 10^{-18}\,\text{s}^{-1}$].

(a) Take the recession speed of NGC 4889 to be $6.48 \times 10^6\,\text{m}\,\text{s}^{-1}$ and calculate two 'extreme' values for the distance to this galaxy. [See question 17.61.]

(b) Express the difference in these distances in light years and comment on the result of your calculations.

17.63 The relation $z \approx v/c$ is only approximate for values of $z < 0.1$, i.e. $v <$ 10% of c.

(a) Up to what distance, to 1 significant figure, will the relation usefully hold when using Hubble's law?

(b) Are the calculations in the previous two questions valid?

17.64* Stars of mass m, in spiral galaxies like the Milky Way and our neighbouring galaxy M31, rotate in a circle of radius r at a constant speed v around the centre of the galaxy. The mass of the galaxy m_G is predicted to be $m_G = rv^2/G$.
Recent measurements predict values for m_G that are as much as twenty times the mass estimated by analysing the matter that can be 'seen'.
Use a search engine to investigate the nature of the 'dark matter' that is responsible for some of this discrepancy.

17.65 The Sun is moving round the centre of our galaxy, the Milky Way, at a speed of $270\,\text{km}\,\text{s}^{-1}$ in a circle of radius 8500 pc.

(a) Calculate a value for m_G, the effective mass of the Milky Way. [Use data.]

(b) Referring to the previous question, what is the mass of 'ordinary matter' we believe exists in the Milky Way?

17.66 A very faint Cepheid star has a mean apparent magnitude of +25 and a period of 51 days.

(a) Use the graph of absolute magnitude against period in question 17.20 to find the absolute magnitude of this star and thus calculate its distance from Earth.

(b) Use the Hubble equation to predict the speed with which this star is receding from us.

(c) Predict what red shift the light from this star will exhibit on Earth.

17.67* Imagine lightly sticking a series of circular pieces of paper of different sizes to the surface of a partly blown up balloon. The pieces of paper are to represent galaxies which move further apart as the balloon is blown up to double its original diameter.

(a) Sketch the balloon at these two sizes. (Remember that the size and relative positions of the pieces of paper will not change.)

(b) Use your sketch to explain to someone interested in the Big Bang hypothesis how our Universe has evolved and is evolving.

17.68* The two full lines in the sketch graph opposite illustrate how the fate of the Universe depends on whether gravitational and other attractive forces will result in it eventually contracting to a 'Big Crunch' or continuing to expand for ever. (The role of 'dark energy' in providing the acceleration in the way galaxies move away from one another is not at present understood.)

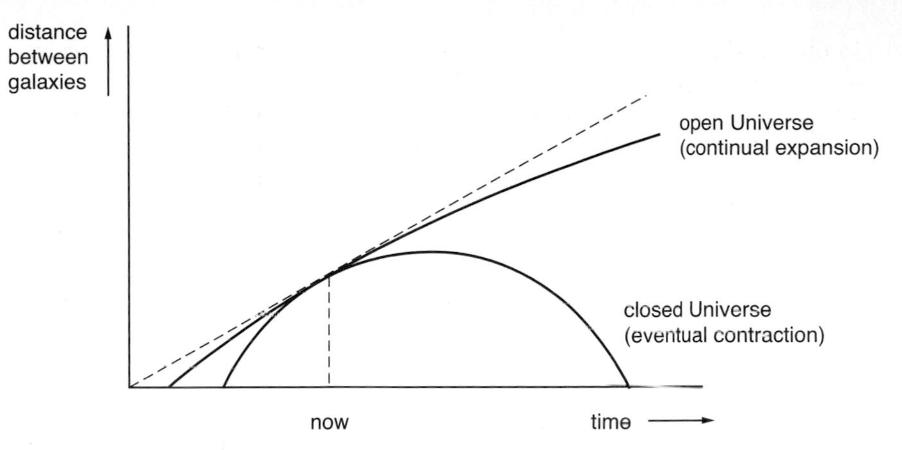

(a) Explain what the diagonal dashed line on the graph represents.
(b) If there were no gravitational forces the time from 'now' back to the Big Bang would equal the reciprocal of the present value of the Hubble constant. Use this to estimate a maximum age for the Universe in years. [Use data.]
(c) What evidence is there that strongly supports the origin of the Universe in a Big Bang?
(d) Use a search engine to learn about the possibility of there being other 'island universes', sometimes called multiverses.

17.69 Somewhere between the two full lines on the graph of the previous question there is a line representing a Universe that will expand until the galaxies come to rest after an infinite time.
(a) How will the line representing this 'critical' condition show up on the graph?
(b) The critical density of the Universe ρ_0 at the present moment is believed to depend on the Hubble constant H_0 and the gravitational constant G according to the relationship

$$\rho_0 = 3H_0^2/8\pi G$$

Calculate this critical density.

17.70 If the present average density of the Universe is 1.0×10^{-26} kg m^{-3}, approximately how many hydrogen atoms per cubic metre does this represent? [$m_p = 1.67 \times 10^{-27}$ kg.]

18 Electric fields

Data: electronic charge $e = 1.60 \times 10^{-19}\,\mathrm{C}$

electric field constant $k = 8.99 \times 10^{9}\,\mathrm{N\,m^2\,C^{-2}}$

permittivity of free space $\varepsilon_0 = 8.85 \times 10^{-12}\,\mathrm{F\,m^{-1}}$

Refer also to data relating to gravitational forces and fields in Chapter 13.

18.1 Electrical forces

In this section you will need to

- use Coulomb's law $F = kq_1q_2/r^2$ for two charges q_1 and q_2 separated by a distance r
- understand that the electric field constant $k = 1/(4\pi\varepsilon_0)$ where ε_0 is the permittivity of free space
- remember that the laws for electrical forces apply to point charges and to charged conducting spheres.

18.1 Two raindrops falling side by side carry electric charges of +4.0 pC and −5.0 pC, respectively.

(a) What electrical force does one drop exert on the other when they are 12 cm apart?

(b) How many times bigger than your answer to **(a)** is the pull of the Earth on each raindrop, if each has a mass of 15 mg?

18.2 A small sphere carrying a charge of +1.0 nC is situated a distance of 180 mm from another small sphere carrying a charge of +4.0 nC.

(a) What is the size of the electrical force F between the two spheres?

(b) What would be the size of this force if they were moved until they were separated by distances of r/mm = 90, 60, 30, 20, 10?

(c) Sketch a graph to show the variation of F with r in this case.

18.3 A uranium nucleus can be regarded as a sphere of radius 8.0×10^{-15} m containing 92 uniformly distributed protons. What is the force on an α-particle just 'touching' the surface of the nucleus? [Use data.]

18.4 The electron and proton in a hydrogen atom are 52.9 pm apart. What force does each exert on the other?

18.5 Two charges, one twice the size of the other, are located 15 cm apart and experience a repulsive force of 95 N. What is the size of the smaller charge?

18.6 A 7.0 μC charge is fixed in position. When a small charged sphere of mass 2.0 g is released 50 cm away, it accelerates towards the charge at $250\,\mathrm{m\,s^{-2}}$. What is the size of the sphere's charge?

18.7 The metal dome of a Van de Graaff machine may be regarded as a sphere of radius 0.15 m. It carries a charge of −4.0 μC. A small polystyrene sphere, of weight 25 mN, supported by a thin fibre thread is held near the dome so that their centres are in the same horizontal plane and a distance 0.20 m apart. In equilibrium the thread hangs at an angle of 45° to the vertical.

(a) Sketch the situation and draw a free-body force diagram for the polystyrene sphere.

(b) What is the electrical force on the polystyrene sphere?

(c) Calculate the charge on the polystyrene sphere.

18.8 Two charges, of 60 μC and Q, are placed as shown in the diagram.
A third charge placed at X, 50 cm from the origin, experiences no force.
Show that the charge Q must be 15 μC.

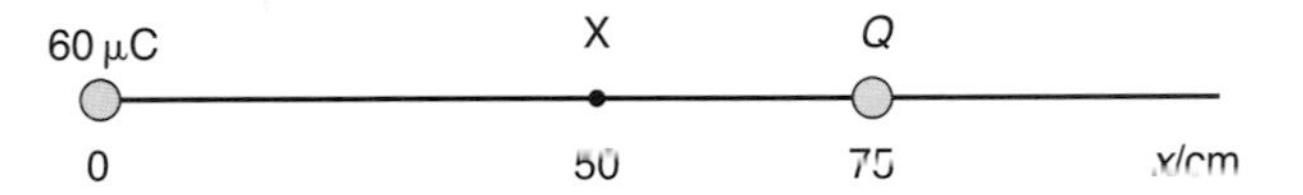

18.9 Express the unit for the permittivity of free space ε_0 (F m^{-1}) in SI base units.

18.10* To test that charges exert forces according to an inverse square law (the $1/r^2$ part of Coulomb's law), the apparatus shown in the diagram can be used.

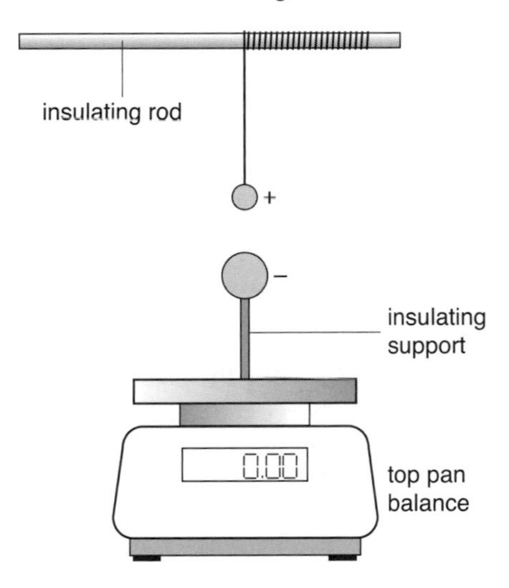

(a) Describe how you would use the apparatus to show how the force between the two charged spheres depends on the distance r between their centres.

(b) If the top pan balance can register vertical forces only to the nearest 0.1 mN, show that the smallest value of q_1q_2, the product of the charges on the spheres, when r = 25 mm, is about 7×10^{-18} C^2.

(c) Suggest what difficulties might arise in using this apparatus to investigate Coulomb's law.

18.11* Explain why a rubber balloon that has been charged by rubbing on a sweater appears to 'stick' to a vertical wall.

18.12 Two small conducting spheres A and B of equal mass are suspended from the same point O by insulating threads of equal lengths. When the spheres are given equal charges they come to rest with both threads inclined at the same angle θ to the vertical.
When half the charge has leaked off *one* sphere, the other remaining fully charged, it is found that both threads are inclined at the *same* smaller angle φ to the vertical.
Explain why the angles must be equal whether or not the charges are equal.

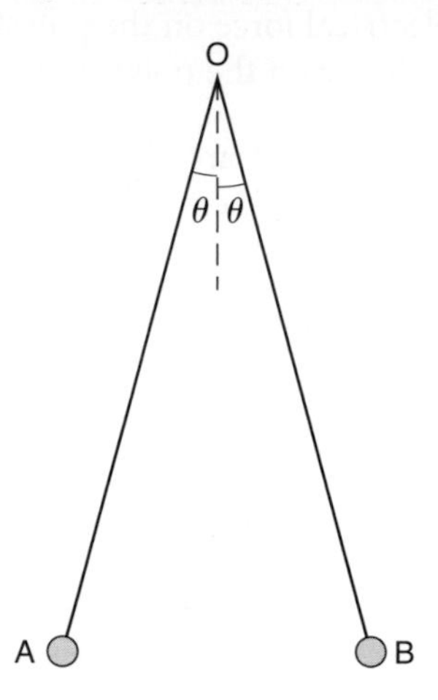

18.13 **(a)** Calculate **(i)** the gravitational force **(ii)** the electrical force between two protons which are 2.0×10^{-10} m apart. Take the mass of a proton to be 1.7×10^{-27} kg. [Use data.]
(b) What is the ratio of the electrical to the gravitational force?

18.2 Uniform electric fields

In this section you will need to

- understand that E is a vector quantity
- use the equation $E = F/q$ for the electric field strength at a point where F is the electrical force on a charge q placed at the point
- remember that the difference in electrical potential energy for charge q between two points in a uniform E-field is $qE\Delta x$ where Δx is measured parallel to the field
- understand the relationship between electrical potential difference (p.d.) and electric field strength and use the equation $E = V/d$ for a uniform E-field
- draw diagrams to show field lines and equipotential surfaces for electric fields.

18.14 By considering the equations defining the units involved, show that $1\,\mathrm{V\,m^{-1}} = 1\,\mathrm{N\,C^{-1}}$.

18.15 The diagram shows a uniform electric field formed between two horizontal charged plates. The strength of the field is $250\,\mathrm{N\,C^{-1}}$.

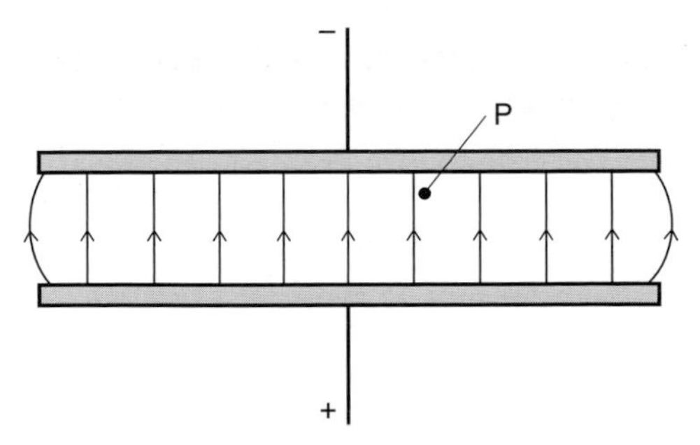

A charged particle at P experiences a downward electric force of 4.0×10^{-11} N.

(a) What is the charge on the particle at P?

(b) What is the force on the particle when, separately, it is moved

(i) vertically downwards towards the positively charged plate

(ii) a short distance horizontally to the left?

18.16 A charged polystyrene ball of mass 0.14 g is suspended by a nylon thread from a fine glass spring. In the absence of any electric field the spring extends by 30 mm. The polystyrene ball is then placed in an electric field that acts vertically upwards, of strength $2.0 \times 10^5\,\mathrm{V\,m^{-1}}$, and the spring extends by a further 6.0 mm. What is the electric charge on the polystyrene ball?

18.17 A charged oil drop of mass 2.0×10^{-15} kg is observed to remain stationary in the space between two horizontal metal plates when the potential difference between them is 245 V and their separation is 8.0 mm. Calculate the charge on the drop and comment on your answer.

18.18 What potential difference would you need to maintain between two horizontal metal plates 6.00 mm apart so that a particle of mass 4.00×10^{-15} kg with three surplus electrons attached to it would remain in equilibrium between them? Which plate would be the positive one? [Use data.]

18.19 Refer to the previous question.

(a) When the potential difference between the plates is 480 V, the particle moves slowly downwards in the space between them. When it has fallen 4.00 mm, what is the work done

(i) by the gravitational force acting on it

(ii) by the electrical force acting on it?

(b) Describe the energy transfers as the particle moves downwards.

18.20* About 100 years ago the American physicist Robert Millikan measured the electric charge on lots of tiny oil droplets. (He altered the charge on the droplets using ionising X-rays.) He deduced that charge was quantised, i.e. always came in multiples of 1.6×10^{-19} C. However, in coming to this conclusion, he ignored the results of at least 25% of his experiments. Use a search engine to learn more about Millikan's oil drop experiment, and comment on his treatment of his results.

18.21 Two parallel vertical plates are placed 40 cm apart in a good vacuum and charged as shown in the diagram. A tiny plastic sphere of mass 1.4×10^{-15} kg is released at P. The sphere carries a charge of $+1.5 \times 10^{-17}$ C.

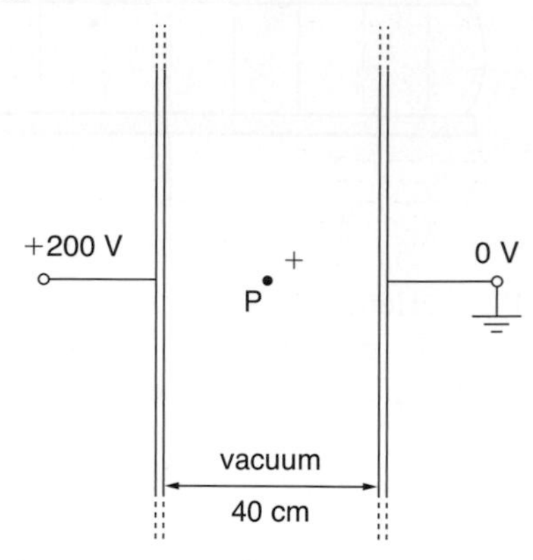

(a) Calculate the size of the two forces acting on the sphere.
(b) Draw a free-body force diagram for the plastic sphere showing the size and direction of these two forces, and deduce the direction in which it will accelerate away from P.

18.22 A vacuum tube contains two parallel electrodes 7.5 mm apart. A p.d. of 150 V is maintained between them.
(a) What is the electric field strength in the gap?
(b) What is the force on an electron in the gap?
(c) How much electric potential energy does an electron gain in crossing the gap?

18.23 The drop generators in inkjet printers can 'fire' over 100 000 droplets per second. Some of these droplets are charged and can be steered to different points on the paper by electric fields. The diagram shows an ink droplet of mass m carrying a charge $-q$ reaching deflecting plates at a speed v. The plates are a distance x apart, have a length ℓ and the p.d. between them is V.

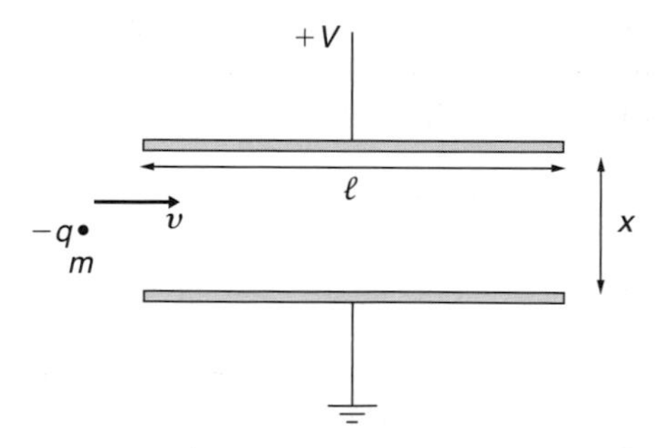

(a) Show that the vertical acceleration of the droplet is qv/mx. Assume that gravitational forces are negligible and state any other assumption made.
(b) Describe the motion of this ink droplet **(i)** as it moves between the deflecting plates **(ii)** as it moves beyond them.
(c) In one printer $\ell = 15$ mm and $v = 200$ m s^{-1}. Calculate the time taken for the droplet to travel through the deflection plates
(d) While the droplet is between the plates it accelerates vertically at 2.2×10^5 m s^{-2}. What is the vertical displacement of the droplet while it moves between the plates?

18.24 The diagram shows some equipotential surfaces between two charged parallel plates which are 0.12 m apart.

(a) Copy the diagram and add values for the electric potential at each of the lines.

(b) Add some electric field lines and calculate a value for the *E*-field between the plates.

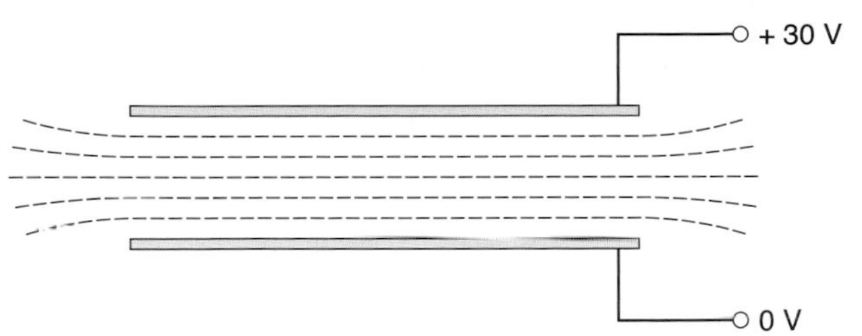

18.25 The simplified diagram shows the inside of a cathode ray oscilloscope (c.r.o).

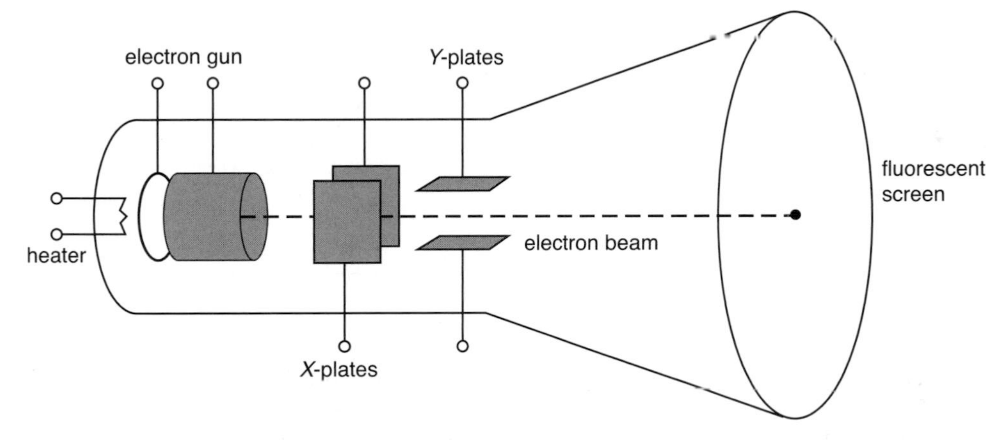

(a) What is the function of the heater?

(b) An alternating p.d. is applied to the *Y*-plates. Draw a sketch to show what would appear on the fluorescent screen.

(c) Research the function of an oscilloscope of this kind. **(i)** Explain the term time base and **(ii)** sketch what would appear on the fluorescent screen when the time base is switched on after an alternating p.d. has been applied to the *Y*-plates.

Questions 6.28 to 6.30 contain calculations on the cathode ray oscilloscope.

18.3 Radial electric fields

In this section you will need to

- use the equation $E = kq/r^2$ for the electric field strength around a point charge q or spherical charged conductor
- use the equation $V = kq/r$ for the electric potential near a point charge q or spherical charged conductor.

18.26 **(a)** Show, by listing five values of the product Er^2, that the graph of E against r in the region of an isolated proton follows an inverse square law.
(b) Calculate a value for the electric field constant k. [Use data.]

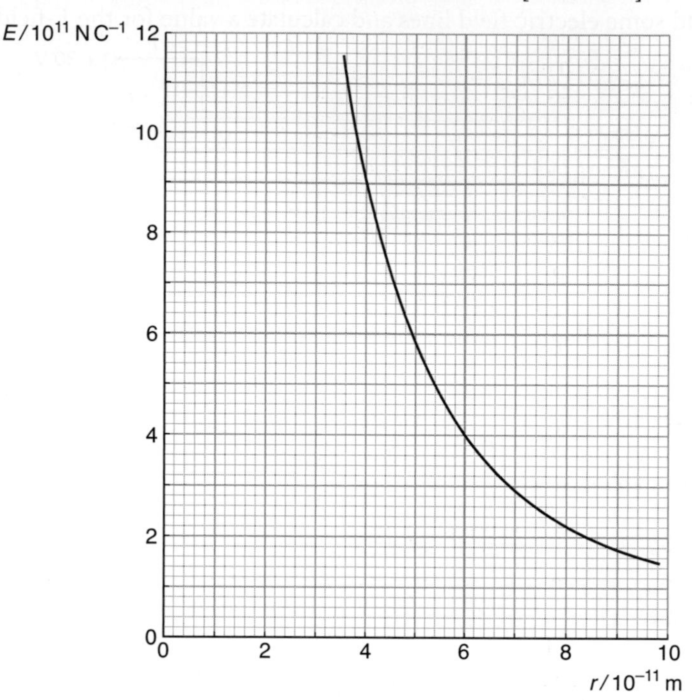

18.27 In a hydrogen atom the average distance apart of the proton and the electron is 5.3×10^{-11} m. Calculate
(a) the electric field at this distance from the proton
(b) the electrical force on the electron
(c) the acceleration such a force would produce on an electron. Take the mass of an electron to be 9.11×10^{-31} kg.

18.28 Here are some data about the electric field near the charged sphere of a Van de Graaff generator:

distance from centre of sphere r/m	0.20	0.25	0.30	0.40	1.00
electric field strength E/kN C^{-1}	75	48	33	19	3.0

(a) Show that this E-field follows an inverse square law by plotting a graph of E against $1/r^2$.
(b) Calculate the charge on the surface of the sphere. [Use data.]

18.29 In a small Van de Graaff machine the metal sphere reaches a potential of 2.4×10^5 V before it discharges by sparking across an air gap.
(a) If the sphere is of radius 0.15 m, calculate the charge on the sphere just before discharge.
(b) What is the electric field at the surface of the sphere just before the discharge takes place?

18.30 The diagram shows two point charges of +8.0 nC and –8.0 nC placed a distance 0.60 m apart.

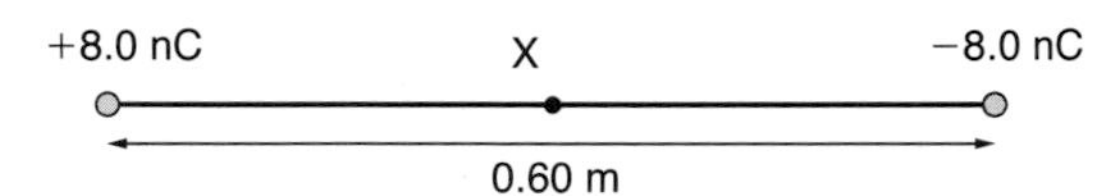

(a) Show that the magnitude of the resultant *E*-field at X, midway between the charges, is 1600 N C^{-1}.

(b) Predict the direction of the resultant *E*-field at Y and draw a vector diagram to support your prediction.

(c) Sketch the shape of the *E*-field around these two charges.

18.31 The Earth's upper atmosphere carries a permanent positive charge. As a result in normal fair weather conditions there is a downward electric field of about 200 N C^{-1} or 200 V m^{-1} at the Earth's surface.

(a) Show that the total charge on the Earth's surface is close to -9×10^5 C. Take the radius of the Earth to be 6400 km.

(b) Calculate the surface density of this charge in C m^{-2}.

18.32 Two charges each of +3.0 μC, separated by a distance of 0.80 m, produce the electric field shown in the diagram.

(a) Calculate the electric potential at A, a distance of 0.50 m from each charge.

(b) Copy the diagram and add dashed lines to show the shapes of the equipotential surfaces in the plane of the paper.

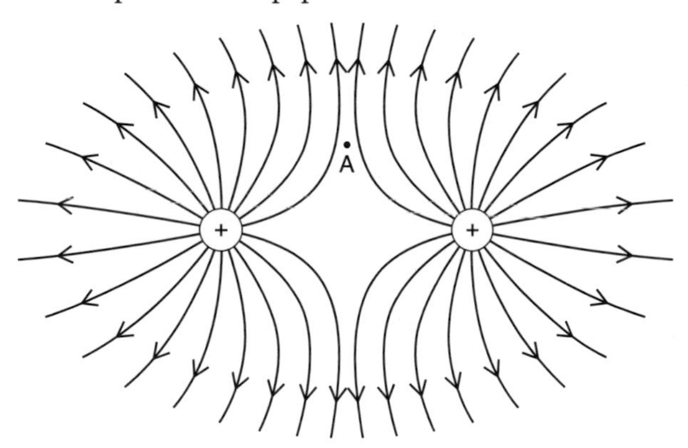

19 Capacitance

Data: electronic charge $e = 1.60 \times 10^{-19}$ C

19.1 Charge and capacitance

In this section you will need to

- understand that when a capacitor stores charge q there is a charge $+q$ on one plate and $-q$ on the other
- use the equation $q = CV$ and remember that the farad, the unit of capacitance, is so large that capacitances are usually measured in μF or pF
- use equations for capacitors in parallel and in series

$$C_{par} = C_1 + C_2 + C_3 \text{ and } \frac{1}{C_{ser}} = \frac{1}{C_1} + \frac{1}{C_2} + \frac{1}{C_3}$$

- describe how to measure capacitance.

19.1 **(a)** A capacitor of capacitance 10 μF is connected to a battery of e.m.f. 12 V. What are the charges on its plates?

(b) What potential difference must be applied between the plates of a 100 μF capacitor for the charges on them to be +0.025 C and −0.025 C?

19.2 The graph shows the result of an experiment in which known charges are transferred to an isolated capacitor and a data logger registers the p.d. across the capacitor after each transfer.

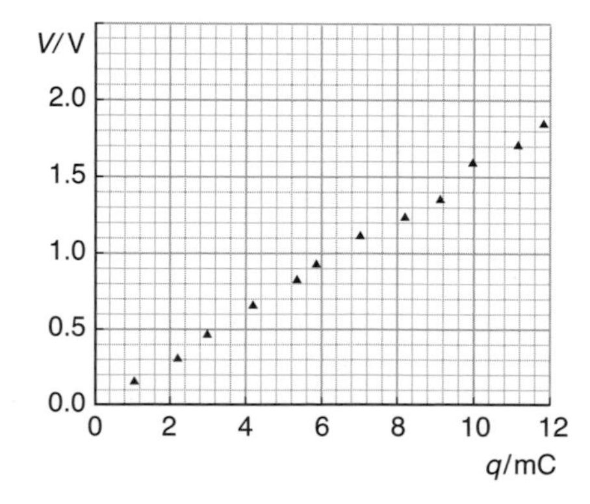

Lay a transparent ruler along the points and use it to determine the capacitance of the capacitor.

19.3 Copy and complete the following table, which gives the charge q on capacitors of capacitance C when there is a p.d. V across them. Express your answers using prefixes (not in standard form).

V	12.0 V		1.5 V		100 V	600 mV
C	2.2 μF	5000 μF	4.7 μF	220 μF		
q		10 mC		66 μC	1.0 nC	280 μC

19.4 A p.d of 12 V displaces 2.5×10^{16} electrons from one plate of a capacitor to the other.

(a) How much charge does this represent?

(b) Calculate the capacitance of the capacitor.

19.5 **(a)** What charge is carried on each of the plates of a 470 μF capacitor when there is a potential difference between the plates of 30 V?

(b) If this capacitor is charged in 4.0 s, what is the average charging current?

19.6 A 33 μF capacitor is charged and then isolated. During the next minute the potential difference between the plates falls by 4.0 V. What is the average leakage current between the plates? Explain your calculation.

19.7 A steady charging current of 50 μA is supplied to the plates of a capacitor and causes the p.d. between the plates to rise from 0 V to 5.0 V in 20 s. What is its capacitance?

19.8 The pair of intersecting rings in the diagram is a symbol for a constant current source, which here produces a current of 0.48 mA.

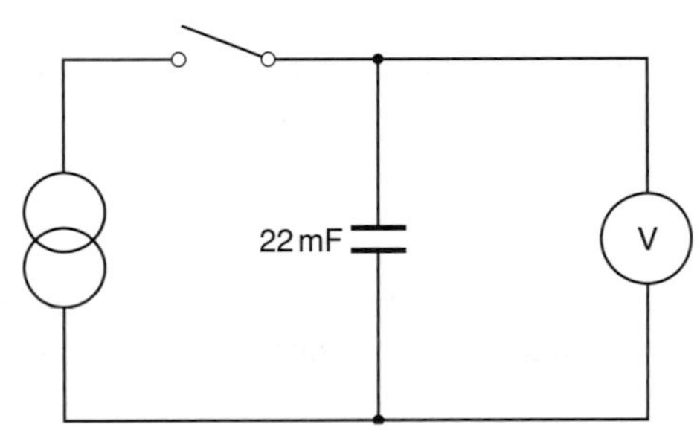

The switch in the circuit is closed for a time and the digital (high-resistance) voltmeter, connected across a 22 mF capacitor, then registers 0.74 V. For how long was the switch closed?

19.9 The diagram shows an RC series circuit with the switch open.

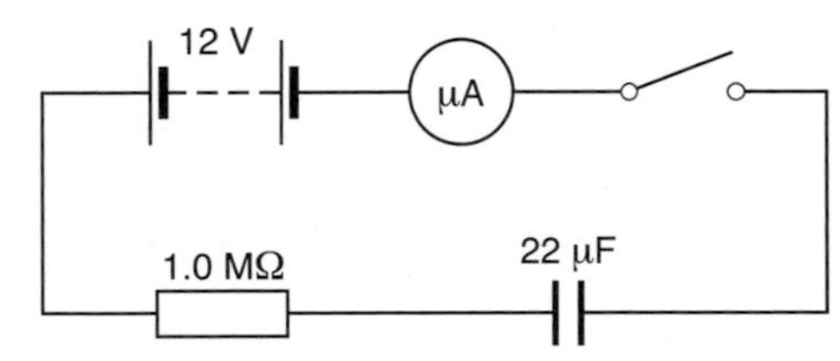

(a) At the instant the switch is first closed, what is

(i) the p.d. across the capacitor

(ii) the p.d. across the resistor

(iii) the current in the circuit

(iv) the charge on each capacitor plate?

(b) What are the same quantities **(i)–(iv)** after the switch has been closed and a steady condition reached?

(c) At a certain instant after the switch has been closed the microammeter reads 9.5 μA. Calculate, for this instant, **(i)** the p.d. across the resistor **(ii)** the p.d. across the capacitor **(iii)** the charge on each capacitor plate.

19.10 A capacitor of capacitance 180 pF is charged to a potential difference of 12 V and then discharged through a microammeter. This sequence of operations is repeated 250 times per second using a reed switch. What is the average current registered by the meter?

19.11 A vibrating reed is used to connect a capacitor alternately to a battery and to a meter as shown in the diagram. In this way the capacitor is fully charged by the battery and fully discharged through the meter 50 times per second.

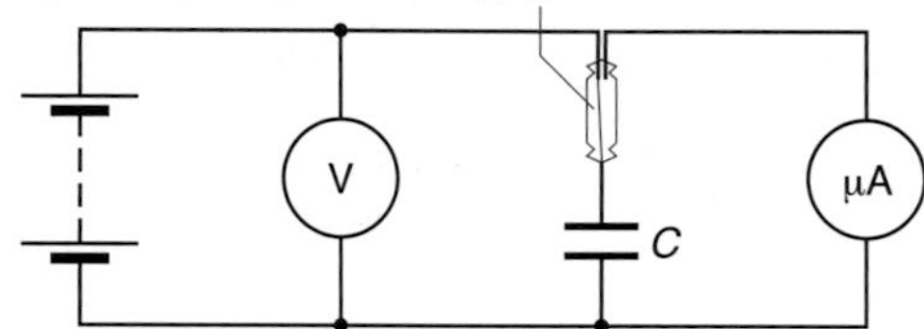

If the e.m.f. of the battery is 12 V and the meter registers an average current of 2.4 mA, what is the capacitance of the capacitor?

19.12* The diagram shows a computer keyboard key. When the key is depressed the separation d of the metal plates decreases from 2.5 mm to 0.50 mm. These plates form a capacitor for which $C \propto 1/d$.
How many times greater is the capacitance of C when the key is depressed? [The computer notes this change and 'knows' which key has been depressed.]

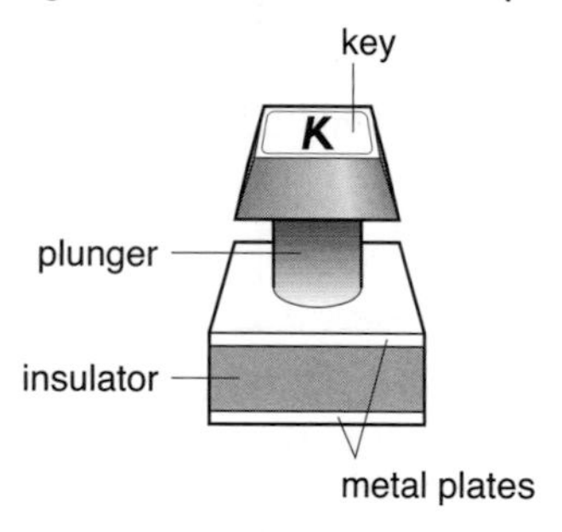

19.13 Three capacitors of capacitances 2.0 μF, 3.0 μF and 6.0 μF are joined **(a)** in parallel **(b)** in series. What is the combined capacitance in each case?

19.14 Two capacitors are arranged in series as in the diagram. A battery is connected between P and Q and it is found that the charge on plate w is +60 μC.

(a) Explain why the charge on plate z of the other capacitor is –60 μC.

(b) What are the charges on plates x and y?

(c) Calculate the p.d. across each capacitor.

(d) What is the p.d. between P and Q?

(e) Calculate the capacitance of a single capacitor placed between P and Q which would store 60 μC of charge when connected to the same battery.

30 μF 60 μF
P Q
w x y z

19.15 Two capacitors of capacitance C_1 and C_2 are connected in series, with a p.d. $V = V_1 + V_2$ across the combination.

(a) Show that the p.d.s V_1 and V_2 across the individual capacitors are

$$V_1 = \frac{C_2 V}{(C_1 + C_2)} \quad \text{and} \quad V_2 = \frac{C_1 V}{(C_1 + C_2)}$$

(b) Do the answers to the previous question support this general conclusion?

19.16 A 22 μF capacitor is connected in series with a 47 μF capacitor to a battery. Say which of the following statements are true, and for each of the true statements explain why it is true.

(a) (i) Each capacitor has the same charge.

(ii) The charge on the 22 μF capacitor is about half the charge on the 47 μF capacitor.

(iii) The charge on the 22 μF capacitor is about double the charge on the 47 μF capacitor.

(b) (i) The p.d. across each capacitor is the same.

(ii) The p.d. across the 22 μF capacitor is about half the p.d. across the 47 μF capacitor.

(iii) The p.d. across the 22 μF capacitor is about double the p.d. across the 47 μF capacitor.

19.17 The diagram shows an arrangement of three capacitors and a 9.0 V battery in a circuit.

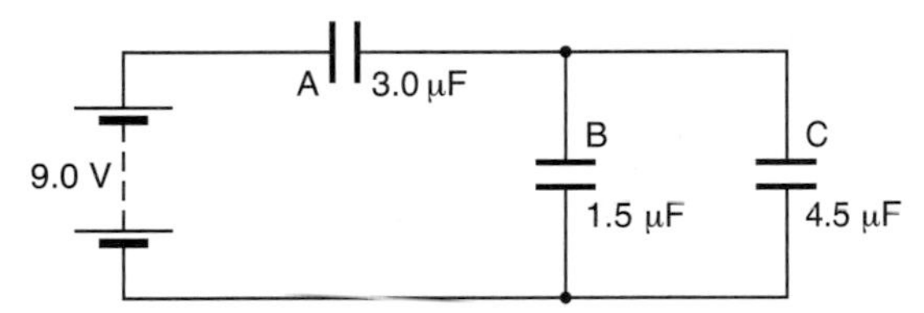

(a) What is the capacitance of the single capacitor D which could be used to replace the capacitors B and C in their positions in this circuit?

(b) Calculate the p.d.s across capacitors A and D.

(c) What is the p.d. across each capacitor A, B and C?

(d) Work out the charge in each capacitor A, B and C.

19.18 If you have three capacitors of capacitance 3.0 μF, 6.0 μF and 8.0 μF, how could you produce a combination of capacitors of capacitance 10 μF?

19.19 A 47 μF capacitor C_1 is charged from a 6.0 V battery.

(a) What is the charge on the capacitor?

It is then disconnected from the battery and connected to an uncharged 22 μF capacitor C_2 as shown in the diagram.

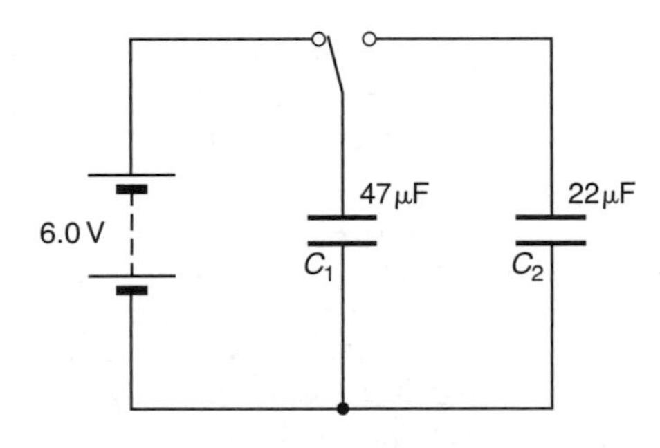

(b) What is the capacitance of these two capacitors connected in parallel?
(c) What is now the p.d. across each capacitor?
(d) Hence calculate the final charges on C_1 and C_2.

19.20 A variable 'trimmer' capacitor used to make fine adjustments has a capacitance range from 10 nF to 30 nF. The trimmer is connected in parallel with a capacitor of capacitance about 1.0 μF.
(a) Draw diagrams showing the 1.0 μF capacitor connected to **(i)** a 10 nF trimmer **(ii)** a 30 nF trimmer.
(b) Calculate the percentage range over which the capacitance of the combination can be varied.

19.21 Capacitors are often marked with a 'nominal' value for the capacitance and a tolerance range within which the actual capacitance lies For example, a 10 μF ± 20% capacitor has capacitance between 8 μF and about 12 μF.
(a) A 0.22 μF ± 20% capacitor is connected in series with a 0.10 μF ± 30% capacitor. Show that the resulting capacitance lies in the range 0.050 μF to 0.087 μF
(b) Express this answer as a capacitance and its associated tolerance.

19.22* The coulombmeters found in schools and colleges measure the charge on, for example, a small capacitor C by transferring the charge to be measured to a 4.7 μF capacitor connected across the input terminals of a digital voltmeter.

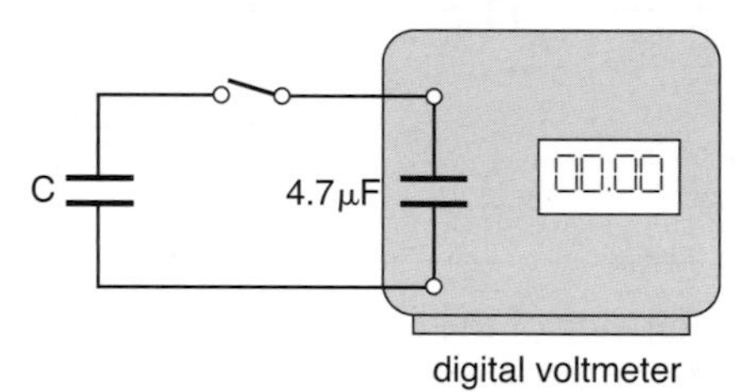

(a) The voltmeter registers 0.25 V. Calculate the initial charge q.
(b) If C has a capacitance of 40 nF, show that the charge remaining on it after the transfer is 10 nC.
(c) Hence calculate the percentage of the charge on C that is transferred to the coulombmeter.

19.2 Energy and capacitors

In this section you will need to

- understand that a capacitor can store energy
- use the equations for the energy stored in a capacitor $E = \frac{1}{2}qV = \frac{1}{2}CV^2$
- understand that there is a useful analogy between a capacitor ($V = \frac{1}{C}q$) and a spring ($F = kx$).

19.23 A capacitor of capacitance 22 μF is charged by connecting it to a 400 V supply, and is then discharged.
(a) Calculate the energy transferred during the discharge.
(b) If the discharge takes 10 μs, what is the average power of the discharge?

19.24 The graph shows how the p.d. across a capacitor changes as it is charged.

(a) What is the capacitance of the capacitor?

(b) (i) Calculate the energy stored in the capacitor when it has a charge of $q/\mu C = 1, 2, 3, 4, 5$.

(ii) Sketch a graph of energy stored against charge.

(iii) What shape would a graph of energy against p.d. have?

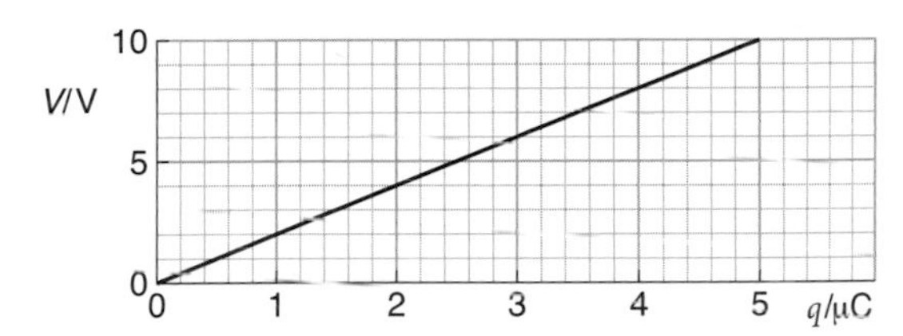

19.25 The table shows part of a spreadsheet for a capacitor of capacitance C. It has been used to calculate the charge q (in column C) on the capacitor and the energy W (in column D) stored in it for different p.d.s V. Column A shows the value of C, and column B shows the chosen values of V. Cells C2 and D2 contain formulae which calculate the values of Q and W when $V = 0$.

	A	B	C	D
1	C/μF	V/V	Q/μC	W/μJ
2	4.70	0.0	0.00	0.00
3	4.70	1.0	4.70	2.35
4	4.70	2.0	9.40	9.40
5	4.70	3.0	14.10	21.15
6	4.70	4.0		
7	4.70	5.0		
8	4.70	6.0		

(a) (i) The formula in cell C2 is =A2*B2. Explain why using this formula gives the value of q in cell C2.

(ii) Cell D2 has the formula =0.5*B2*C2. Explain why using this formula gives the value of W in cell D2.

(b) Create your own spreadsheet by filling in the headings and columns A and B, and then typing in cells C2 and D2 the two equations shown in part **(a)**.

(c) You should find that when you select cells C2 and D2 and drag downwards, the columns C and D are completed. Your results should agree with the results already shown. What are the quantities in cells C8 and D8? *(Save this spreadsheet to use in the next question.)*

19.26 Open the spreadsheet you used in the previous question.

(a) Type the formula =0.5*A2*B2^2 into cell E2. You should see the value 0 or 0.00.

(b) Express this formula as an equation using the symbols W, C and V.

(c) Select cell E2 and drag downwards so that the column E is completed. Comment on your results.

(d) Type the formula =0.5*(C2^2)/A2 into cell F2. You should see the value 0 or 0.00.

(e) Express this formula as an equation using the symbols W, Q and C.

(f) Select cell F2 and drag downwards so that the column F is completed. Comment on your results.

19.27* Capacitors of high capacitance can be used to provide a back-up energy source for a solar-powered wristwatch. The basic circuit is shown in the diagram.

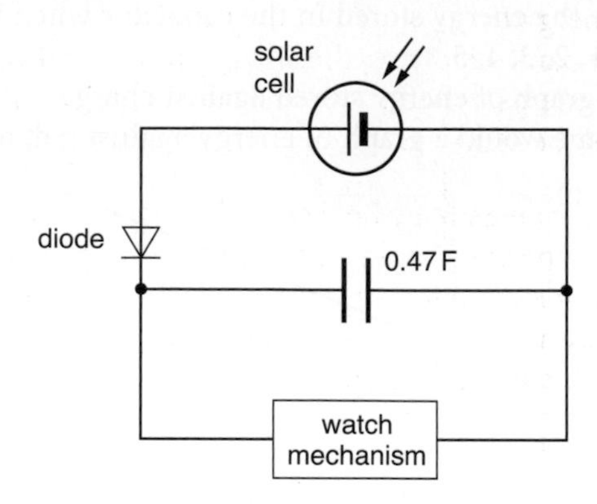

(a) In direct sunlight the solar cell powers the watch and charges the capacitor to a p.d. of 2.3 V. Calculate **(i)** the charge **(ii)** the energy then stored in the capacitor.
(b) In darkness, the p.d. produced by the solar cell drops to zero and the capacitor takes over as the power supply. Calculate the charge that will have flowed through the watch mechanism as the p.d. across the capacitor falls from 2.3 V to 1.0 V. (The watch mechanism ceases to function if the p.d. across it falls below 1.0 V.)
(c) As the p.d. falls from 2.3 V to 1.0 V, the watch draws a *constant* current of 1.0 μA from the capacitor. **(i)** For how long, in hours, can the back-up capacitor run the watch? **(ii)** What average power does it deliver to the watch during this back-up period?

19.28* Use a search engine to look up 'super-capacitors', and the uses to which they can be put. (Some have capacitances of over 3000 F!)

19.29 In the circuit shown, calculate the energy stored in the 4.0 μF capacitor
(a) with the switch S closed
(b) with the switch S open.

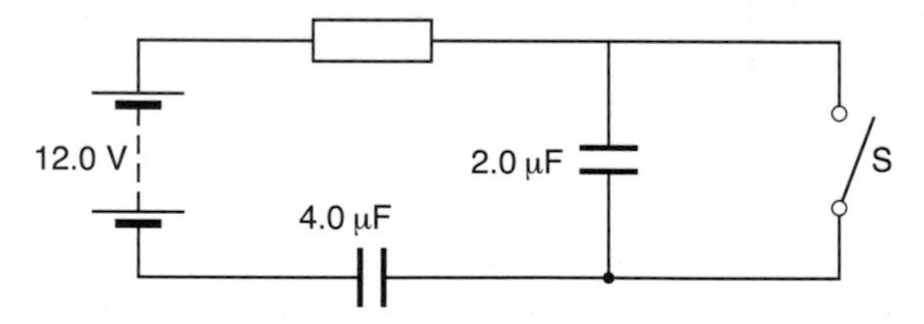

19.30 A capacitor of capacitance 22 μF is charged to a p.d. of 200 V and then isolated from the supply.
(a) What is the energy stored in it?
(b) If an identical capacitor, initially uncharged, is joined across it, what is the energy now stored in the pair of capacitors? Comment on your answer.

19.31 An 8.0 μF capacitor is charged by joining it to a 500 V supply through a resistor.
(a) What charge flows through the supply and the resistor?
(b) How much electrical energy is taken from the supply?
(c) How much electrical energy is stored in the capacitor?
(d) How do you account for the difference between these two amounts?

19.32 A laser fusion experiment in the Lawrence Livermore Laboratory in California can deliver light energy at a rate of 10^{14} W. But the laser pulse lasts for only 10^{-9} s.

(a) How much energy is delivered in one pulse?

(b) The capacitor bank supplying this energy has a total capacitance of 0.26 F. Only 0.17% of this energy actually appears as light. To what approximate p.d. must the capacitor bank be charged?

19.33 It is possible to draw an analogy between what happens when a force is exerted on a spring and what happens when a p.d. is applied to a capacitor.

(a) For a spring $F = kx$ and for a capacitor $q = CV$, where the symbols have their usual meanings, copy and complete the table to show which quantities for a capacitor are *analogous* to F, k and x for a spring.

spring	capacitor
F	
k	
x	

(b) Write down an expression for the energy stored in a stretched spring in terms of k and x.

(c) Write down an analogous expression for the energy stored in a charged capacitor.

19.3 Capacitor discharge

In this section you will need to

- draw graphs showing the variation of charge and potential difference against time for the discharge of a capacitor through a resistor
- use the equation $I = \Delta q/\Delta t = C\Delta V/\Delta t$ for the rate of flow of charge from a capacitor
- understand the equation $dq/dt = -q/RC$ for the discharge of a capacitor through a resistor, the solution of which is called an exponential decay curve
- use the equations $q = q_0 e^{-t/RC}$, $I = I_0 e^{-t/RC}$ and $V = V_0 e^{-t/RC}$ for capacitor discharge
- remember that the product RC is called the time constant of the circuit and that the time for the charge to halve is $RC\ln 2 \approx 0.7RC$.

19.34 The diagram shows a capacitor of capacitance 0.61 mF that is charged from a 6.0 V battery and then discharged through a variable 100 kΩ resistor set at a resistance R.

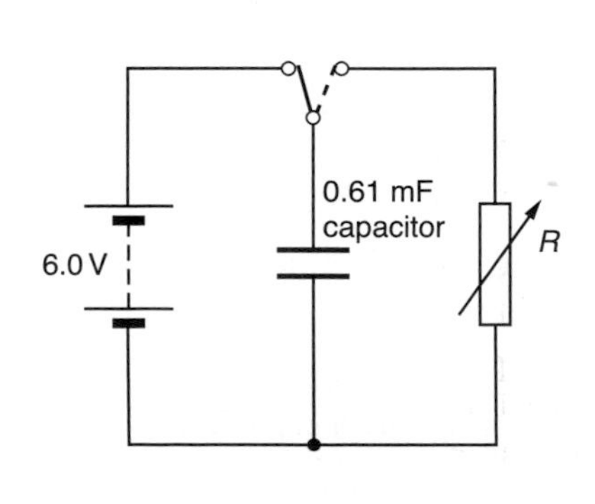

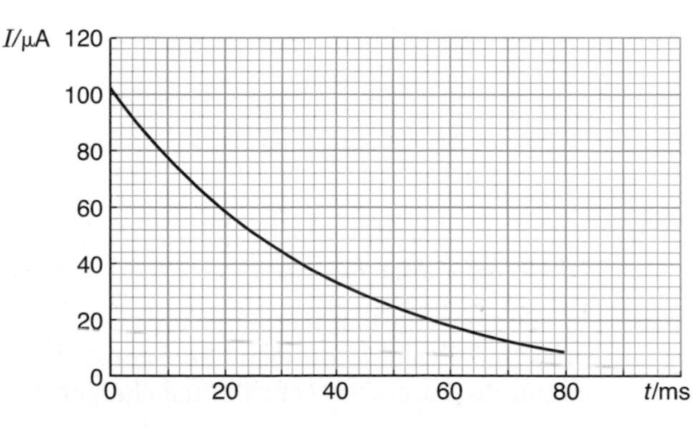

The graph on the previous page shows how the current in the capacitor–resistor circuit varies with time as the capacitor discharges through the resistor.

(a) Determine the half-life of this decay, taking *three* independent readings from the graph. Hence calculate the time constant of the circuit.

(b) Show that the variable resistor was set at $R = 59\,\text{k}\Omega$ during this decay.

(c) The initial current in the circuit was $102\,\mu\text{A}$. Explain how this is consistent with the value $R = 59\,\text{k}\Omega$.

19.35 Show that the unit of the time constant RC is the second.

19.36 A $47\,\mu\text{F}$ capacitor is charged to 6.0 V and then discharged through a $1.0\,\text{M}\Omega$ resistor.

(a) What is the initial charge stored in the capacitor?

(b) Calculate the initial current and hence find the time t_d the capacitor would take to discharge fully if this discharging current were to remain constant.

(c) What is the product RC, the time constant, for this circuit?

19.37 The diagram shows a capacitor and a resistor connected in series with a switch and a 1.5 V cell. At time $t = 0$ the switch is closed and the graph shows how the p.d. across the resistor then varies with time. During this process the p.d. across the resistor and the p.d. across the capacitor are related by the equation $V_R + V_C = 1.5\,\text{V}$.
Sketch the variation with time of **(a)** the p.d. across the capacitor **(b)** the charge on the capacitor **(c)** the current in the resistor.

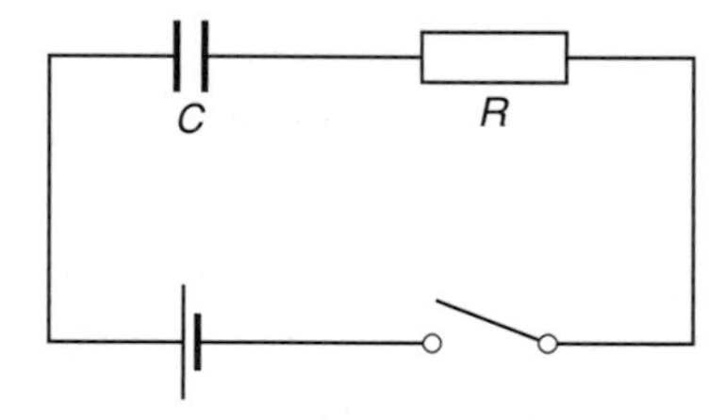

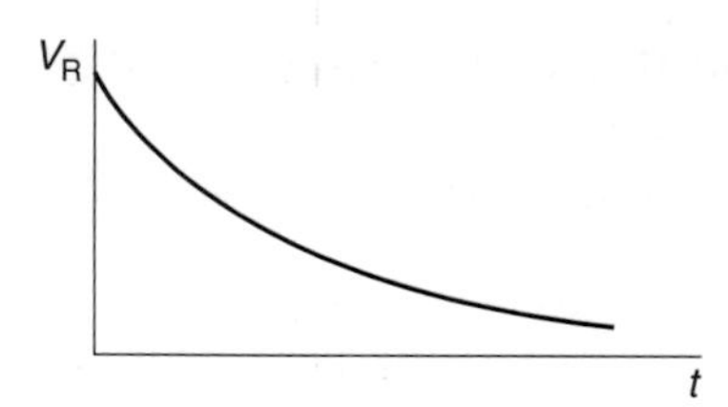

19.38 The graph shows the charge on a capacitor as it discharges through a resistor.
What is the current in the resistor **(a)** at $t = 0$ **(b)** at $t = 20\,\text{s}$? In each case show how you obtain your answer.

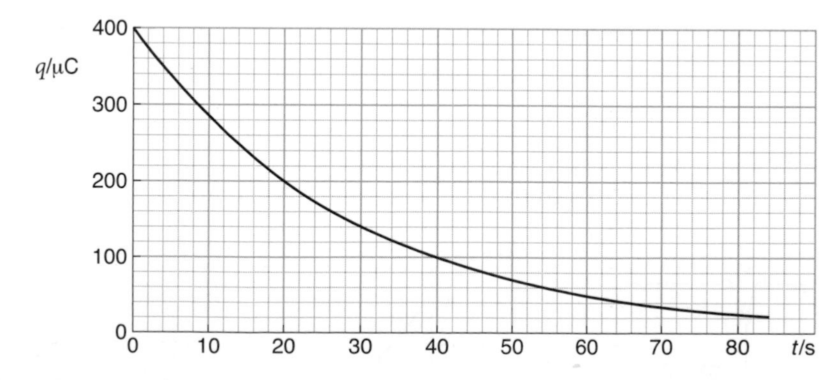

19.39 Check that the graph in the previous question is an exponential decay graph by showing that $q_0/q_{10\,\text{s}} = q_{10\,\text{s}}/q_{20\,\text{s}} = q_{20\,\text{s}}/q_{30\,\text{s}} = q_{30\,\text{s}}/q_{40\,\text{s}}$, etc.

19.40 In one time constant the charge on a capacitor discharging through a resistor falls to 0.37 of its initial value. How many time constants must elapse before a capacitor in an RC circuit discharges 99% of its initial charge?

19.41 A charged capacitor is connected to a resistor. Sketch graphs to show the variation with time, as the capacitor discharges, of

(a) the charge on the capacitor
(b) the p.d. across the capacitor
(c) the current in the resistor.

19.42 In one method for measuring the speed v of a rifle bullet, the bullet is made to break strips of conducting foil first at A and then at B, where the distance AB is d as shown in the diagram.

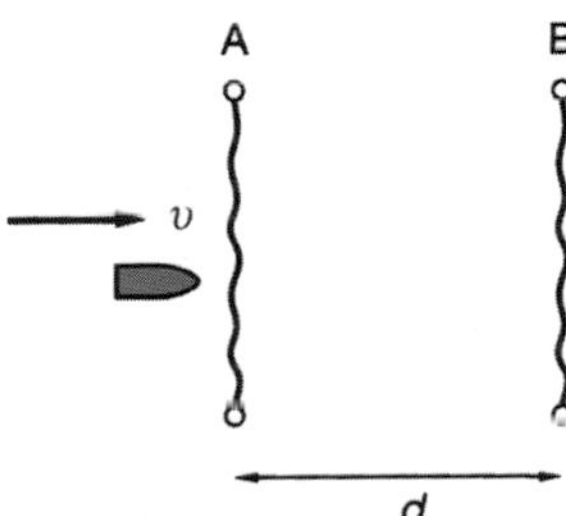

(a) Design an RC circuit in which opening a switch as A is broken starts the discharge of a capacitor and opening a switch as B is broken stops the discharge. Add a voltmeter which would enable you, if R and C are known, to find the time taken by the bullet to get from A to B.
(b) If v is about 300 m s^{-1}, suggest suitable values for d, R and C which would enable v to be measured. [Hint: You would wish C to lose about half its charge during the experiment.]

19.43* Often, when a car door is opened, a light comes on inside the car and fades away after a few seconds, a time that can be adjusted by the car owner. Suggest how this might be achieved with an RC circuit and suggest values for R and C.

19.44* The table shows part of a spreadsheet which has been created to model the decrease with time t of p.d. V in an RC circuit. Cells A2, B2, C2 and D2 show the initial values of t and V and the values of R and C.

	A	B	C	D
1	t/s	V/V	R/Ω	C/F
2	0	12	2.00E+05	1.00E-04
3	1	11.41		
4	2	10.86		
5	3	10.33		
6	4	9.82		
7	5	9.35		
8	6	8.89		

(a) Cells C2 and D2 contain the numbers 2.00E+05 and 1.00E–04. Express these values of resistance and capacitance in standard form.
(b) The formula in cell B3 which is used to calculate the p.d. after 1 s is
=B$2*EXP(–A3/(C$2*D$2)).
Explain how this relates to the equation $V = V_0 e^{-t/RC}$ for the exponential decay of p.d. V on a capacitor.

(c) The formula in cell B4 has been obtained by selecting cell B3 and dragging down the table. What is the formula in cell B4?

(d) Set up your own version of this spreadsheet for $V_0 = 10\,\text{V}$, $R = 100\,\text{k}\Omega$, $C = 47\,\mu\text{F}$ and use it to find **(i)** the values of V for $t/\text{s} = 1, 2, 3, 4$ and **(ii)** the last whole-number value of t for which V is greater than 0.50 V.

19.45 The p.d. across a capacitor of capacitance 220 μF is monitored as it discharges through a resistor. Use the graph shown to calculate

(a) **(i)** the initial rate of change of p.d. across the capacitor

(ii) the initial rate of flow of charge, i.e. the current, through the resistor.

(b) What is the current in the resistor after 4.0 s?

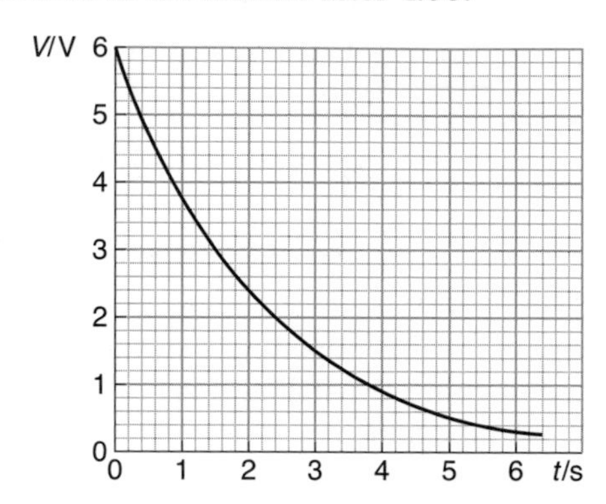

19.46 The graph used in the previous question is one showing exponential decay. What is **(a)** the half-life **(b)** the time constant, for this circuit?

19.47 A capacitor of capacitance 22 μF is charged to 12 V and isolated. The charge leaks away through a resistance of 10 MΩ until the p.d. across the capacitor is 8.0 V.

(a) **(i)** Calculate the charge which leaks off the capacitor.

(ii) Assuming that the average p.d. driving the discharge is 10 V, what is the average leakage current?

(iii) Deduce how long the charge took to leak away.

(b) In reality the p.d. falls exponentially according to the equation $V = V_0 e^{-t/RC}$. Use this equation to calculate how long the charge took to leak away.

(c) What is the percentage difference between your answers to **(a)** and **(b)**?

19.48* A charged capacitor is discharged through a resistor and a student plots a graph of the current in the circuit during the first 40 s of the discharge. The student then analyses graph A and produces a second graph B showing the charge on the capacitor at each stage of the discharge.

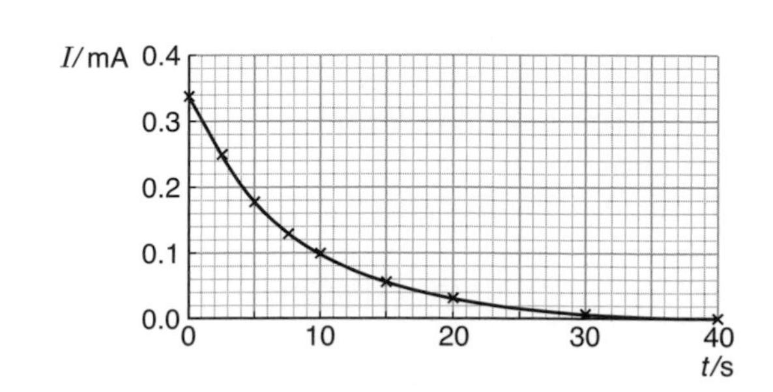

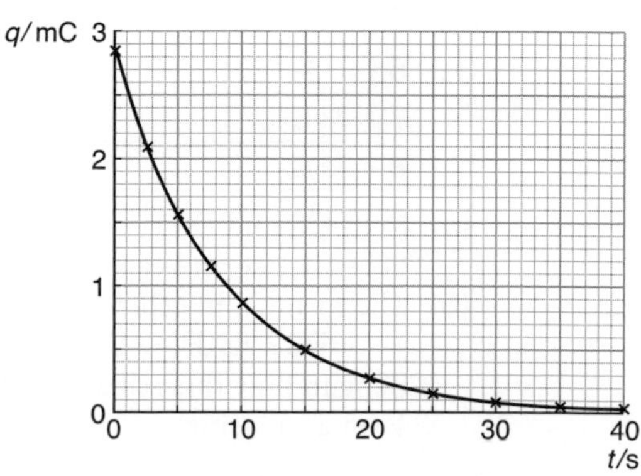

Graphs A and B are related in several ways.

(a) Explain how the student was able to produce graph B from graph A.

(b) Could graph A have been produced from graph B if graph B had been drawn first? Explain.

(c) What other information about the circuit can be deduced from either or both graphs?

19.49 A capacitor was joined across a digital voltmeter and charged to a potential difference of 1.00 V. The p.d. V was then measured at 20 s intervals and at time $t = 30$ s a resistor of resistance $R = 1.5\,\text{M}\Omega$ was connected across the capacitor. The following results were obtained:

t/s	0	20	40	60	80	100	120
V/V	1.00	1.00	0.81	0.54	0.35	0.23	0.15

(a) Plot a graph of ln (V/V) against t to demonstrate the exponential fall of p.d. with time.

(b) Use your graph to show that the time constant RC of this decay process is 48 s.

(c) Deduce the capacitance of the capacitor.

20 Electromagnetism

Data: gravitational field strength at Earth's surface $g_0 = 9.81\,\text{N kg}^{-1}$

electronic charge $e = 1.60 \times 10^{-19}\,\text{C}$

rest mass of electron $m_e = 9.11 \times 10^{-31}\,\text{kg}$

rest mass of proton $m_p = 1.67 \times 10^{-27}\,\text{kg}$

magnetic field constant $\mu_0 = 4\pi \times 10^{-7}\,\text{N A}^{-2}$

speed of electromagnetic radiation $c = 3.00 \times 10^{8}\,\text{m s}^{-1}$

20.1 Magnetic forces

In this section you will need to

- remember the rule for predicting the direction of the magnetic force on a current-carrying wire placed in a magnetic field
- draw magnetic field lines which show the direction of fields into and out of the plane of a diagram as × and · respectively
- use the equation $F = B_{\perp}I\ell$ for the force on a current-carrying wire (where $B_{\perp}$ is the resolved part of B at right angles to I and ℓ)
- use the equation $F = B_{\perp}qv$ for the force on a charged particle moving in a magnetic field (where $B_{\perp}$ is the resolved part of B at right angles to v)
- understand that a charged particle deflected by a magnetic field will move in a circular path
- use the expression v^2/r for centripetal acceleration.

20.1 A horizontal conductor of length 50 mm carrying a current of 3.0 A lies at right angles to a horizontal magnetic field of flux density 0.50 T.
(a) What is the size of the magnetic force on it?
(b) Draw a diagram to show the directions of the current, the field and the force.

20.2 What is the unit of magnetic flux density in terms of
(a) N, A and m
(b) the base units of the SI?

20.3* A horizontal length of copper wire is supported perpendicular to the magnetic field produced by two ceramic magnets on an iron 'yoke' (see diagram).
The U-magnet rests on an electronic balance which registers changes in the vertical force on it when there is an electric current in the wire.

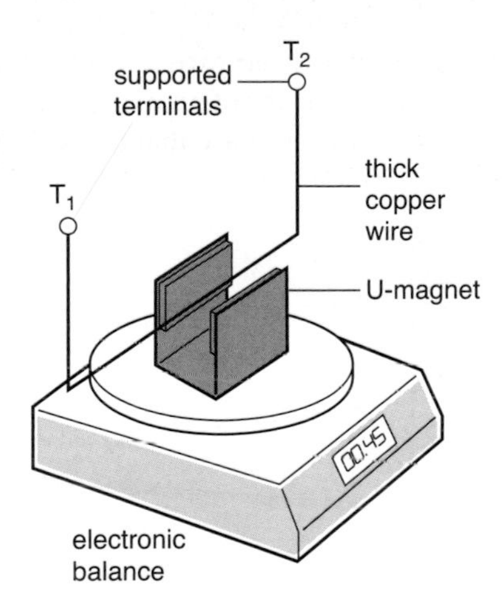

(a) Sketch a circuit diagram to show how the current between the terminals T_1 and T_2 can be varied and measured.

(b) The table shows the balance readings Δm as the current I in the circuit is varied:

I/A	1.1	2.3	3.3	4.5	5.1
Δm/g	0.60	1.30	1.85	2.50	2.85

Plot a graph of Δm against I and deduce a value for B, the magnetic flux density between the poles of the magnet. Take the length of the current-carrying wire in this B-field to be 5.0 cm.

20.4 A power cable crosses a gap of 80 m between two pylons and carries a current of 2500 A. The Earth's magnetic field in the region of the pylons is 48 μT at an angle of 20° to the vertical. If the cable is perpendicular to the Earth's field calculate

(a) the force on the cable

(b) the vertical force on the cable.

20.5 There is a current of 5.0 A through each of the conductors OA, OB, OC and OD shown in the diagram. The conductors are in a magnetic field of flux density 0.15 T parallel to the plane of the diagram. What is the size and direction of the magnetic force on each conductor?

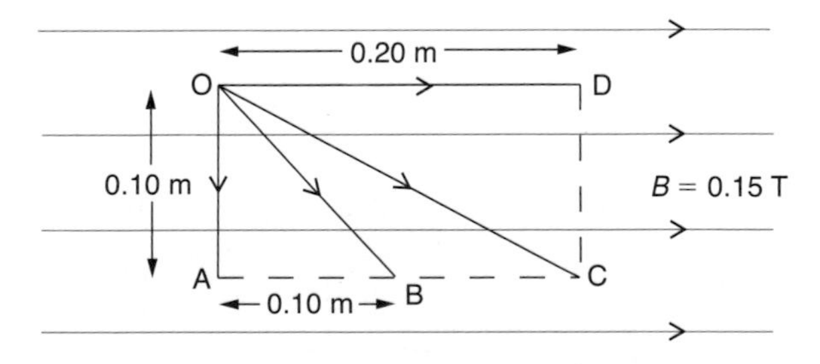

20.6 A piece of stiff wire AB of mass 0.40 g and length 0.35 m lies on a laboratory bench. A current I is arranged to pass in the wire and is slowly increased until the wire just rises from the bench, when I = 0.20 A. Describe the magnitude and orientation of a magnetic field that causes the wire to rise.

20.7 The diagram shows a rectangular three-turn coil PQRS in which there is a steady current $I = 0.40\,A$. The coil is free to rotate about an axis perpendicular to a magnetic field of flux density $B = 0.15\,T$. On the right are diagrams showing, end-on, the coil in two positions as it rotates.

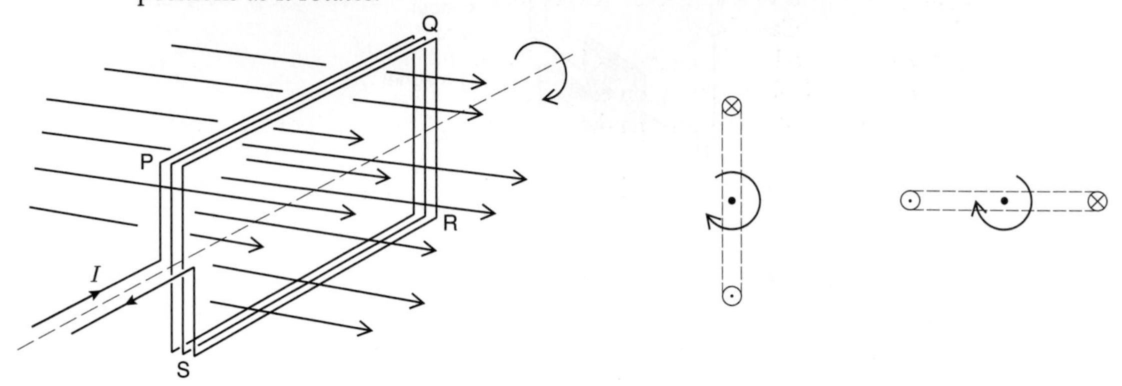

(a) Copy the two end-on diagrams and add the magnetic forces acting on the top and bottom of the coil in each case.

(b) If PQ = RS = 24 cm and PS = QR = 15 cm, calculate **(i)** the size of each of these magnetic forces **(ii)** the turning effect or torque which they produce on the coil in each case.

(c) Discuss the effect, if any, which the magnetic forces acting on the sides PS and QR have on the coil as it rotates.

(d) Explain why, if the coil is to continue to rotate, i.e. the system is to act as a d.c. motor, the current must be arranged always to enter the upper side of the coil.

20.8* An early demonstration that electromagnetic forces can produce movement (the basis of simple electric motors) was the 'jumping wire' demonstration shown in the diagram.

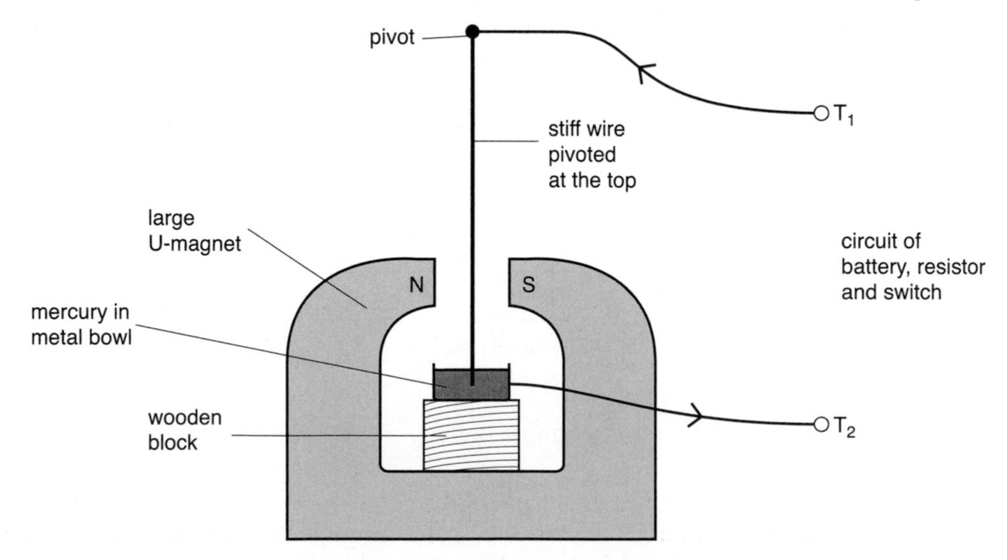

When the switch is closed the current down the stiff wire pivoted at its top causes its lower end to 'jump' out of the page and to break contact with the mercury in the metal bowl. It then falls back and the motion continues.

(a) Explain the physics behind this demonstration.

(b) Use a search engine to discover how Michael Faraday first produced continuous circular motion of a pivoted wire.

20.9 An electric current of 5.0 A is passing through a copper wire of radius 0.50 mm which lies perpendicular to a magnetic field of 0.25 T. The density of free electrons in the wire is $1.0 \times 10^{29}\,\text{m}^{-3}$. Calculate

(a) the drift speed of the electrons [use $I = nAqv$]

(b) the magnetic force acting on each electron

(c) the number of free electrons there are in a 0.40 m length of the wire

(d) the force on the wire.

Compare your answer to **(d)** with the force F given by the expression $F = BI\ell$ for this length of wire.

20.10 When a current is switched on in a solenoid, do the coils attract or repel each other? Describe how, given any apparatus you might reasonably expect to find in your laboratory, you could investigate this effect quantitatively.

20.11 **(a)** What is the size of the magnetic force on an electron moving with a velocity of $2.0 \times 10^7\,\text{m s}^{-1}$ at right angles to a uniform magnetic field of flux density 15 mT?

(b) Show on a diagram the directions of the field, the velocity and the force.

20.12 A stream of helium nuclei each carrying a charge of 3.2×10^{-19} C is travelling with a speed of $1.5 \times 10^7\,\text{m s}^{-1}$ at right angles to a magnetic field of 2.0 T.

(a) What is the magnetic force on each particle?

(b) What effect does this force have on the path followed by the nuclei?

(c) If the mass of each nucleus is 6.6×10^{-27} kg, calculate the acceleration of each nucleus.

20.13 The diagram shows two charged plates creating an electric field between them. The plates are a distance d apart and have a potential difference V between them. There is also a uniform magnetic field of magnetic flux B directed into the paper in the region marked by the small crosses. An electron, of mass m_e and charge e, enters these crossed E- and B-fields from the left. The electron is moving at a speed v.

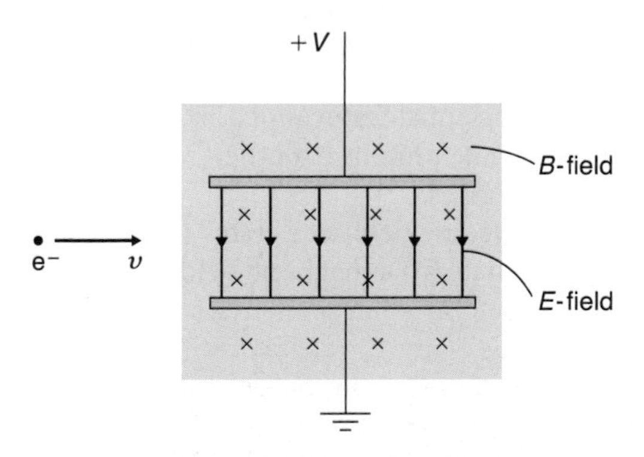

(a) What is **(i)** the magnitude and direction of the electric force F_e on the electron **(ii)** the magnitude and direction of the magnetic force F_m on the electron?

(b) Hence show that, when $F_e = F_m$, i.e. when the electron moves through the crossed fields in a straight line, $v = V/Bd$, and confirm that V/Bd has the unit m s^{-1}. [This arrangement is sometimes called a *velocity selector*.]

(c) With $B = 2.4$ mT and $d = 25$ mm, what value of V 'selects' those electrons that go straight through the fields at $7.0 \times 10^6\,\text{m s}^{-1}$?

20.14 A particle with a charge q and mass m is moving with speed v at right angles to a magnetic field of flux density B, and moves in a circular path of radius r. Derive an expression

(a) for r in terms of q, m, v and B

(b) for the period T in terms of B, q and m.

(c) Check that the units of the right-hand side in the above expressions are metre and second, respectively.

20.15 In one type of mass spectrometer two beams of singly ionised neon atoms are fired perpendicular to a magnetic field B. The two beams have the same speed v but are of different isotopes of neon with masses $20m$ and $22m$. What is the ratio of the radii in which the isotopes move? State which has the greater radius.

20.16 The diagram shows a uniform magnetic field: its direction is into the paper and its strength is 2.5 mT. An electron at P is fired in the direction shown at a speed of $2.0 \times 10^{7}\,\mathrm{m\,s^{-1}}$.

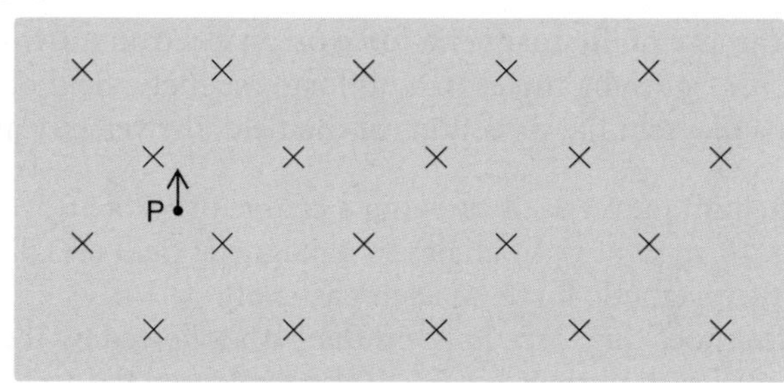

(a) On a copy of the diagram mark the direction of the magnetic force on the electron and calculate **(i)** the radius r of the electron's circular path **(ii)** the time T taken for the electron to make one complete circular orbit. [Use data.]

(b) A second electron is fired from P in the same direction with twice the speed. How long will this second electron take to complete one circular orbit? Comment on your answer.

(c) Show that $T = 2\pi m/Be$, i.e. that the time to complete one circular orbit is independent of the speed v with which the electron is projected.

20.17 The diagram shows the essential features of a cyclotron that can accelerate protons (or other positive ions) to high energies.

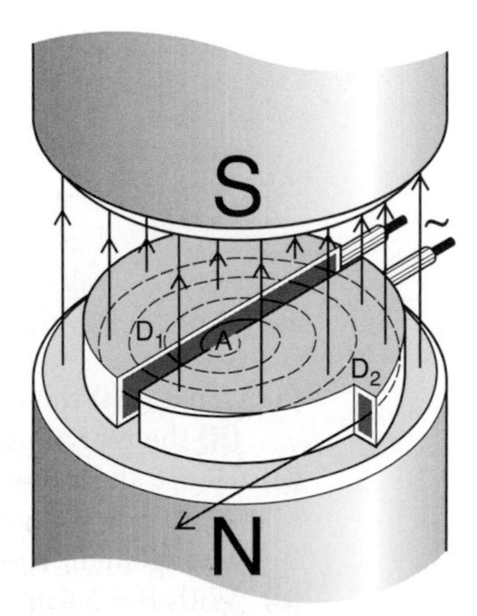

(a) Explain how **(i)** an electric field accelerates the protons every half circle **(ii)** a magnetic field confines the protons to move in semicircles at a constant speed.

(b) A particular cyclotron has an alternating p.d. of 10 kV and a magnetic field of magnetic flux density 1.3 T. Use the result of question 20.16(c) to calculate the time T taken by the accelerating protons to complete a circle and hence the frequency f of the alternating p.d. [Use data.]

(c) How many circles does a proton make in reaching a final energy of 10 MeV? Hence calculate the time a proton takes from injection at A to its ejection from the dees.

(d) Show that a proton given 10 MeV of energy will show a relativistic mass increase of 1.8×10^{-29} kg.

(e) What percentage of the rest mass of a proton does this represent? [Use data.]

20.18 A cyclotron is rather like a 'coiled up' linear accelerator (linac).

(a) What is it that 'coils up' the linac to form a cyclotron?

(b) Draw the first four drift tubes of a typical linac and compare how charged particles are accelerated in **(i)** a linac **(ii)** a cyclotron – you may refer to the cyclotron diagram in question 20.17.

(c) List any other features common to both accelerators.

20.19 The diagram, taken from a photograph and half real size, shows the path of an electron of rest mass m_e spiralling in a magnetic field B that is into the page.

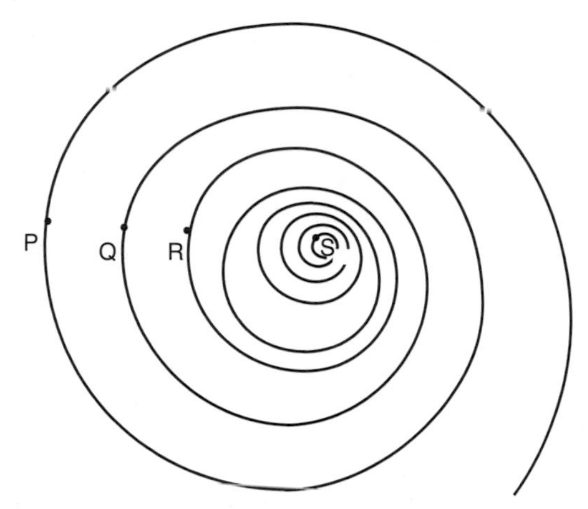

(a) Show that the momentum p of an electron with charge of magnitude e moving in a circular path of radius r is given by $p = Ber$.

(b) In the photograph there was a magnetic field of flux density 1.2 T into the page. By carefully measuring the radii PS, QS and RS, calculate the momentum of the electron at P, Q and R. (Remember the diagram is half real size.)

(c) The speed of the electron is known to be 3.0×10^8 m s^{-1} at each position. Deduce the effective mass of the electron at each position and comment on your answers.

20.20* This famous cloud chamber photograph shows an energetic cosmic ray particle. It was possible to deduce that the rest mass of the particle was the same as the electron. There is a magnetic field of flux density 1.5 T into the photo and the lines across the centre are the edges of a lead plate that was 6 mm across.

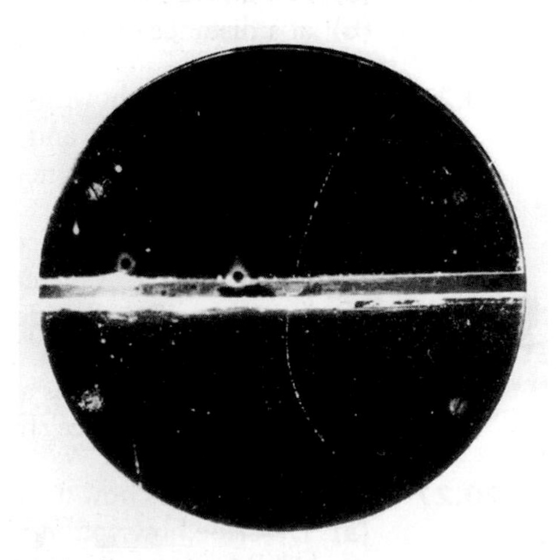

(a) Explain how the photograph tells you that the particle which leaves the track passing downwards through the lead plate was positively charged.

(b) The photograph was taken by Carl Anderson in 1932 and was the first evidence of the positron – the antiparticle to the electron. Use a search engine to learn more about Anderson and the positron.

20.2 Magnetic field patterns

In this section you will need to

- draw, in two dimensions, magnetic field (flux) patterns near a long straight wire, a flat coil and a long solenoid
- remember a rule for predicting the relation between the sense of a current and that of the field lines
- understand that magnetic flux density B is a vector quantity
- use the equation $B = \mu_0 I/2\pi r$ for a straight wire and $B = \mu_0 nI$ for a solenoid.

20.21 **(a)** The diagram shows wires carrying currents into and out of the paper. Copy the diagram, leaving some space above and below it, and sketch the separate magnetic field patterns in the plane of the paper caused by these currents.
(b) On a second copy of the diagram sketch the resultant magnetic field pattern. [As is usual, the lines should be close together where the field is stronger.]
(c) Sketch the magnetic field pattern in and around a current-carrying solenoid.

20.22 At a latitude of 55° the Earth's magnetic flux density is 50×10^{-6} T or 50 μT. In the northern hemisphere this field is directed downwards into the Earth at an angle of 65° to the horizontal. Calculate
(a) the horizontal component B_h
(b) the vertical component B_v of this field.

20.23 A long straight wire carries a current of 10 A. Draw a graph with labelled axes to show how the magnetic flux density varies with distance r from the wire for values of r from 5 mm to 50 mm.

20.24 The magnetic flux density at a distance of 10 mm from a long straight wire carrying a current of 10 A is 0.20 mT. What is the magnetic flux density
(a) at a distance of 6.0 mm, with the current still 10 A
(b) at a distance of 10 mm, if the current is 3.0 A?

20.25 Two parallel straight wires A and B carry currents in opposite directions: the currents are, respectively, 2.0 A and 3.0 A. The wires are 0.10 m apart. At what distance from wire A will the resultant magnetic flux density be zero?

20.26* The cross-channel direct current cable between the UK and France carries a maximum current of 14.8 kA. A typical depth below the surface of the sea is 50 m.
(a) What magnetic flux density is created at the surface of the sea by the maximum current?
(b) Your answer to **(a)** is a significant fraction of the strength of the Earth's magnetic field but in practice ships need make no allowance for it in navigating. Why not?

20.27 The graph shows how the magnetic flux density B varies along the axis of a solenoid.
(a) Describe how you would produce the experimental data needed to plot a graph of this kind.

(b) Express the magnetic flux density at the ends of the solenoid as a fraction of that at its centre.

(c) By considering the magnetic flux density at the join when two such solenoids are placed end-to-end, argue that your answer to **(b)** is theoretically predictable.

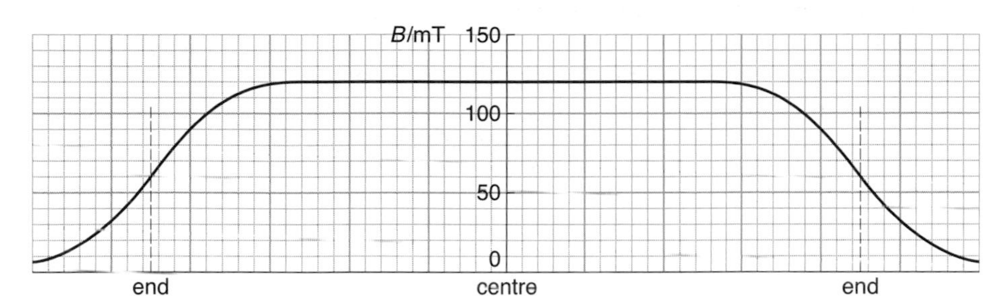

20.28 A solenoid used in magnetic resonance imaging (MRI) is wound from a niobium–titanium superconducting wire 2.0 mm in diameter, with adjacent turns separated by an insulating layer of negligible thickness. (The solenoid is 2.4 m long and 95 cm in diameter.) Calculate the current in the superconducting wires necessary to produce a magnetic field of flux density 1.5 T at the centre of the solenoid.

20.29 Show that the units of the equations $B = \mu_0 nI$ and $B = \mu_0 I/2\pi r$ are consistent.

20.3 Electromagnetic induction

In this section you will need to

- use the equation $\mathcal{E} = B_{\perp}\ell v$ for the e.m.f. produced across a conductor moving in a magnetic field
- use the equation $\Phi = B_{\perp}A$ which defines magnetic flux
- understand that electromagnetic induction involves the transfer of mechanical energy to electrical energy
- understand that e.m.f.s are induced when the magnetic flux linking a circuit, e.g. a loop of wire, is changing
- use Faraday's law of electromagnetic induction: $\mathcal{E} = N(\Delta\Phi/\Delta t)$, i.e. the induced e.m.f. is equal to the rate of change of flux linkage
- remember that the direction of an induced e.m.f. is such as to oppose the change causing it.

20.30 A teacher demonstrates that an e.m.f. is produced when a wire is moving between the poles of a strong U-shaped magnet. The diagram (on the next page) shows the direction of the field and the way in which the wire is moved. The circle shows an enlargement of part of the wire. On a copy of the inset mark

(a) the directions of the magnetic forces on the positive ion and the electron

(b) the positive and negative ends of this part of the wire

(c) the direction a current would have if the ends of the wire were connected by a circuit, none of which passed through the magnetic field.

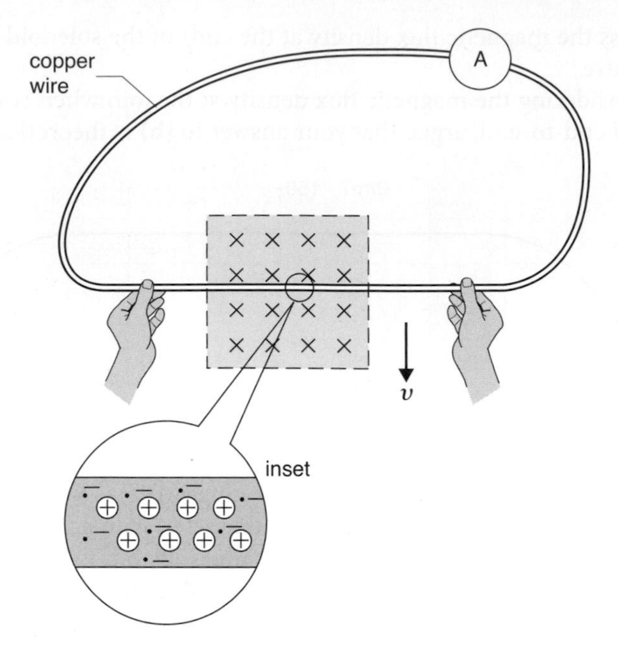

20.31 **(a)** If, in the previous question, there is an induced current in the wire, in which direction is the magnetic force produced by this current?
(b) Is this direction the same as, or opposite to, the direction in which the wire was being moved? Comment on your answer.

20.32 If, in question 20.30, the magnetic field is of flux density 0.12 T, the length of the wire in the field is 25 mm, and the teacher moves the wire at a speed of $1.5\,\text{m}\,\text{s}^{-1}$, what is the e.m.f. produced?

20.33 Show that the unit of $B\ell v$ is the volt.

20.34 The wingspan of Concorde, a supersonic aircraft, was 25.6 m. It cruised at 50 000 ft and travelled at $550\,\text{m}\,\text{s}^{-1}$ (Mach 2 at that altitude).
(a) Explain whether the size of the e.m.f. induced between its wingtips depended on the direction in which Concorde was flying at a particular place.
(b) Calculate the size of this e.m.f. at a place where the vertical component of the Earth's magnetic field is $50\,\mu\text{T}$.
(c) Where might Concorde have been flying if the induced e.m.f. was zero?

20.35 A metal scaffolding pole, 5.0 m long, falls from the top of a high building. The pole is aligned east–west while locally the Earth's magnetic field points north–south and has a horizontal component of $18\,\mu\text{T}$. Calculate the induced e.m.f. between the ends of the pole 3.0 s after it started to fall.

20.36 The diagram opposite shows two railway lines, of negligible resistance, joined at the left by a conducting link of resistance R. An axle, of resistance r, rolls along the rails at a speed v. The vertical component of the Earth's magnetic field B acts downwards as shown.
(a) Explain why an e.m.f. is induced in the circuit ABCD.
(b) Given that $B = 50\,\mu\text{T}$ and $v = 4.0\,\text{m}\,\text{s}^{-1}$, estimate a value for d, the separation of the railway lines, to calculate a value for this e.m.f.
(c) What value of $(R + r)$ would give rise to a current of 60 mA in the circuit?

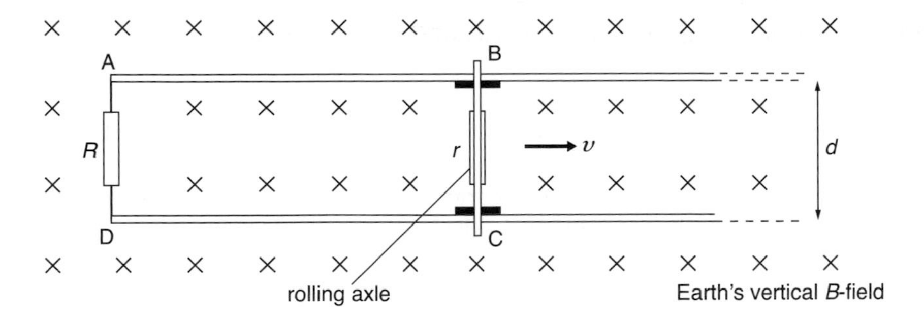

20.37 Calculate the magnetic flux through a football pitch which measures 110 m by 70 m at a place where the flux density of the Earth's magnetic field is 49 μT or 49 μWb m^{-2} in a direction making an angle of 22° with the vertical.

20.38 The diagram shows a wire loop ABCD whose centre is O. It is placed at right angles to a uniform magnetic field of flux density 0.50 T. The dimensions of the loop and the space covered by the field are shown on the diagram.

(a) What is the flux through the loop?

(b) What is the size of the change of flux through the loop when the loop

(i) is moved, in its own plane, 0.20 m to the right

(ii) is moved, in its own plane, 0.30 m to the right

(iii) is rotated, in its own plane, through 90°, about its centre O

(iv) is rotated, about the line BC, through 90°

(v) is rotated, about the line BC, through 180°

(vi) is raised 0.10 m, parallel to the field (i.e. out of the paper).

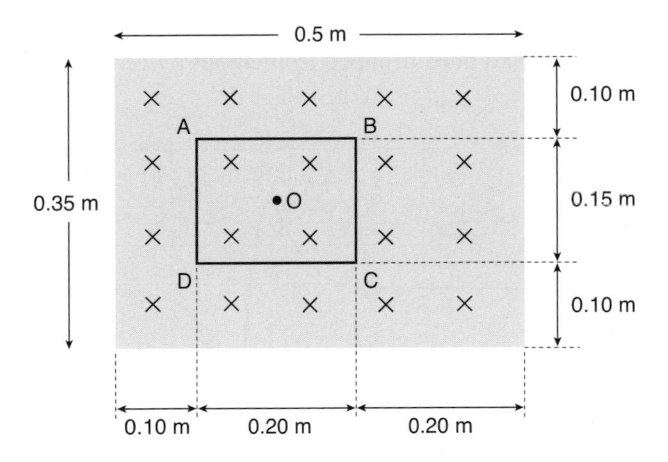

20.39 A large electromagnet has circular pole pieces of diameter 0.20 m. The total flux produced by the magnet is 0.050 Wb. Calculate the average flux density between the pole pieces.

20.40 Refer to question 20.28 about an MRI scanner. Take the diameter of the 'tube' into which a patient can be placed for a full body scan to be 0.70 m. Assuming the magnetic flux density to be uniform across the tube, calculate the magnetic flux through the tube.

20.41 A bar magnet is moved towards a circular coil as shown in the diagram. An e.m.f. is induced in the coil.

(a) Explain why an e.m.f. is induced.

(b) By considering the direction of the induced current, deduce the direction of the induced e.m.f.

(c) What is the source of the energy which the induced current transfers to the coil?

(d) State the effect of **(i)** moving the magnet more quickly **(ii)** moving the magnet away from the coil.

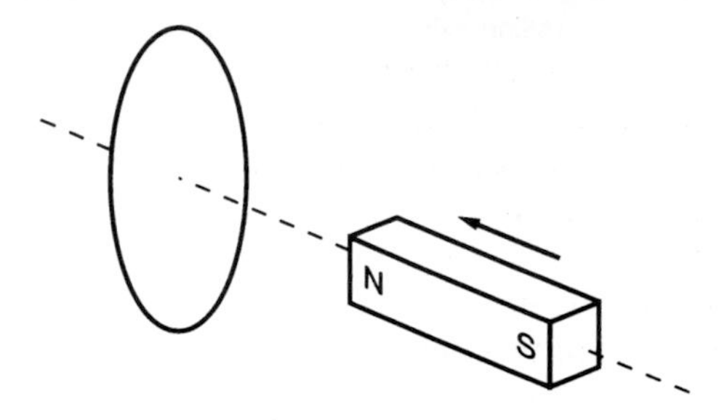

20.42 A circular coil is placed with its axis vertical and a bar magnet, with its axis aligned with the axis of the coil, is held above the coil and then dropped. A data logger connected to the coil records the e.m.f. induced in the coil at short time intervals and later draws a graph to show how the e.m.f. varies with time.

(a) The diagram shows the graph obtained as the magnet falls through the coil. Explain the shape of the graph.

(b) Copy the graph, and on the same axes sketch the graphs which would have been obtained if, separately (with the data logger again connected)

(i) the coil had been replaced with one with twice the number of turns

(ii) the magnet had been dropped from about twice the height.

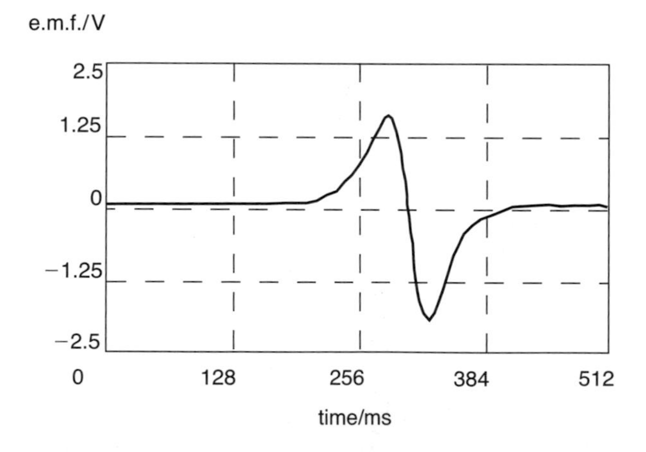

20.43* Refer to the previous question. Explain why the magnet would take longer to fall through the coil if the data logger were removed and, instead, the ends of the coil were connected together:

(a) using an argument based on forces

(b) using an argument based on energy.

20.44 A large U-magnet is placed on a bench so that the magnetic field between its poles is horizontal. A long wire is connected to a data logging device.
Draw sketch graphs of the induced e.m.f. $\mathcal{E}$ against time t as recorded by the data logger when the wire is moved

(a) at the same steady speed vertically downwards and then vertically upwards between the poles of the magnet
(b) slowly downwards and then quickly upwards between the poles of the magnet
(c) at a steady speed horizontally from one pole of the magnet to the other.

20.45 A metal framed window, which measures 0.80 m by 1.2 m, pivots about a vertical side and faces south. When closed its plane is vertical and at right angles to the Earth's magnetic field which has a local horizontal flux density of 18 μT. The window is opened through 90° in 1.5 s. Calculate

(a) the initial flux through the window
(b) the change of flux through it in 1.5 s
(c) the e.m.f. induced in the window frame.

20.46 A ring of copper wire of area 100 cm^2 and electrical resistance 0.012 Ω is placed with its plane at right angles to a uniform magnetic field. If the field is changing at the rate of 2.0 T s^{-1}, calculate

(a) the e.m.f. induced in the ring
(b) the induced current in the ring.

20.47 The diagram shows a bar magnet that swings to-and-fro above a vertical coil of wire. The coil is connected to an oscilloscope with its time base switched on.

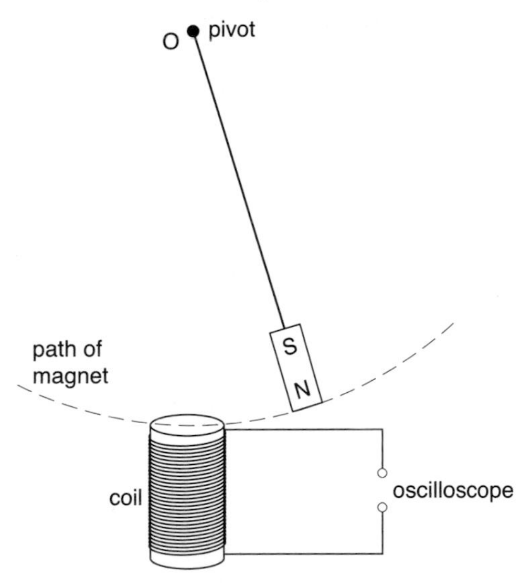

(a) Sketch a graph to show the trace on the oscilloscope during two complete swings of the magnet.
(b) The maximum e.m.f. is 4.0 mV. Calculate the rate of change of magnetic flux needed to induce this e.m.f. in a coil of 400 turns.

20.48 A circular coil and a copper ring are placed, as in the diagram, flat on a horizontal table.

(a) What is the direction of the current in the ring if the current in the coil is
(i) clockwise and increasing **(ii)** clockwise and decreasing
(iii) anticlockwise and increasing **(iv)** anticlockwise and decreasing?

(b) The current in the coil is increased from zero to 4.0 A in 2.0 s. How will the current in the ring be affected if the current in the coil is increased from zero
(i) to 2.0 A in 2.0 s **(ii)** to 2.0 A in 1.0 s **(iii)** to 4.0 A in 1.0 s?

(c) How will the energy converted in the ring be affected if the three processes given in **(b)** occur?

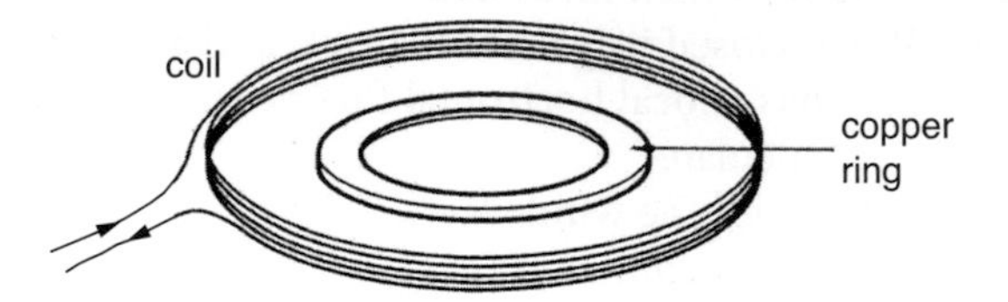

20.49* The diagram shows a bicycle dynamo in which a permanent magnet is made to rotate close to a fixed coil of wire of many turns. From the dynamo, leads connect to a lamp fixed to the handlebars of the bike.

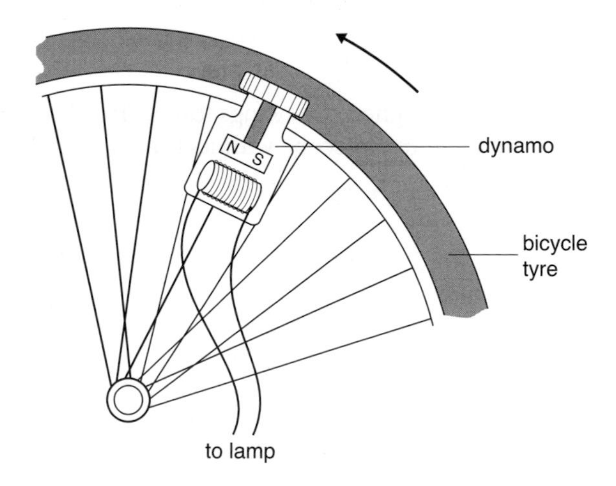

Explain how the dynamo lights the lamp when the cyclist is moving at a steady speed and state how the brightness of the lamp varies with the speed of the bike.

20.50 A rigid metal frame ABCD of side 40 cm moves horizontally at a constant speed of $5.0\,\text{cm}\,\text{s}^{-1}$ through the shaded region, perpendicular to which there is a uniform magnetic field of flux density 25 mT acting downwards.

(a) For each position 1, 2 and 3, calculate the e.m.f. induced in the frame, in each case stating the sense, clockwise or anticlockwise, in which the e.m.f. would drive a current round the frame.

(b) If the resistance of the metal frame is $2.0\,\text{m}\Omega$, describe how the current round the frame ABCD varies as the frame crosses the region of the magnetic field.

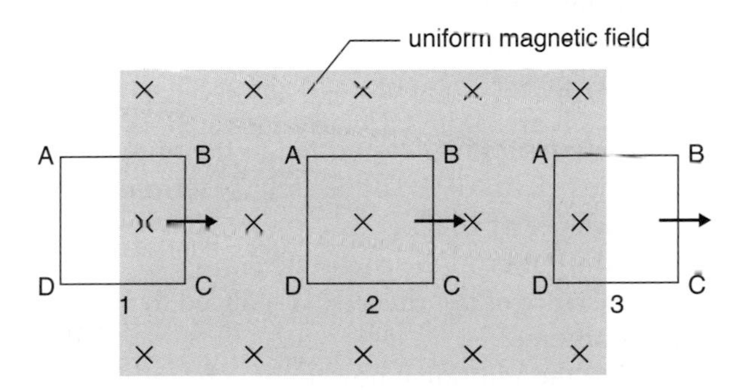

20.51 A coil with 5000 turns of mean area $1.0\,\text{cm}^2$ – a search coil – is placed in a magnetic field. The flux density of the field varies as shown in the graph below.

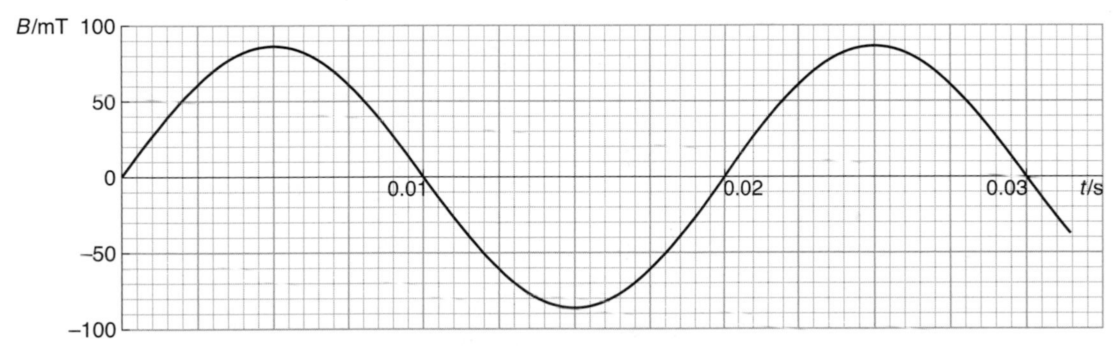

(a) Explain at what time or times the e.m.f. induced in the coil will be a maximum.

(b) What is the maximum rate of change of the magnetic field in $\text{T}\,\text{s}^{-1}$?

(c) Estimate the maximum e.m.f. induced in the coil.

(d) Assuming that the induced e.m.f. is sinusoidal, sketch its variation with time and comment on the relationship between your sketch and the graph in the diagram.

20.4 The transformer

In this section you will need to

- understand the principle of the transformer
- use the ideal transformer equation $V_P/V_S = N_P/N_S$ for primary and secondary voltages and turns.

20.52 The diagram gives information about an ideal transformer. Explain as fully as possible, supporting your explanation with calculations, why this transformer is said to be ideal.

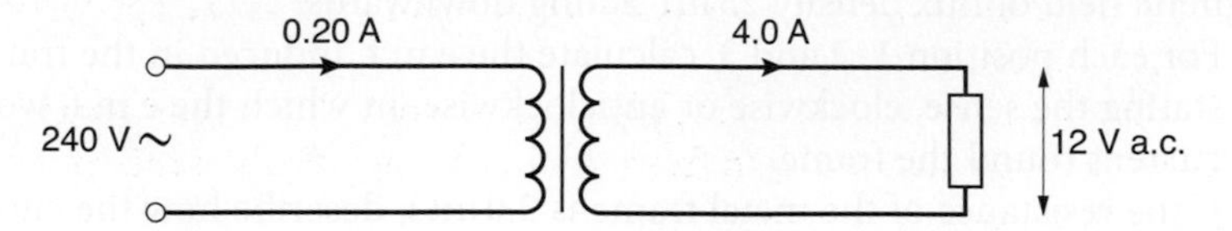

20.53 An ideal transformer has 4800 primary turns and 120 secondary turns. It is used to light a 6.0 V 24 W lamp. What is **(a)** the primary voltage **(b)** the primary current **(c)** the primary power?

20.54 A 12 V alternating supply is connected to the primary terminal of a transformer which steps down the p.d. to 6.0 V. The secondary terminals are connected to a rheostat which is set to a resistance of 10 Ω.

(a) What is the power in the rheostat?

(b) If the resistance of the rheostat is reduced to 5.0 Ω, what does the power in the rheostat become?

(c) Where has the additional power come from?

20.55 The diagram shows a laboratory demonstration of electromagnetic induction involving a 400 turn coil connected to a 50 Hz 12 V supply and an aluminium ring.

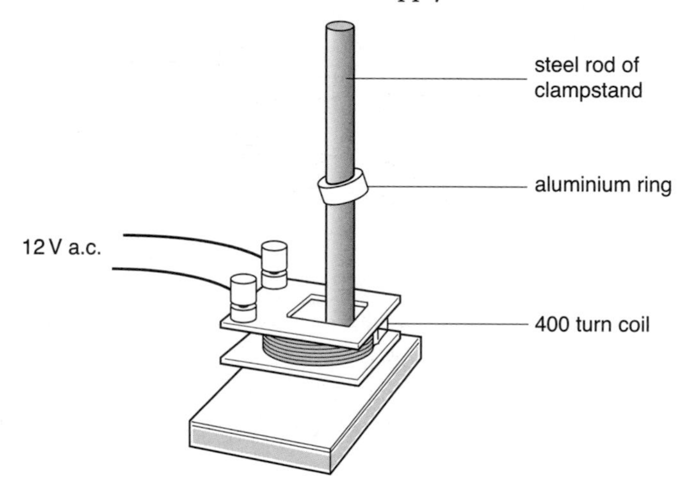

(a) Switching on the alternating current in the 400 turn coil causes the aluminium ring to rise up the steel rod and 'float' there. Explain the origin of the upward force supporting the ring in this position.

(b) The coil and ring act as a step-down transformer. However, it is not an ideal transformer, because the ring is so far from the coil. Assume its efficiency is 2.0%. If the p.d. supplied to the coil is 12 V, and the current in the coil is 2.0 A, what is the power transferred in the ring?

(c) The resistance of the aluminium ring is 85 μΩ. How big is the induced current in the ring?

(d) The aluminium ring becomes hot. If the initial rise of temperature of the ring is 11 K min^{-1}, calculate the mass of the ring. The s.h.c. of aluminium is 880 J kg^{-1} K^{-1}.

20.56* A building site has a transformer which steps down the incoming p.d. from 11 kV to 415 V. Assume that the transformers in this question are 100% efficient.

(a) If the number of turns on the primary is 3000, what is the approximate number of turns on the secondary?

(b) The p.d. of 415 V is used to supply a crane which has a maximum power of 60 kW. Calculate the current drawn from the 11 kV supply when this crane is working at maximum power.

The site has a second transformer which, for safety reasons, steps down the p.d. of 415 V to 25 V for some 100 W hand-held lamps.

(c) When five of these hand-held lamps are in use, what current is drawn from the 11 kV supply?

20.57 A laboratory power supply contains a transformer so that the mains p.d. of 240 V may be stepped down to low p.d.s increasing in steps of 2 V from zero to 12 V. The primary coil has 1200 turns. Explain how the secondary p.d.s may be obtained.

20.58* The diagram shows how sparks are produced in the ignition system of a petrol-driven car. The result of breaking the contact at C in the primary circuit (which has tens of turns) is to induce a huge voltage surge in the secondary coil (which has thousands of turns).

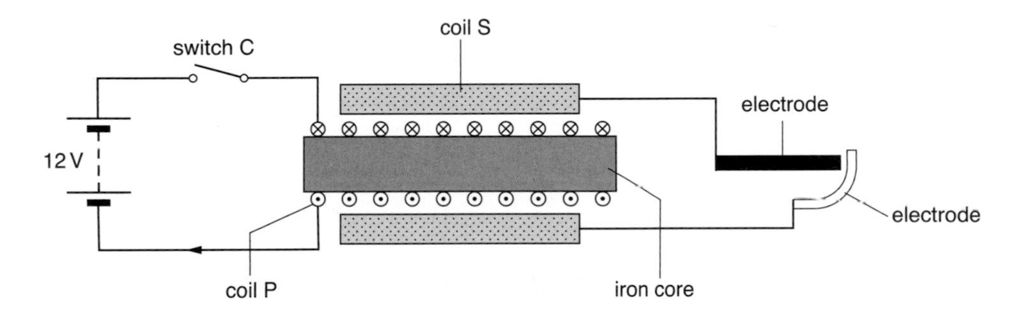

(a) Describe in detail the physics which leads to there being a spark between the electrodes each time the contact at C is broken.

(b) A spark in air occurs when the E-field between the electrodes exceeds 3.2 MV m^{-1}. Suggest a possible value for the p.d. required between the electrodes when the gap between them is 0.90 mm.

Answers

1 Mechanics: linear motion

1.1 **(a)** 3.3 m s $^{-1}$
(b) zero
1.3 **(a) (i)** 0.16 mm
(ii) 0.018 mm
(b) (i) 0.13 mm s^{-1}
(ii) 0.018 mm s^{-1} from A towards B
1.4 **(a)** +9.0 m s^{-1}
(b) −21 m s^{-1}
(c) −12 m s^{-1}
(d) 20 m s^{-1} west
1.5 **(a)** 3.03 m s^{-1}
(b) 10.6 m s^{-1}
1.6 310 s (≈ 5 minutes)
1.7 **(a) (i)** 1.5 m s^{-1}
(ii) about 0.9 m s^{-1}
(iii) about 0.6 m s^{-1}
(b) 0.90 m s^{-1}
1.8 0.3 s (3% at least)
1.9 **(b)** 6.7 m
1.10 **(b) (i)** about 23 m s^{-1}
(ii) about 11 m s^{-1}
(c) about 35 m s^{-1}
1.11 **(a)** P takes 50 s, Q takes 33 s, R takes 20 s
(b) 1.0 m s^{-1}
1.13 **(a)** 15.40 s
(b) 64.92 m s^{-1}
1.14 **(a)** 72 m **(b)** 72 m
1.15 **(b)** 1.8 m
1.18 **(a)** 8.4 km N 17° W
(b) 29 m s^{-1} S 79° W
1.19 **(a)** 3.0 m s^{-1}
(b) 6.0 m s^{-1}
(c) 1.2 m s^{-2}
1.20 **(a)** 57 m s^{-2}
(b) −190 m s^{-2}
1.21 **(a)** +2.5 m s^{-2}, zero, −5.0 m s^{-2}
1.22 **(a)** 33 s **(b)** 1.9 km
1.24 **(a)** about 4 m s^{-2}
(b) about 35 m
1.25 400 m
1.26 **(b)** $d_{\text{think}} \propto v$
(c) 27 m
1.27 **(b)** k is numerically 0.015
(c) (i) −6.5 m s^{-2}
(ii) −6.5 m s^{-2}
1.28 dry 81 mph
wet 68 mph
1.29 **(a)** 800 m s^{-2}
(b) (i) 0.07 s
(ii) 30 mm
1.30 **(a)** 9.0×10^{12} m s^{-2}
(b) 2.3 m
1.31 **(a)** 9.5 m s^{-1}
(b) (i) 8 m s^{-2}
(ii) about 1.5 m s^{-2}
1.32 **(a)** $v^2 = u^2 + 2as$
(b) $s = ut + \frac{1}{2}at^2$
1.33 21 m s^{-1}
1.34 55 cm
1.35 220 m s^{-2}
1.36 **(a)**, **(b)** and **(c)** all 600 m
1.37 **(a)** 9.8 m s^{-1}, 0, −9.8 m s^{-1}, −19.6 m s^{-1}
(b) 19.6 m, 0
(c) 39.2 m
1.38 **(a) (i)** 0.45 s
(ii) 0.64 s
1.40 **(a)** 1.51 m s^{-1}
(b) 3.03 m s^{-1}
(c) 10.1 m s^{-2}
1.41 1.8 m
1.42 44 m
1.43 7.0 m s^{-1}
1.45 **(a) (i)** 1.2 m
(ii) 4.9 m s^{-1}
(b) 1.4°
1.46 **(b)** 10.0 m s^{-2} each time
(c) 0.80 m
(i) 0.45 m
(ii) 0.29 m
1.47 **(b) (i)** 27 m s^{-1}
(ii) 22 m s^{-1}
1.50 **(a)** 9.4 m s^{-1}
(c) 16°
1.51 **(b)** 8.65 m

2 Balanced and unbalanced forces

2.1 35 N
2.2 **(a)** $U = 22$ N, $P = 150$ N, $T = 29$ N
2.4 **(a)** 50 N
(c) 850 N
2.5 **(b) (i)** 700 N
(ii) 100 N
(c) 710 N
2.6 **(b)** $P = 11000$ N, $Q = 2000$ N
2.7 **(a)** 19 N
(b) 5 N
2.8 **(a)** 10 N
(b) 17 N
(c) 17 N
2.9 **(b)** $W = 49$ N, $F = 17$ N, $R = 46$ N
2.10 **(a)** 5.2°
2.11 **(b)** 12°
2.12 **(a)** 43°
(b) 30 N
2.13 **(a)** 0.71 kN
(b) 69 kg
2.14 19°
2.15 **(a)** 1200 N m anticlockwise
(b) zero
(c) 600 N downwards
2.16 **(a)** 2000 N m and 600 N m, both anticlockwise
(b) 1300 N
2.17 **(a)** 140 N
(b) 38 N
2.18 **(a)** $N + Y = W$
(b) 10 N
2.20 **(b)** 1950 N down, 3100 N up
2.21 **(a)** 0.12 MN
(b) 11 m from B
2.22 **(a)** 3.3 N m clockwise
(b) 5.0 cm
(c) 67 N
(d) 50 N
2.23 **(b)** $M = 17$ N

(d) (i) 148 N
(ii) 6.5°
2.24 **(b)** 115 N
(c) $X = 58$ N, $Y = 100$ N
2.25 **(a)** 5.0 m s^{-2}
(b) 4.0 m s^{-2}
(c) 7.5 m s^{-2}
(d) 1.2 m s^{-2}
(e) 5.0 m s^{-2}
2.26 **(a)** −36 N
(b) −10 N
(c) −2.2 N
(d) +26 kN
(e) -1.8×10^{-9} N
2.27 0.55 m s^{-2}
2.28 **(a) (i)** 21 kN
(ii) 8.0 kN
(b) both 26 times
2.29 1200 N
2.30 **(a)** −3000 m s^{-2}
(b) −600 MN
2.31 2.0×10^4 N
2.32 **(a)** 5.0 m s^{-2}
(b) 210 kN
2.34 **(a) (i)** 340 N
(ii) 87 N
(b) (i) 5.2 m s^{-2}
(ii) 1.3 m s^{-2}
2.35 **(a) (i)** 6.0 kN
(ii) 5.2 kN
2.37 **(a) (i)** 0.83 kN, 0.67 kN, 0.34 kN, 0.085 kN
(ii) zero, 0.16 kN, 0.49 kN, 0.75 kN
2.38 **(a)** 12 kN
(b) 14 kN
(c) 8.2 kN
(d) 15 kN
2.39 **(b) (i)** 2.7 kN
(ii) 1.7 kN
2.40 2.9 kN
2.41 **(a) (i)** 32 N
(ii) 6.4 N
2.42 **(a)** 11 kN
2.44 0.44 kN

3 Work and energy

3.1 **(a)** 8000 J
(b) −8000 J
3.2 **(a) (i)** 2.3 kJ
(ii) 1.4 kJ
3.3 2.5 kJ
3.4 **(a)** parallel 35 kN, perpendicular 8.1 kN
(b) (i) 140 MJ
(ii) zero
3.5 **(a)** 10 J **(b)** 38 J **(c)** 22 J
3.6 **(a) (i)** 0.80 J
(ii) 3.2 J
3.8 **(a)** 9.3 kJ, zero, −7.7 kJ, −1.6 kJ
3.9 5.9 MN
3.10 **(a)** 35 MJ
(b) (ii) 190p
3.11 50 kN
3.12 zero
3.13 about 6.3 GJ
3.14 **(a)** about 10 MJ
(b) about 1.2 kW
3.15 both kg m^2 s^{-3}
3.16 **(a)** 27 W
(b) 7.1 kW
3.17 **(a) (i)** A £9.00, B £29.10
(ii) A £108.00, B £19.80
(b) at these prices and costs A is 2.4 times as expensive as B
3.18 98 W
3.19 **(a)** 44 s
(b) 180 kJ
3.20 **(a) (i)** 79%
(ii) 100%
(b) (i) 330 N
(ii) 250 N
3.21 **(a)** about 2.7%
3.22 **(a)** 50×10^6 m^2 or 50 km^2
(b) 0.017%
(c) 1.4%
3.25 **(a)** 8.5 mm
(b) 4.0×10^{26} J
3.26 **(a)** 310 kJ
(b) 1.7 kJ
3.27 **(a)** 1.2 kJ
(b) about 3.5 kJ
(c) about 10 MJ
(d) about 1.5 kJ
3.28 93 μW
3.29 **(a)** 150 kJ
(b) 830 W
3.30 **(a)** 9.0 kJ **(b)** 1.5 kJ
3.31 **(a) (i)** 23 m
(ii) 46 m
(b) 103 m
3.32 **(a)** 3.2 J **(b)** 2.7 J **(d)** 0.5 J
3.34 **(a)** 72 W **(b)** 56 W
3.35 **(a) (i)** g.p.e. = 13(.7) kJ, k.e. zero
(ii) g.p.e. = 2.7 kJ, k.e. = 11 kJ
(iii) g.p.e. zero, k.e. zero
(b) (i) 13(.7) kJ
(ii) 3.4 kN
3.36 0.88 mW
3.37 gain of k.e. = 71 kJ
loss of g.p.e. = 103 kJ
average push = 210 N
3.38 **(a)** 55–58 J
(b) about 37 m s^{-1}
(c) 69 m
3.39 **(a) (i)** 1.6 J
(ii) 1.6 J
(b) speed is 4.4 m s^{-1}, it breaks
3.40 **(a)** 1.1×10^9 kg (about a million tonnes)
3.41 **(a)** 46 s
(b) 36 ms^{-1}
(c) 3.5 kN
3.44 **(a) (ii)** (e.g.) k.e. to e.p.e. (elastic energy) plus some g.p.e.
(b) about 5.0 m

4 Electricity: charge and energy

4.2 **(a)** 0.20 A **(b)** 0.20 A
(c) 0.60 A **(d)** 0.40 A
4.3 5.0 A
4.4 0.50 kA
4.5 20 kA
4.6 C 0.20 A, D 0.05 A
4.7 **(a)** 144 C
(b) 9.0×10^{20}
4.8 100 s
4.9 72 mC
4.10 **(a)** 30 C
(b) 10 mC
(c) 20 μC
4.11 **(c)** 0.36 kC

(d) 6.3 kC
4.12 (b) (i) 6.0 kC
(ii) 0.60 kC
(iii) 13.2 kC
4.13 290 kC
4.15 (a) (i) 2.8 C (ii) 3.0 C
4.16 (b) (i) 1.7 A
(ii) 0.57 A
4.17 17 μA
4.18 8.2 mm s^{-1}
4.20 about 15 hours
4.22 6.9 mm s^{-1}
4.23 1.7 mA
4.24 $0.6 \times 10^{16} s^{-1}$
4.25 (a) 1 : 4
(b) the thinner wire
4.26 (a) from A to D
(b) (i) from A to D
(ii) from D to A
(c) $2.4 \times 10^{-7} m s^{-1}$
4.27 31 m s^{-1}
4.28 (a) 15 J (b) 45 J (c) 30 J
4.29 (a) (i) 4.5 J
(ii) 9.0 J
(iii) 18 J
(b) 6.0 V
4.30 (a) 7.5 V
(b) 4.5 V
(c) 1.5 V
4.31 (a) 12 J (b) 15 J (c) 27 J
4.33 0.54 kJ
4.34 (a) 3.6 kJ (b) 25 h
4.35 (a) 36 kC (b) 54 kJ
4.36 68p, 17p and 11p
4.37 (a) 28 kJ, 43 kJ, 86 kJ
(b) 360 J g^{-1}, 430 J g^{-1}, 660 J g^{-1}
4.40 all except D & E and H & A
4.41 (a) 4.0 V (b) 4.0 C
(c) 16 J (d) 4.0 V
(e) 50 mA (f) 0.20 A
(g) 0.80 W
4.42 (a) 0.60 kC
(b) 7.2 kJ
(c) 12 V
4.43 all 2.89 V
4.44 (a) (i) both 3.0 V
(ii) all 2.0 V
(iii) 2.0 V, 4.0 V
(iv) 3.0 V, 1.5 V, 1.5 V
4.45 (a) (i) +1.0 V
(ii) zero
(iii) +3.0 V
(iv) −3.0 V
(b) (i) *a* and *b*, *b* and *c*, *k* and *l*, *l* and *n*
(ii) *e* and *g*, *i* and *j*
(iii) *b* and *m*
(iv) *g* and *e*, *j* and *i*
4.46 (a) 0.15 m
(b) 0.45 m
4.48 (a) (i) 4.0 V, 6.0 V, 2.0 V
(ii) each 2.4 V
4.49 (a) (i) +0.8 V (ii) −0.4 V
(b) (i) *w* and *x*
(ii) *q* and *v*
4.50 (a) +6.0 V, +4.0 V, 0.0 V
(b) 0.0 V, −2.0 V, −6.0 V
4.51 (a) 4×10^9 J
(b) about 11 000
4.53 0.12 kW
4.54 4
4.55 0.14 MJ
4.56 (a) (ii) 17.3p
(b) (i) 3.24×10^6 J
(ii) 5.3p
4.57 (a) 11 W: 48 mA, 60 W: 260 mA
(b) 11 W: 160 MJ, 60 W: 860 MJ
4.58 (a) E: £15.95, F: £62.50
4.60 54 kJ
4.61 (a) 2.5 mC
(b) 0.25 kA
(c) 60 kW
4.62 (d) 24 W
(e) 10 A
4.63 over a gigawatt!
4.64 (a) 20 kA
(b) 4.3×10^{13} J
(c) 2.6×10^{13} J
4.65 18 m^2

5 Electrical resistance

5.1 25 Ω
5.2 8.6×10^{10} Ω
5.3 50 V
5.4 (a) 0.20 Ω
(b) 0.12 V
(c) 0.14 W
5.5 (a) 91 kΩ
(b) 100 kΩ
(c) 85 kΩ
(d) metal
5.6 (a) (i) 100, 100
(ii) 25, 100
(iii) 50, 50
(iv) 75, 25, 100
(v) 100, 0, 100
(b) $R_{AC} + R_{BC} = R_{AB}$
5.7 (a) 0.71 kW
(b) 74 Ω
5.8 (a) 0.12 A
(b) 40 Ω, 0.15 A
(c) 30 Ω, 0.20A; 20 Ω, 0.30 A; 10 Ω, 0.60 A
5.10 (a) second, 2.6 W
(b) first, 11 W
5.11 (a) 2.4 Ω
(b) 15 W
(c) resistance decreases, power increases
5.12 0.51 m
5.13 (b) (i) 12 V
(ii) 0.10 A
(iii) 0.10 A
5.14 (a) 25 Ω
(b) 6.0 Ω
5.15 (i) 2.50 Ω (ii) 5.00 Ω
(iii) 8.33 Ω (iv) 6.67 Ω
5.16 (a) (i) 6.0 V
(ii) 3.0 V
(iii) 2.4 V
(iv) 3.0 V
(b) (ii) zero
(iii) 3.6 V
5.17 0.042 Ω
5.18 (a) 50 Ω
(b) 150 Ω
(c) 50 Ω
(d) 25 Ω
(e) 125 Ω
(f) 25 Ω
5.19 (a) 132 mA
5.20 (a) <1.00 Ω
(b) <1.00 Ω
5.21 from the top, 0.2 A, 0.3 A, 0.6 A
5.22 (a) 6.3 V
(b) 9.5 V
5.25 0.14 kW
5.26 (a) 2.3 V

(b) 7.4 V
(c) 23 V
5.27 **(b)** 68 Ω
5.28 **(a) (i)** infinite
(ii) 29 Ω
(iii) 4 Ω
(b) (i) infinite
(ii) almost zero
5.30 0.16 kW at 40 kV but 2.5 kW at 10 kV
5.31 $kg\,m\,s^{-3}\,A^{-2}$
5.32 Ω m
5.33 0.11 Ω
5.34 **(b)** second wire, twice
5.35 **(a)** 240 mm
(b) 16
5.36 **(a)** $13\,mm^2$
(b) 2.5 Ω
(c) 0.42 Ω
5.37 **(a)** all numerically about 0.425
5.38 2.5 m
5.39 **(a)** 8.0 MΩ
(b) the same
5.40 **(a) (i)** 4.27 Ω
(ii) 21.3 Ω
(b) 95 cm
5.41 **(a)** 0.32 kΩ
(b) 12
5.42 **(a)** 0.10 A
(b) 1.4 V
(c) 1.4 V
5.43 **(a)** 3.00 V, 1.00 Ω
(b) 1.50 V, 0.25 Ω
5.44 **(a)** 0.27 kJ
(b) 0.25 kJ
5.45 **(b)** 1.5 Ω
(c) (i) 0.30 W
(ii) 0.24 W
(iii) 0.060 W
5.46 **(a)** 0.52 Ω
(b) 1.50 V
5.47 **(a)** 8.3 Ω
(c) 1.7 Ω
5.48 **(b)** $60\mathcal{E}$
(c) 15%
5.49 **(a)** 12.0 V
(b) 11.9 V
(c) 8.0 V
5.52 **(b)** $47\,J\,s^{-1}$
5.53 **(a)** 40 μA
(b) 120 μA
5.56 5.2 V
5.57 **(b)** both 1.50 V
(c) 0.58 A, 0.64 Ω
(d) 1.10 V, 0.75 Ω

6 Electrical circuits

6.1 **(a)** 2.7 V **(b)** 230 Ω
6.2 **(a)** 1.25 V **(b)** 4.7 V
6.3 **(a)** A 4.8 V, B 1.2 V, C 1.2 V
(b) A 120 mA, B 80 mA, C 40 mA
6.4 **(a)** 6.0 Ω
(b) (i) 2.5 V **(ii)** 3.2 V
(c) (i) 0.53 A
(ii) 0.32 A
(iii) 0.21 A
6.5 **(a)** is 4 times greater than **(b)**
6.6 **(b) (i)** 920 Ω, 460 Ω and 230 Ω
(ii) 0.25 A, 0.50 A and 1.0 A
(iii) 58 W, 115 W and 230 W
6.7 **(a) (i)** 1.2 V **(ii)** 1.7 V
(b) 55 mA
(c) 17/12
(d) 0.19 A
6.9 **(b)** 0.12 kΩ
6.10 **(a) (i)** 2.4 V
(ii) 20 mA
(iii) 220 mA
6.11 [R_e = 4.8 kΩ, R_h = 2.4 Ω]
6.12 **(a)** 12.5 Ω **(b)** 48 mA
6.15 **(a)** 3.0 V
(b) (i) 6.0 V
(ii) 5.0 V
(iii) 5.7 V
6.16 **(a) (i)** 6.0 V **(ii)** 0 V
6.17 **(b) (i)** 11 Ω **(ii)** 0.58 V
6.18 0.23 MΩ and 1.9 kΩ
6.19 0.27 Ω
6.20 **(c)** 2.1 V **(d)** 0.13 A
(e) (i) 46 mA **(ii)** 86 mA
6.21 **(a)** V_c/V = 1.5, 2.4, 3.5, 4.1
6.22 **(a) (i)** 0.13 V **(ii)** no
(b) 86 kΩ
6.23 **(a)** V/V = 0.83, 1.2, 1.7, 2.1
6.24 **(a) (i)** both 3.0 V
(ii) both 3.0 V
(iii) both 4.0 V
(b) $P/R = Q/S$
(d) 46 Ω
6.25 **(c)** from X to Y
6.26 **(a)** 0.20 A, 4.0 V
(b) 0.20 A, 6.0 V
6.27 **(b)** 28 mV **(c)** 80 mA
6.28 **(a)** 40 Hz
(b) 33 Hz
(c) 29 Hz
6.29 **(a)** 1.3 V
(b) 125 Hz
(c) 2.8 V
(d) 41 mA, 15 mA
6.30 **(b)** 6.8 mV **(c)** $1\,mV\,div^{-1}$
(d) 0.25 V, $100\,mV\,div^{-1}$
(e) 63 mV **(f)** 6.3 mA

7 Physics of materials

7.1 **(a)** $2.7 \times 10^3\,kg\,m^{-3}$
(b) $3.5 \times 10^2\,kg$
(c) $7.8 \times 10^{-3}\,m^3$
7.2 $5.5 \times 10^3\,kg\,m^{-3}$
7.3 17 kg
7.4 **(a)** $3.9 \times 10^5\,N\,m^{-2}$
(b) $0.28\,m^2$
(c) $1.3\,m^2$
(d) $40\,N\,m^{-2}$
(e) $4.0 \times 10^2\,N$
7.5 1.1 kPa, 1.6 kPa, 3.3 kPa
7.6 **(a)** 27 000 kPa
(b) 530 kPa
(c) 16 kPa
7.9 **(b)** 101 kPa
7.10 **(a)** 16 kPa, 11 kPa
(b) 100 mmHg
(c) from top, 56, 100, 144, 188 mmHg
(d) 208 mmHg
7.12 **(a)** 1.75 MN
(b) (i) 0.119 MN
(ii) 0.238 MN
(c) (i) 1.63 MN
(ii) 1.51 MN
7.15 **(a)** $W - U = F\,(6\pi r\eta v)$
(b) $39\,mm\,s^{-1}$
7.16 **(b)** $6.7 \times 10^{-6}\,m$
(c) $9.9 \times 10^{-12}\,kg$
7.19 0.25 MN
7.20 **(b)** 5000 N

7.21 **(a)** 59 N **(b)** 59 N
7.22 **(a)** 59 N **(b)** 69 N **(c)** 64 N
7.23 **(a)** 100 N m^{-1}, 40 N m^{-1}, 10 N m^{-1}
(b) **(i)** 12 N
(ii) 4.8 N
(iii) 1.2 N
(c) **(i)** 0.16 m **(ii)** 0.40 m
7.24 **(a)** 6.0 N **(b)** 12 N
7.25 **(a)** 10 N m^{-1}
(b) 60 N m^{-1}
(c) 60 N m^{-1}
7.26 **(a)** $2k$ **(b)** $\frac{1}{2}k$
7.27 **(a)** **(i)** 6–7 cm
(ii) about 8 N
(iii) about 10
7.28 **(a)** 5.0×10^{-5} **(b)** 2.0
7.29 **(a)** 2.0×10^{-3}
(b) 2.5
(c) 30 mm
(d) 18 mm
(e) 3.6 m
7.30 **(a)** about 8 kN m^{-1}
(b) about 20%
7.31 no
7.32 **(a)** 1.0×10^{-7} m^2
(b) 3.4×10^7 N m^{-2}
7.33 **(a)** 2.4×10^7 N m^{-2}
(b) 1.6×10^8 N m^{-2}
(c) 1.1×10^7 N m^{-2}
7.34 **(a)** 1.01×10^5 N m^{-2}
(b) 2.7×10^5 N m^{-2}
(c) 3.5×10^7 N m^{-2}
(d) 2.8×10^9 N m^{-2}
(e) 2.35×10^{11} N m^{-2}
7.35 **(a)** 60 MPa
(b) 6.0 GPa
(c) 2.0 kN
(d) 5.0 N
(e) 0.50 mm^2
7.37 **(a)** 83 GPa
(b) 2.0 GPa
(c) 12 MPa
(d) 1.0×10^{-3}
(e) 1.0×10^{-2}
7.38 **(a)** **(i)** no **(ii)** yes
(iii) yes **(iv)** yes
(b) stress and strain
7.39 **(a)** 100 MPa
(b) 7.7×10^{-4}
(c) 0.92 mm
7.40 **(a)** nickel
(b) Ni 200 GPa, Cu 130 GPa, Al 70 GPa
7.41 **(a)** **(i)** 8 **(ii)** 8
(iii) 4 **(iv)** 2
7.43 **(a)** 37 kN
(b) 0.12 GPa
(c) 5.9×10^{-4}
7.44 **(a)** tungsten **(b)** copper
7.46 **(b)** 55 MPa
(c) 0.41 MN
7.48 **(a)** 18 GPa
(b) 0.15 GPa
(c) 8.5×10^{-3}
7.50 **(a)** 0.15 J
(b) 0.60 J
(c) 1.4 J
7.51 **(b)** **(i)** 2.2 J **(ii)** 3.8 J
(d) 1.7 J
7.52 **(a)** 1.7 mm **(b)** 39 mJ
7.53 **(a)** about 3 J
(b) 2 J
7.54 **(a)** just more than 20 mJ
(b) 180 mJ
(d) same shape
(e) 16 MPa, 0.005
7.55 **(a)** high tensile steel
(b) 2.5 J, 8.1 J
(c) mild steel
7.58 **(a)** **(i)** ≈ 5 MPa
(ii) > 3 MPa
(b) **(i)** 25×10^4 J m^{-3}
(ii) 2.0×10^6 J m^{-3}
7.59 **(b)** 7 GPa
7.60 **(a)** 5.6×10^9 Pa
(b) **(i)** yes **(ii)** 2.7 mN

8 Waves

8.1 **(b)** 47 ms, 53 ms
8.2 1.5×10^8 m
8.3 **(a)** 20 m
(b) 67 h
8.5 15 mm to 6 mm
8.6 **(b)** **(i)** P & T, Q & U
(ii) P & R, Q & S, R & T, S & U
8.7 **(a)** 0.13 m
(b) 2.7 kHz
8.9 **(a)** **(i)** 0.20 m **(ii)** 2.0 m
(b) **(i)** 20 m s^{-1}
(ii) 10 Hz
(iii) 0.10 s
8.11 **(b)** 23 m
8.12 **(b)** 9.8 N
8.21 **(a)** 0.11 mm to 0.023 mm
(b) 0.17 mm
8.22 **(a)** 4.5 μs
(b) 7.2 mm
8.24 **(a)** 5.3 Hz
8.27 For $\theta_a = 20°$: $\theta_a = 13.0°$, $\varphi = 7.0°$
8.28 **(a)** **(i)** 41.5° **(ii)** 39.6°
8.29 4.3°
8.30 **(a)** 1.1 m
8.31 **(a)** **(i)** 49.8° **(ii)** 40.4°
(b) 2.42
8.32 **(a)** 0.49°
(b) red light
8.33 **(a)** 4935 ns, 4945 ns
(b) 10 ns
8.34 **(b)** 9.4°
8.35 **(a)** 81.28°
(b) 14.73 μs
(c) 14.91 μs
8.36 **(b)** 44 kHz
8.46 **(a)** about 520 nm
8.47 **(a)** green 0.59 mm, orange 0.71 mm
8.48 **(a)** 350 mm and 316 mm
(b) 34 mm
8.49 **(a)** **(i)** 800.0049 mm, 800.0064 mm
(ii) 1.5 μm
(b) dark
8.50 **(a)** $1\frac{1}{2}\lambda$
8.51 **(a)** 400 nm
(b) 100 nm
8.53 **(b)** 44 m s^{-1}
8.54 **(b)** **(i)** 330 Hz
(ii) 220 Hz
(iii) 170 Hz
8.56 **(b)** 0.40 m s^{-1}
8.57 **(b)** 0.43 kHz
8.59 **(a)** 104 m
8.60 **(a)** 32 mm
8.61 **(b)** 28 mm
8.66 **(a)** **(i)** 1.57 m
(ii) 636 mm^{-1}
(b) 646 nm
8.67 **(a)** 10.5°

9 Photons and electrons

9.1 **(a)** 3.0×10^{-19} J
(b) 4.6×10^{-19} J
9.2 **(a)** f/Hz $= 3 \times 10^{20}$, 3×10^{17}, etc.
(b) E/J $= 2 \times 10^{-13}$, 2×10^{-10}, etc.
9.3 **(a)** 1.33×10^{-19} J
(b) 3.64×10^{-19} J
(c) 5.45×10^{-19} J
(d) 1.29×10^{-15} J
(e) 8.6×10^{-14} J
9.4 **(a)** **(i)** 2.41×10^{14} Hz
(ii) 1.24×10^{-6} m
(b) infrared
(c) **(i)** 4.83×10^{14} Hz
(ii) 6.22×10^{-7} m
(d) red
9.6 **(a)** **(i)** lights
(ii) radiators
(iii) radios
(b) **(i)** 2.5 eV
(ii) 0.06 eV
(iii) 4×10^{-6} eV
9.7 **(a)** 5.18 **(b)** 3.44 **(c)** 2.15
9.8 **(a)** 3.14×10^{-19} J
(b) $7.1 \times 10^{14}\,s^{-1}$
9.9 **(a)** 4.6×10^{19} Hz
(b) 0.19 MeV
9.10 **(a)** 6.0 J
(b) 1.6×10^{-19} J
9.11 **(a)** 2.0 eV **(b)** 20 eV
(c) 1.0 MeV **(d)** 5.0 MeV
9.12 **(a)** 1.3 kV
9.14 **(a)** 3.2×10^{-19} J
(b) 3.2×10^{-18} J
(c) 1.6×10^{-13} J
(d) 8.0×10^{-13} J
9.15 **(a)** 1.0 eV
(b) 2.0 eV
(c) 2.0 MeV
9.16 **(a)** **(i)** $3.6 \times 10^{6}\,m\,s^{-1}$
(ii) $4.2 \times 10^{6}\,m\,s^{-1}$
(b) **(i)** $8.5 \times 10^{4}\,m\,s^{-1}$
(ii) $9.8 \times 10^{4}\,m\,s^{-1}$
9.17 **(a)** 200 keV
(b) 3.2×10^{-14} J
9.18 **(a)** $\frac{1}{2}$
(b) $\sqrt{2}$
9.19 **(a)** **(i)** $0.14\,mW\,m^{-2}$
(ii) $0.035\,mW\,m^{-2}$
(iii) $0.016\,mW\,m^{-2}$
(b) $1 : \frac{1}{4} : \frac{1}{9}$
9.20 32 m
9.21 **(a)** about $40\,mW\,m^{-2}$
(b) 3.6×10^{-19} J
(c) about $1.8 \times 10^{15}\,s^{-1}$
9.22 **(a)** **(i)** 100 keV
(ii) 1.6×10^{-14} J
(b) **(i)** 2.4×10^{19} Hz
(ii) 1.2×10^{-11} m
9.23 **(a)** 5.89×10^{-7} m
(b) 3.88×10^{-7} m, 1.14×10^{-6} m
(c) 388 nm is just ultraviolet, 1.14 μm is in the infrared
9.24 **(a)** **(i)** 10.19 eV
(ii) 1.63×10^{-18} J
(iii) 2.46×10^{15} Hz
(iv) 1.22×10^{-7} m
(v) ultraviolet
(b) **(i)** 1.90 eV
(ii) 3.04×10^{-19} J
(iii) 4.59×10^{14} Hz
(iv) 6.54×10^{-7} m
(v) red
(c) **(i)** 0.66 eV
(ii) 1.1×10^{-19} J
(iii) 1.6×10^{14} Hz
(iv) 1.9×10^{-6} m
(v) infrared
9.25 six
9.26 **(a)** 13.6 eV
(b) 4.09×10^{-6} m
(c) 9.75×10^{-8} m
9.27 **(b)** $f_2 - f_1$ or $f_1 - f_2$
9.28 **(a)** $f_1 = 2.63 \times 10^{14}$ Hz, $f_2 = 5.09 \times 10^{14}$ Hz, $f_3 = 7.72 \times 10^{14}$ Hz
(b) yes
9.30 **(a)** $2.19 \times 10^{6}\,m\,s^{-1}$
(b) 91.4 nm
9.31 **(b)** Paschen
(c) Lyman 122 nm, Balmer 657 nm, Paschen 1877 nm
9.37 **(a)** **(i)** 8.3×10^{-10} eV
(ii) 0.025 eV
(ii) 6.2×10^{5} eV
9.38 **(a)** **(i)** 4×10^{-5} m
(ii) 1×10^{-13} m
(ii) 6×10^{-22} m
9.39 **(a)** 1.2×10^{-11} m
(b) 2.9×10^{-13} m
(c) 1.4×10^{-13} m
9.43 **(a)** 2×10^{-11} m
(b) about 4 kV
9.45 **(a)** 2.3 eV and 4.3 eV
(b) yes, sodium
9.46 **(a)** 2.9 eV
(b) 6.6×10^{-20} J
(c) $3.8 \times 10^{5}\,m\,s^{-1}$
9.47 **(a)** 1.00×10^{15} Hz, not visible
(b) 0.813 eV
9.49 **(a)** about 4.2×10^{-15} V s
(b) 6.7×10^{-34} J s
(c) 2.3 eV
9.50 **(b)** 11 kV

10 Practising calculations

10.1 **(a)** 3.4×10^{-2}
(b) 5.3×10^{-3}
(c) 1.5×10^{-1}
(d) 1.5×10^{-2}
(e) 6.7×10^{2}
10.2 **(a)** 2 **(b)** 3 **(c)** 2
10.3 **(a)** 10^{9} **(b)** 10^{6} **(c)** 10^{9}
(d) 10^{-9} **(e)** 10^{3} **(f)** 10^{6}
(g) 10^{7} **(h)** 10^{-3} **(i)** 10^{6}
10.4 **(a)** 10^{-2} **(b)** 10^{-4} **(c)** 10^{2}
(d) 10^{3} **(e)** 10^{-8}
10.5 **(a)** 1.4×10^{6}
(b) 6.4×10^{3}
(c) 6.8×10^{-42}
10.6 **(a)** 1.3×10^{2}
(b) 3.6×10^{7}
(c) 1.2×10^{-19}
10.7 **(a)** 1.73×10^{1}
(b) 1.05×10^{-4}
(c) 2.8
10.8 **(a)** 0.20 **(b)** 0.020
(c) 0.000 40 **(d)** 3.0
10.9 **(a)** 100% **(b)** 200%
10.10 **(a)** 90 **(b)** 120
(c) 132 **(d)** 360
10.11 **(a)** 10% **(b)** 87%
(c) 230% **(d)** 0.20%
10.12 **(a)** 124 **(b)** 244 **(c)** 1.30
10.13 $v/m\,s^{-1}$ = 30.1, 32.7, 29.9

10.14 **(a)** 0.49 **(b)** 0.069 **(c)** 0.023 **(d)** 12 **(e)** 1.3 **(f)** 9.2×10^2 **(g)** 0.66 **(h)** 0.063 **(i)** 11 **(j)** 0.99

10.15 **(a)** 6.3×10^{-2} m
(b) 1.2×10^{-2} m
(c) 8.3×10^5 m
(d) 5.5×10^{-7} m
(e) 5.3×10^{-2} kg
(f) 5.0×10^5 kg
(g) 1.2×10^{-4} kg
(h) 2.3×10^{-6} kg
(i) 1.8×10^3 s
(j) 2.3×10^{-2} s
(k) $8.6 \times 10^4\,\Omega$
(l) $4.5 \times 10^{-2}\,\text{m}^3$

10.16 **(a)** $1.6 \times 10^{-4}\,\text{m}^2$
(b) $5.3 \times 10^{-6}\,\text{m}^2$
(c) $1.7 \times 10^{-6}\,\text{m}^2$
(d) $7.8 \times 10^{-6}\,\text{m}^3$
(e) $3.4 \times 10^{-8}\,\text{m}^3$

10.17 **(a)** $4.9\,\text{m}^2$
(b) $0.23\,\text{m}^2$
(c) $5.3 \times 10^{-6}\,\text{m}^2$

10.18 **(a)** 4.7×10^{-10} F
(b) 1.5×10^3 V
(c) 5.0×10^7 W
(d) 4.0×10^{-8} s

10.19 1.26×10^5 N

10.20 $9.8 \times 10^{-4}\,\text{m}^3$

10.21 **(a)** $v = x/t$
(b) $h = V/bd$
(c) $r = \sqrt{(A/\pi)}$
(d) $h = V/\pi r^2$
(e) $V = m/\rho$
(f) $a = (v - u)/t$
(g) $a = (v^2 - u^2)/2x$
(h) $T_1 = T_2/(1 - \eta)$
(i) $I = \sqrt{(P/R)}$
(j) $R = V^2/P$
(k) $f = 1/T$
(l) $l = gT^2/4\pi^2$
(m) $r = (E - V)/I$
(n) $r = (3V/4\pi)^{1/3}$

10.22 5.49×10^{14} Hz

10.23 2.0×10^{-2} m

10.24 9.1×10^{-2} m

10.25 9.0×10^{-3} m

10.26 5.06×10^3 s

10.27 **(a)** m **(b)** kg **(c)** s **(d)** A **(e)** K

10.28 $1\,\text{N} = 1\,\text{kg}\,\text{m}\,\text{s}^{-2}$

10.29 $1\,\text{Pa} = 1\,\text{kg}\,\text{m}^{-1}\,\text{s}^{-2}$

10.30 $1\,\text{J} = 1\,\text{kg}\,\text{m}^2\,\text{s}^{-2}$

10.31 $1\,\text{W} = 1\,\text{kg}\,\text{m}^2\,\text{s}^{-3}$

10.32 $1\,\text{C} = 1\,\text{A}\,\text{s}$

10.33 $1\,\text{V} = 1\,\text{kg}\,\text{m}^2\,\text{s}^{-3}\,\text{A}^{-1}$

10.34 $1\,\text{Hz} = 1\,\text{s}^{-1}$

10.35 $1\,\text{F} = 1\,\text{kg}^{-1}\,\text{m}^{-2}\,\text{s}^4\,\text{A}^2$

10.36 **(a)** $30\,\text{m}\,\text{s}^{-1}$, speed
(b) 1.0×10^{-3} J, work or energy
(c) 29 Pa, pressure
(d) $76.2\,\text{m}^3$, volume

10.37 **(a)** 7.1×10^{-7} C
(b) 2.0 J
(c) 2.8×10^6 V

10.39 $1.4 \times 10^3\,\text{W}\,\text{m}^{-2}$

10.40 $0.21\,\text{MN}\,\text{m}^{-2}$

10.41 2.47 acres

10.42 £1.37

10.43 **(a)** t is greater than 10 s
(b) t is greater than 20 s and less than, or equal to, 40 s
(c) A is proportional to the square of r
(d) the change in x is 0.35 m

10.44 **(a)** $R \propto \ell/A$ **(b)** $F \propto v^2/r$

10.45 **(a)** $\Delta v = +1.4\,\text{m}\,\text{s}^{-1}$
(b) $\Delta t = +1.64$ s
(c) $\Delta V = -47\,\text{cm}^3$

10.46 If the current remains constant, and the temperature rises by 10 K, the potential difference falls by 2.0 V.

10.47 **(a)** yes **(b)** yes **(c)** yes **(d)** yes **(e)** yes **(f)** no **(g)** yes **(h)** yes **(i)** no **(j)** yes

10.48 **(a)** all **(b)** B

10.49 **(a)** all **(b)** B, C

10.50 **(a)** $2.0\,\text{m}\,\text{s}^{-1}$
(b) $70\,\text{m}\,\text{s}^{-2}$
(c) $0.40\,\Omega$
(d) $5.0 \times 10^{-4}\,\text{C}\,\text{V}^{-1}$ or 5.0×10^{-4} F
(e) $4.0 \times 10^{14}\,\text{m}^3\,\text{s}^{-2}$
(f) 3.8×10^{15} V s

10.51 **(a)** 200 m **(b)** 100 m **(c)** 0.15 J

10.52 **(a)** 62 m **(b)** 5.8×10^{-3} C

10.53 **(a)** W, x^2, $\frac{1}{2}k$
(b) E, $1/d$, V
(c) F, $1/r^2$, Gm_1m_2
(d) f, $1/\lambda$, c or λ, $1/f$, c
(e) C, A, $\varepsilon_r\varepsilon_0/d$
(f) C, $1/d$, $\varepsilon_r\varepsilon_0A$
(g) T, $\sqrt{\ell}$, $2\pi/\sqrt{g}$
(h) T, $1/\sqrt{k}$, $2\pi\sqrt{m}$
(i) V_s, f, h/e

10.54 **(b)** yes **(c)** yes
(d) $6.5 \times 10^3\,\text{N}\,\text{m}^{-1}$
(e) $1.3 \times 10^{11}\,\text{N}\,\text{m}^{-2}$

10.55 **(b)** yes
(c) no
(d) $-0.60\,\Omega$
(e) $m = -0.60\,\Omega$, $c = 1.52$ V

10.56 **(c)** $0.34\,\text{km}\,\text{s}^{-1}$

10.57 **(a)** 0.427
(b) 0.904
(c) 0.473

10.58 **(a)** 0 **(b)** 1 **(c)** 0

10.59 **(a)** 1 **(b)** 0

10.60 because tan 90° is infinite.

10.61 **(a)** 0.5938 **(b)** 36.43°

10.62 **(a)** 0.8046, 36.43°
(b) 0.7380, 36.43°

10.63 **(a)** 11.2 m **(b)** 5.00 m

10.64 0.75 m

10.65 **(a)** **(i)** 60° **(ii)** 50° **(iii)** 65° **(iv)** 40°
(b) **(i)** x: $(2.6\,\text{m}\,\text{s}^{-1})\cos 30°$; y: $(2.6\,\text{m}\,\text{s}^{-1})\cos 60°$
(ii) x: $(5.4\,\text{N})\cos 50°$; y: $(5.4\,\text{N})\cos 40°$
(iii) x: $(2.6\,\text{N}\,\text{s})\cos 25°$; y: $(2.6\,\text{N}\,\text{s})\cos 65°$
(iv) x: $(0.35\,\text{T})\cos 40°$; y: $(0.35\,\text{T})\cos 50°$

10.66 **(a)** yes **(b)** yes **(c)** no **(d)** yes **(e)** yes

10.67 **(a)** 7.57 m **(b)** 5.54 m

10.68 **(a)** 6.33 cm **(b)** 0.622 cm

10.69 **(a)** 25 N **(b)** 16°

10.70 $354\,\text{m}\,\text{s}^{-1}$ in a direction N 8.13° E

10.71 **(a)** 5.73° **(b)** 57.3° **(c)** 115° **(d)** 360°

10.72 **(a)** $\frac{1}{2}\pi$ rad
(b) $\frac{1}{3}\pi$ rad

(c) $\frac{2}{3}\pi$ rad
10.73 (a) 1.50 m
(b) 3.00 m
(c) $s = r\theta$
10.74 (a) 1.05 rad (b) 2.62 m
10.75 0.660 m
10.76 (a) (i) 3.636×10^{-3}, 0.2083°, 3.636×10^{-3} rad
(ii) 3.636×10^{-3} rad
10.77 19 m
10.78 (a) 2.98×10^{-5} rad
(b) 0.102 minutes
10.79 (a) 3.09×10^{16} m
(b) 3.67×10^{-6} rad = 0.758 seconds
10.80 (a) (i) 0.767 rad = 43.9°
(ii) 41.9°
(b) +4.7%
10.81 (c) +1, −1
(d) 0°, 180°, 360°
10.82 (c) +1, −1 (d) 90°, 270°
10.83 −1.2 m
10.84 −1.5 m
10.85 (b) 4.0 s (c) velocity
(d) t = 1.00 s (e) 1.9 m s^{-1}
(f) (i) 3.8 m s^{-1}
(ii) 0.94 m s^{-1}
10.86 (a) 1.4 m s^{-1} (b) 3.8 km s^{-2}
10.87 (a) 3.1×10^{6} V s^{-1}
(b) 5.0 μs
(c) ±0.62 V
10.88 (a) 2×10^{-5} Hz
(b) 2×10^{-3} m s^{-1}
10.89 (a) 1.40 (b) 2.40 (c) 3.40
10.90 (a) 3.22 (c) 5.52 (d) 7.82
10.92 (a) 1.00 (b) 2.00 (c) 3.00
10.93 (a) 1.00 (b) 2.00 (c) 3.00
10.94 (a) 2.00 (b) 3.00 (c) 2.50
10.95 (a) 2.00 (b) 3.00 (c) 2.50
10.96 (a) 6.7×10^{-2}
(b) 0.13
(c) 0.62
10.97 (b) 1.5
10.98 (a) 40 y^{-1} (b) 100 y^{-1}
10.100 (a) (i) 640 (ii) 512
(b) 1250
10.101 (a) (i) 147 (ii) 89 (iii) 54
(b) 1.39 s (c) 9.21 s
10.102 (b) 31 s; the half-life
(c) 62 s
(d) 45 s
10.103 (a) 10.0, 8.19, 6.70, 5.49, 4.49, 3.68
10.104 (a) s
(b) 0.33 s
(c) (i) 0.33 s
(ii) 0.66 s
(iii) 0.99 s
(d) 0.23 s
(e) (i) 0.46 s (ii) 0.69 s
10.105 (d) 0.43 h^{-1} (e) 1.6 h
10.107 (a) 1050 s^{-1} (b) 7 (c) yes
(d) number too small for laws to hold
10.108 (a) 0.14 h^{-1} (b) 1.3×10^{10}
10.109 (a) 1.25×10^{-4} m^{-1}
(c) (i) 29 kPa (ii) 8.2 kPa
10.111 (c) 0.44 d^{-1} (d) 1.6 d
10.112 (a) second
(b) (i) 1.97 V (ii) 3.16 V
(c) 2.41 s
10.114 (ii) and (iii) are correct
10.115 (a) 3.16×10^{7}
(b) 1.26×10^{7}

11 Thermal physics

11.1 (a) (i) 310 K, 77 K
(ii) −269 °C, −39 °C
11.8 (a) +0.20 MJ
(b) −0.08 MJ
(c) +0.12 MJ
11.9 (a) +, −, +
(b) 0, −, +
(c) ΔW is electrical work
11.11 (a) 0.45 (b) 0.54 (c) 0.60
11.12 (b) 0.38
11.13 (b) 5.0 MJ (c) 1.5
11.15 (b) £576
(c) 40 kW
(e) £384
11.17 3.2 GJ (3.2×10^{9} J)
11.18 £2
11.19 10 s
11.20 8.1 minutes
11.21 (a) 3.2 litres min^{-1}
(b) 3.7 litres min^{-1}
11.22 about 6 kJ must be removed
11.23 (b) 4.8 K min^{-1}, 22 W or 22 J s^{-1}
(c) 1.1 J s^{-1}
11.24 (a) 35 J
(b) about 2.2 K
11.25 about 4 minutes
11.26 (a) 108 000 kg
(b) 2.0 K
11.27 0.30 K
11.29 about 130 J kg^{-1} K^{-1}
11.32 (a) 4.5 MJ (b) 0.17 MJ
11.33 (a) 38 min (b) 30 min
11.35 15.4 W
11.36 24 g
11.37 (b) 840 J kg^{-1} K^{-1}
(d) about 60 kJ kg^{-1}
11.40 (a) 0.12 K min^{-1}
(b) 13 g min^{-1}
11.41 (a) 3.8
11.43 (a) 1.7×10^{-3} m^{3}
(b) 6.5×10^{-6} m^{3}
(c) 3.3×10^{-9} m^{3}
11.44 (a) 85 kPa (b) 0.16 MPa
(c) 0.16 MPa (d) 0.19 MPa
(e) 45 kPa
11.47 (a) 1.0×10^{25}
(b) 17 mol
11.49 39.9 °C
11.50 0.364 m^{3}
11.51 (a) 8.0 J mol^{-1} K^{-1}
11.52 (a) 5.7 MPa
11.53 (a) 300 N m
(b) 0.12 mol
(c) 773 K
11.54 (a) 2.4 mol
(b) 200 kPa
(c) 12×10^{-2} m^{3}
11.56 (a) 2.5×10^{22}
(b) 4.0×10^{-26} m^{3}
(d) about 3000
11.57 (a) 432 m s^{-1}
(b) 432 m s^{-1}
11.58 (a) 33.3 m s^{-1}
(b) 33.7 m s^{-1}
11.59 (a) 300 K
(b) 600 K
11.60 (b) (i) 400 (ii) 1250

12 Linear momentum

12.1 (a) (i) 150 kg m s^{-1} north
(ii) 150 kg m s^{-1} south
(b) 2.0×10^{4} kg m s^{-1} east
(c) 5.0×10^{9} kg m s^{-1} west
12.4 (a) 0.80 m s^{-1} to right

(b) 1.0 m s^{-1} to right

12.5 **(a)** 3.0 m s^{-1} away from bank
(b) no

12.6 **(a)** 0.28 kg

12.7 **(b)** 0.96 J

12.8 0.52 m s^{-1} north

12.9 9.5 kg

12.11 **(a)** 2.2 m s^{-1} north
(b) 0.15 MJ

12.12 **(a) (i)** -3.0×10^4 kg m s^{-1}
(ii) $+3.0 \times 10^4$ kg m s^{-1}
(b) (i) A: +3.0 m s^{-1}
B: +0.50 m s^{-1}
(ii) A and B: +1.5 m s^{-1}
(c) 75 000 kg m s^{-1} at all times

12.13 **(a)** −0.67 m s^{-1}

12.14 **(a)** 8.7 m s^{-1} along the line
(b) 370 J

12.15 **(b)** 9/25 or 0.36
(c) 9/25 or 0.36
(d) 1/4 or 0.25
(e) 56, iron

12.16 **(a)** 4.1×10^5 m s^{-1}
(b) 2.8%

12.19 **(b)** 0.99 m s^{-1}
(c) 250 m s^{-1}

12.20 $h/4$

12.24 **(a)** 3.0×10^3 N s east
(b) 100 N s down
(c) 0.48 N s down

12.25 **(a)** 3.2 N s
(b) 3.2 N s

12.26 **(a)** 40 N s
(b) 140 N s
(c) 75 N s

12.27 2.7 m s^{-1}

12.28 650 N

12.29 **(a)** 7.4 kN

12.30 **(b)** 90 N s; 1.1 m s^{-1}

12.31 **(a)** acceleration
(b) 83 kN
(c) 83 kN

12.33 **(a)** 30 N **(b)** 20 N

12.34 **(a)** 1.2 m s^{-1}
(c) 8.0 kN, 8.0 kN

12.35 50 ms

12.36 $\Delta v = \dfrac{h}{m\lambda}$

12.37 **(a)** 0.15 N m^{-2}
(b) 0.016 mm

12.38 1400 m s^{-1}; 1000 kg s^{-1}; 3.4×10^7 N

12.39 **(a)** 4.2×10^4 N s
(b) 4.2×10^4 N
(c) 1.7×10^5 N

12.40 **(a)** 5.3×10^{-23} N s
(b) 1.9×10^{21}
(c) 100 kPa

13 Circular motion and gravitation

13.1 **(a) (i)** 15° s^{-1}
(ii) 0.26 rad s^{-1}
(b) (i) 0.10° s^{-1}
(ii) 1.7×10^{-3} rad s^{-1}

13.2 **(b)** 463 m s^{-1}
(c) 288 m s^{-1}

13.3 12.5 m

13.4 **(a)** 21.7 rad s^{-1}, 207 rev min^{-1}
(b) 51.0 rad s^{-1}, 487 rev min^{-1}

13.5 **(a)** 10 m s^{-1}
(b) 10 m s^{-1}
(c) zero
(d) 6.4 m s^{-1} S
(e) 9.1 m s^{-1} SE
(f) 20 m s^{-1} W
(g) 14 m s^{-1} SW

13.6 **(a)** about 75 m
(b) less than 0.001 m s^{-2}

13.7 **(a)** 7.3×10^{-5} rad s^{-1}
(b) 3.1 km s^{-1}
(c) 0.22 m s^{-2}

13.8 29 m s^{-2} (almost 3g)

13.9 **(a)** 21 m s^{-1}
(b) about 1700 m s^{-2}

13.10 **(a)** 13 s

13.11 18 m s^{-1}

13.12 **(b) (i)** 11 m s^{-2} upwards
(ii) 0.63 kN

13.13 **(a)** 4.3×10^2 m s^{-2}
(b) 3.2 kN

13.14 **(b)** 0.62 kN
(c) (i) 0.62 kN
(ii) 1.1 kN

13.15 **(a)** 733.5 N
(c) 731.0 N
(d) 731.0 N

13.16 **(a)** 12 m s^{-1}
(b) (ii) each 16 kN

13.18 **(a) (i)** 0.91 m s^{-1}
(ii) 5.7 m s^{-2}

13.20 **(a)** 220 N
(b) 58°
(c) 0.96 rev s^{-1}
(d) 1.9 rev s^{-1}

13.23 0.12 kN

13.24 **(a)** 5.97×10^{24} kg
(b) 5.52×10^3 kg m^{-3}

13.26 **(a)** 74 N, 0.18 kN, 0.46 kN
(b) 0.12 kJ, 0.29 kJ, 0.74 kJ
(c) 5.9 J kg^{-1}, 14 J kg^{-1}, 37 J kg^{-1}

13.27 **(a) (i)** +120 kJ **(ii)** −82 kJ
(b) 29 J kg^{-1} etc.

13.28 **(a)** 10 m
(c) 5.9 kJ

13.31 **(a)** the two quantities are linearly related
(b) −2.0 to 2 sig.fig.
(c) $g \propto r^{-2}$

13.32 $m_E = 81 m_M$

13.33 5.9×10^{13} N kg^{-1}

13.34 **(a)** 9.81 N kg^{-1}, 9.72 N kg^{-1}
(b) ≈ 60 km

13.35 **(a)** 1.5×10^{10} J
(b) (i) 1.2×10^{10} J
(ii) 2.7×10^{10} J
(iii) 6.0 km s^{-1}

13.36 **(a)** $v_e = \sqrt{(2GM_E/r_E)}$
(b) 11 km s^{-1}

13.37 **(a)** 2.4 km s^{-1}

13.40 **(a) (i)** 8.8 N kg^{-1}
(ii) 8.8 m s^{-2}
(c) (i) 7.7 km s^{-1}
(ii) 91 minutes

13.41 **(a) (i)** 3.1 km s^{-1}
(ii) 0.22 m s^{-2}
(b) 0.22 N kg^{-1}

13.43 **(b)** 2.0×10^{30} kg

13.44 **(b)** 380×10^3 km
(c) 8.7×10^{25} kg

13.45 **(a)** 4.2×10^7 m, about 7
(c) about 40°

13.46 **(a)** 8.3 N kg^{-1}
(b) (i) 7.6 km s^{-1}
(ii) 15 times
(c) 2.0 m^2

14 Oscillations

14.4 **(a) (i)** 40 mm **(ii)** 0.36 Hz

(b) (ii) about 0.09 m s^{-1}

14.5 **(b)** possible answers are
(i) 3.8 ms
(ii) 0.95 s
(iii) 20 μHz

14.6 **(a)** 8.3 Hz
(b) 0.42 m s^{-1}

14.7 **(a)** s^{-2} **(b)** $f_A = 3f_B$

14.8 **(a)** 16.0 cm **(b)** −16.0 cm
(c) zero **(d)** 11.3 cm

14.9 **(a)** 4.0 cm, 2.8 Hz
(b) 0.70 m s^{-1}

14.10 **(a) (i)** 1.6 m **(ii)** 0.67 s
(b) (i) zero
(ii) 15 m s^{-1}
(iii) zero

14.11 **(a) (i)** $-kx$ to the left
(ii) kx/m
(b) $2\pi\sqrt{(m/k)}$

14.12 **(a) (i)** increased
(ii) decreased
(iii) decreased

14.14 **(a)** 0.13 m s^{-1} **(b)** 29 m s^{-2}

14.15 0.17 km s^{-1}

14.16 460 Hz

14.17 0.995 m

14.18 **(a) (i)** 98 mm **(ii)** 0.50 Hz
(b) 0.31 m s^{-1}

14.20 0.79 s

14.22 **(a)** 720 N m^{-1} **(b)** 3.1 Hz

14.23 72 kg

14.24 13 kN m^{-1}

14.25 **(a)** 2
(b) $4m$

14.26 **(b)** 3.5 kN

14.27 **(a)** 0.38 kN m^{-1}
(b) 0.94 s
(c) 0.53 m s^{-1}

14.31 **(b) (ii)** 2.0 m s^{-1}, 1.7 m s^{-1}, zero

14.32 **(a)** 0.13 kN m^{-1}
(b) about 5 s

14.35 **(a)** $\sqrt{2}$ **(b)** $1/\sqrt{2}$
(c) 2 **(d)** $\sqrt{2}$

15 Radioactivity

15.2 **(a)** about 2000
(b) about 300 000

15.3 **(a)** 12, 13, 14
(b) 2.32×10^{-26} kg

15.4 7.6×10^{22}

15.7 **(a)** 2.1×10^{-4}
(b) 2.4×10^{17} kg m^{-3}

15.9 **(a)** 20 mm
(b) about 300 m

15.10 **(a)** A
(b) Am
(c) $4\pi A r_0^3/3$
(d) $3m/4\pi r_0^3$
(f) 4×10^{17} kg m^{-3}

15.11 $\approx 6 \times 10^{17}$ kg m^{-3}

15.14 23.0 g

15.15 1.00, 1.14, 1.27, 1.37, 1.54

15.17 neodymium-143

15.18 **(b)** 27, 29, 27
(c) $^{56}_{26}$Fe

15.19 **(a)** $^{140}_{57}\text{La} \rightarrow {}^{136}_{55}\text{Cs} + {}^{4}_{2}\text{He}$

15.21 8 α-particles, 16 β$^-$-particles

15.22 **(a)** $^{26}_{13}\text{Al} + {}^{0}_{-1}\text{e} \rightarrow {}^{26}_{12}\text{Mg}$

15.25 **(b)** γ_1 2.65×10^{-15} m, γ_2 3.01×10^{-15} m

15.26 **(c)** 3.0×10^{-14} J

15.30 **(a)** 2.1×10^{5}

15.31 **(b)** it has no charge
(e) $^{35}_{16}\text{S} \rightarrow {}^{35}_{17}\text{Cl} + {}^{0}_{-1}\beta + \bar{\nu}_e$

15.32 **(b)** 0.011 cm^2 mg^{-1}

15.35 **(a)** 18
(b) 1

15.36 **(a)** about 4 mm
(b) 3.8 mm

15.37 **(b) (i)** 33 ns **(ii)** 0.21 ms

15.38 **(a)** 768 kBq
(b) 15.0 h
(c) 1.51×10^{10}
(d) 193 kBq

15.39 **(a)** 1.7×10^{-17} s^{-1}
(b) 3.6×10^{20}
(c) 6.1 kBq

15.40 **(b)** 3 alpha, 2 beta
(c) 54 s

15.41 **(a)** 0.898
(b) (i) 527 **(ii)** 473
(c) 729

15.42 2.7×10^{14}

15.43 **(a) (iii)** 56 s

15.45 **(a)** 3.14×10^{11}
(b) 0.72 kBq
(c) 2700 y

15.46 1.0×10^{4} y

15.47 **(a)** 200 litres

15.49 **(a)** $^{99}_{42}\text{Mo} \rightarrow {}^{99\text{m}}_{43}\text{Tc} + {}^{0}_{-1}\beta$
(b) 0.29

15.51 **(a)** 3.2×10^{-5} s^{-1}

15.53 **(a)** about 13%
(c) (i) 6% **(ii)** 37%

15.55 **(a) (i)** 0.087 d^{-1}
(ii) 0.046 d^{-1}
(b) 0.133 d^{-1}
(c) 5.2 days

15.56 **(a)** 1.5×10^{-5} s^{-1}
(b) 3.4×10^{10}
(c) 6.9×10^{-6} μg

15.57 **(a)** 1.9 litres

15.59 **(a)** 8.0×10^{10} Bq
(b) 3.2×10^{20}
(c) 0.13 g
(d) over 40 years

15.60 6.9×10^{9} Bq

15.61 **(b)** 2.0×10^{15}, 0.43 μg

16 Nuclear physics

16.1 **(a)** 1.008 67 u
(b) 0.000 548 u
(c) 12.000 0 u

16.2 **(a)** $1.672\,51 \times 10^{-27}$ kg
(b) 1.51×10^{-10} J
(c) 941 MeV

16.4 **(a)** 0.005 88 u
(b) 0.15%

16.5 **(a)** 0.005 88 u
(b) 11 times

16.6 **(a)** 1.94×10^{-8} kg
(b) 3.9×10^{-12}

16.7 **(a)** 15.125 u
(b) 0.114 40 u
(c) 107 MeV

16.8 **(b)** 112 MeV, 107 MeV
(c) nitrogen-15

16.9 **(b)** $^{1}_{1}\text{H} + {}^{7}_{3}\text{Li} \rightarrow {}^{4}_{2}\text{He} + {}^{4}_{2}\text{He}$

16.10 **(a)** about 60
(b) 530 MeV
(c) 9.4×10^{-28} kg

16.11 **(a)** 5.79 MeV
(b) the recoiling Rn nucleus

16.12 **(b)** 1.02 MeV or 1.64×10^{-13} J
(d) 2.43×10^{-12} m

16.13 **(a)** $^{15}_{8}\text{O} \rightarrow {}^{15}_{7}\text{N} + {}^{0}_{1}\beta + \nu_e$
(b) 90 mm

16.14 β$^-$

16.15 **(a)** **(i)** 1.64×10^{-13} J,
(ii) 1.21×10^{-12} m
(b) 4.24×10^{7} m s^{-1}
(c) upper: positron
lower: electron

16.16 **(a)** −0.005 27 u
(b) 4.91 MeV

16.17 **(b)** 1.63×10^{7} m s^{-1}; assume that no energy goes to the recoiling Ra nucleus

16.18 **(a)** 0.81 MeV
(b) the chlorine atom recoils

16.19 31.9739 u

16.20 **(a)** 1.7×10^{-27} kg
(c) 0.41

16.21 **(b)** 0.006 876 u
(e) 6.405 MeV
(f) 6.29 MeV

16.22 **(a)** 7.7 MeV, 8.7 MeV and 8.4 MeV
(b) 1810 MeV and (792 + 1184) MeV
(c) 166 MeV

16.24 **(a)** 0.1860 u
(b) 173 MeV
(c) 4.25 mol
(d) 2.56×10^{24}
(e) 4.44×10^{26} MeV
(f) **(i)** 7.10×10^{13} J
(ii) 1.97×10^{7} kW h
(g) 19.7 h

16.25 **(a)** Ba 1.52, Kr 1.56
(c) 3, 4

16.27 **(b)** **(ii)** ${}^{238}_{92}\mathrm{U} + {}^{1}_{0}\mathrm{n} \rightarrow$
${}^{239}_{93}\mathrm{Np} + {}^{0}_{-1}\mathrm{e} + \bar{\nu}_e$ and
${}^{239}_{93}\mathrm{Np} \rightarrow {}^{239}_{94}\mathrm{Np} + {}^{0}_{-1}\mathrm{e}$
$+ \bar{\nu}_e$

16.29 **(a)** 2500 MW
(b) 2.81×10^{23}
(c) 0.467 mol
(d) 0.11 kg

16.30 **(b)** 1.6
(c) 8%

16.31 **(b)** **(i)** 0.003 50 u
(ii) 3.26 MeV
(c) 3.83×10^{21} s^{-1}
(d) 2.2 kg

16.32 **(b)** **(i)** 0.026 51 u
(ii) 24.7 MeV

16.33 0.511 MeV

16.34 **(b)** 6.4×10^{9} K

16.35 about 10^{38} s^{-1}

16.36 **(a)** $+\frac{2}{3}e$, $-\frac{1}{3}e$ where $e = 1.6 \times 10^{-19}$ C
(b) **(i)** $+e$ **(ii)** 0
(c) $\bar{u}\bar{u}\bar{d}$

16.38 **(a)** 1.5031×10^{-10} J

16.39 **(b)** **(i)** +1 **(ii)** −1
(c) the strong (nuclear) force

16.40 yes to all

16.43 **(a)** uud, udd, $\bar{u}\bar{u}\bar{d}$, $\bar{u}\bar{u}\bar{d}$
(b) uuu and 15 other possibilities besides the four listed in **(a)**

16.44 **(a)** baryons
(b) mesons
(c) hadrons

16.45 $u\bar{d}$, $+e$, 0, π^+

16.46 **(b)** **(ii)** 0 **(iii)** 0

16.47 **(b)** 1.6×10^{-28} kg
(c) 170 GeV/c^2

16.48 **(a)** yes, for **(i)**, **(ii)** and **(iv)**

16.50 **(a)** left-hand side is $u\bar{s}$ (meson) + uud (baryon)

16.51 **(a)** e^- and $\bar{\nu}_e$
(b) udd changes to uud
(c) yes

16.52 **(b)** **(i)** $n \rightarrow p + e^- + \bar{\nu}_e$

16.54 e.g. strong nuclear: gluon with a range $\approx 10^{-18}$ m

16.55 **(a)** (e.g.) leptons, quarks

16.56 **(b)** X is uds (and is called a neutral lambda particle Λ^0)

16.57 **(c)** and **(d)** only

16.58 cdd, charge 0

16.59 at A: two + charged particles and at least one neutral particle emerge from the reaction.
at B: a neutral particle decays into a pair of + and − charged particles

16.60 **(a)** **(i)** negative
(ii) it carries no charge
(iii) it has *very* high momentum
(b) p, because the radius of curvature of its path is greater than that of q

16.61 **(b)** $K^- + p \rightarrow \Sigma^+ + \pi^-$
(c) $(+e, 1, -1)$
(d) **(i)** principle of conservation of linear momentum
(ii) it has no charge
(e) neutron

16.62 **(a)** $K^- + p \rightarrow \Sigma^- + \pi^+ + \pi^+ + \pi^-$
(b) uus, dds

16.63 **(a)** both possible
(b) **(i)** decreases by 1
(ii) increases by 1
(iii) increases by 1

17 Astrophysics

17.2 500 s or about 8 minutes

17.5 206 264.8

17.9 **(a)** 2.6×10^{5} m s^{-1}
(b) 2.6×10^{-10} m s^{-2}

17.12 **(a)** **(i)** 4.22 ly **(ii)** 8.60 ly
(b) **(i)** 0.008″

17.13 **(b)** **(i)** 9.7×10^{-9} radians
(ii) 1.5×10^{19} m, 0.50 kpc

17.14 **(b)** **(ii)** 11.4 ly

17.15 about 4×10^{27} W

17.16 1.1×10^{18} m

17.18 $M_{\text{Sirius}} = +1.5$,
$M_{\text{Procyon A}} = +2.7$;
Sirius is brighter

17.20 **(a)** −3.1
(b) 100 pc

17.21 3×10^{22} m

17.22 **(a)** **(i)** −5.2 **(ii)** −3.8

17.24 **(a)** **(i)** $L \approx 7 \times 10^{35}$ W
(ii) $M \approx -18$

17.26 **(a)** 1.4 kW m^{-2}

17.27 **(a)** 6.9×10^{11} m, about 4.6 AU
(b) **(i)** 9.8×10^{-8} W m^{-2}
(ii) 9.8×10^{-5} mW m^{-2}
(iii) nearly 1 mW cm^{-2}

17.29 5800 K

17.30 **(b)** 2.76 K

17.31 **(i)** 3.4×10^{31} W
(ii) $8.6 \times 10^{4} L_{\odot}$

17.32 $A = 1.3 \times 10^{17}$ m^2;
$r = 1.0 \times 10^{8}$ m

17.33 7.0×10^{8} m

17.34 **(a)** about 1.7 AU
(b) 4300 K, reddish

17.35 **(a)** 5.1

(b) (i) about 21
(ii) about 4, yes

17.36 yes; ratio of $\lambda_{max} \approx 1.5$

17.38 **(a)** Betelgeuse $1.9 \times 10^7\,W\,m^{-2}$, Bellatrix $3.5 \times 10^{10}\,W\,m^{-2}$
(b) Betelgeuse reddish, Bellatrix bluish

17.39 **(b)** 4.2 nm

17.40 $4.74 \times 10^7\,m\,s^{-1}$, about 15% of c

17.41 **(a)** $3.0 \times 10^5\,m\,s^{-1}$

17.42 **(a) (i)** $11\,km\,s^{-1}$
(ii) $19\,km\,s^{-1}$
(b) $30\,km\,s^{-1}$, 1 AU = $1.5 \times 10^{11}\,m$

17.43 **(a)** $I_{Vega}/I_{Deneb} = 2.8$

17.45 **(a) (i)** $4.3 \times 10^9\,kg\,s^{-1}$
(ii) 1.5×10^{12} years
(b) 4 to 5 billion years

17.47 **(a)** dwarf – blue
(b) supergiant – bluish
(c) giant – reddish
(d) Main Sequence – white
(e) supergiant – red

17.48 **(a) (i)** 11 **(ii)** 280

17.49 **(a) (i)** red giant
(ii) white dwarf
(b) becomes a cold black dwarf

17.51 about $1 \times 10^9\,kg\,m^{-3}$

17.52 just over 10 km

17.53 3.6%

17.54 **(c)** more massive stars spend less time on the MS

17.55 8.9 mm

17.56 13 km, perhaps a neutron star or a black hole

17.59 **(b)** $500\,km\,s^{-1}\,Mpc^{-1}$ [$1.6 \times 10^{-17}\,s^{-1}$]

17.60 **(a)** $74 \pm 4\,km\,s^{-1}\,Mpc^{-1}$ [$2.4 \pm 0.1 \times 10^{-18}\,s^{-1}$]

17.61 **(a)** 0.0216
(b) $6.48 \times 10^6\,m\,s^{-1}$
(c) 88 Mpc [$2.7 \times 10^{24}\,m$]

17.62 **(a)** larger H_0: 83 Mpc [$2.6 \times 10^{24}\,m$]; smaller H_0: 93 Mpc [$2.8 \times 10^{24}\,m$]
(b) about 30 *million* ly

17.63 **(a)** 400 Mpc [$1 \times 10^{25}\,m$]
(b) yes

17.65 **(a)** about $3 \times 10^{41}\,kg$
(b) about $1.5 \times 10^{40}\,kg$

17.66 **(a)** 17 Mpc
(b) $1.2 \times 10^6\,km\,s^{-1}$
(c) $z = 0.0041$

17.69 **(a)** $1.0 \times 10^{-26}\,kg\,m^{-3}$

17.70 6

18 Electric fields

18.1 **(a)** $1.2 \times 10^{-11}\,N$
(b) 1.2×10^7 times

18.2 **(a)** $1.1 \times 10^{-6}\,N$
(b) at $r = 30\,mm$, $F = 4.0 \times 10^{-5}\,N$

18.3 0.66 kN

18.4 $8.2 \times 10^{-8}\,N$

18.5 11 μC

18.6 2.0 μC

18.7 **(b)** 25 mN
(c) 2.8 nC

18.9 $kg^{-1}\,m^{-3}\,s^4\,A^2$

18.13 **(a) (i)** $4.8 \times 10^{-45}\,N$
(ii) $5.8 \times 10^{-9}\,N$
(b) 1.2×10^{36}

18.15 **(a)** +0.16 pC
(b) both $4.0 \times 10^{-11}\,N$

18.16 −1.4 nC

18.17 $6.4 \times 10^{-19}\,C$, $4e$

18.18 0.49 kV, upper

18.19 **(a) (i)** $+1.57 \times 10^{-16}\,J$
(ii) $-1.54 \times 10^{-16}\,J$
(b) some work is done by air friction forces

18.21 **(a)** $1.4 \times 10^{-14}\,N$, $7.5 \times 10^{-15}\,N$
(b) 29° with the vertical

18.22 **(a)** $2.0 \times 10^4\,N\,C^{-1}$
(b) $3.2 \times 10^{-15}\,N$
(c) $2.4 \times 10^{-17}\,J$

18.23 **(c)** 75 μs
(d) 0.62 mm

18.24 **(b)** $0.25\,kV\,m^{-1}$

18.25 **(b)** a vertical line

18.26 **(b)** $k = 9.1 \times 10^9\,N\,m^2\,C^{-2}$

18.27 **(a)** $5.1 \times 10^{11}\,N\,C^{-1}$
(b) $8.2 \times 10^{-8}\,N$
(c) $9.0 \times 10^{22}\,m\,s^{-2}$

18.28 0.33 μC

18.29 **(a)** 4.0 μC **(b)** $1.6\,MV\,m^{-1}$

18.31 **(b)** $-1.8\,nC\,m^{-2}$

18.32 **(a)** $1.1 \times 10^5\,V$

19 Capacitance

19.1 **(a)** ±0.12 C
(b) 0.25 kV

19.2 6.5 mF

19.3 2.0 V, 0.30 V; 10 pF, 0.47 F; 26 μC, 7.1 μC

19.4 **(a)** 4.0 mC
(b) 330 μF

19.5 **(a)** ±14 mC
(b) 3.5 mA

19.6 2.2 μA

19.7 200 μF

19.8 34 s

19.9 **(a) (i)** 0 **(ii)** 12 V
(iii) 12 μA **(iv)** 0
(b) (i) 12 V **(ii)** 0
(iii) 0 **(iv)** 26 mC
(c) (i) 9.5 V
(ii) 2.5 V
(iii) ± 55 μC

19.10 0.54 μA

19.11 4.0 μF

19.12 ×5

19.13 **(a)** 11 μF **(b)** 1.0 μF

19.14 **(b)** −60 μC, +60 μC
(c) 2.0 V, 1.0 V
(d) 3.0 V **(e)** 20 μF

19.17 **(a)** 6.0 μF
(b) A: 6.0 V, D: 3.0 V
(c) for B: 3.0 V, 4.5 μC

19.19 **(a)** 0.28 mC
(b) 69 μF
(c) 4.1 V
(d) 0.19 mC, 0.090 mC

19.20 **(b)** ±1%

19.21 **(b)** 0.069 μF ± 27%

19.22 **(a)** 1.2 μC
(c) 99%

19.23 **(a)** 1.8 J
(b) $1.8 \times 10^5\,W$

19.24 **(a)** 0.50 μF
(b) for $Q = 3\,\mu C$, $W = 9.0\,\mu J$

19.25 **(c)** 28 μC, 85 μJ

19.27 **(a) (i)** 1.1 C **(ii)** 1.2 J
(b) 0.61 C
(c) (i) 170 hours

(ii) 1.7 μW
19.29 **(a)** 0.29 mJ **(b)** 0.032 mJ
19.30 **(a)** 0.44 J **(b)** 0.22 J
19.31 **(a)** 4.0 mC
(b) 2.0 J
(c) 1.0 J
19.32 **(a)** 100 kJ **(b)** about 20 kV
19.34 **(a)** $t_{1/2}$ = 25 s, λ = 36 s
19.36 **(a)** 0.28 mC
(b) 6.0 μA, 47 s
(c) 47 s
19.38 **(a)** 13 μA **(b)** 6.7 μA
19.40 5
19.44 **(c)** =B$2*EXP(–A4/(C$2*D$2))
(d) **(i)** 8.08V, 6.53V, 5.28V, 4.27V
(ii) 14 s
19.45 **(a)** **(i)** 3.0 V s^{-1}
(ii) 0.66 mA
(b) 0.11 mA
19.46 **(a)** 1.5 s **(b)** 2.2 s
19.47 **(a)** **(i)** 88 μC
(ii) 1.0 μA
(iii) 88 s
(b) 89 s **(c)** just over 1%
19.49 **(c)** 32 μF

20 Electromagnetism

20.1 **(a)** 75 mN
20.2 **(b)** kg s^{-2} A^{-1}
20.3 0.11 T
20.4 **(a)** 9.6 N **(b)** 3.3 N
20.5 e.g. F_{OB} = 75 mN
20.6 56 mT, horizontal and perpendicular to AB
20.7 **(b)** **(i)** $F_{PQ} = F_{RS}$ = 14 mN, $F_{PS} = F_{QR}$ = 9.0 mN;
(ii) torque from forces on PQ and RS = 2.2 N m, torque from forces on PS and QR = 0
20.9 **(a)** 0.40 mm s^{-1}
(b) 1.6 × 10^{-23} N
(c) 3.1 × 10^{22}
(d) 0.50 N
20.11 **(a)** 4.8 × 10^{-14} N
20.12 **(a)** 9.6 × 10^{-12} N
(b) circular arc
(c) 1.5 × 10^{15} m s^{-2}
20.13 **(a)** **(i)** eV/d **(ii)** Bev
(c) 420 V
20.14 **(a)** $r = mv/Bq$
(b) $T = 2\pi m/Bq$
20.15 1.1, 22m
20.16 **(a)** **(i)** 46 mm **(ii)** 14 ns
(b) 14 ns
20.17 **(b)** 20 MHz
(c) 500, 25 μs
(e) about 1%
20.18 **(c)** e.g. both operate in a good vacuum
20.19 **(b)** at Q: p = 8.8 × 10^{-21} N s
(c) at Q: $m = 32m_e$
20.22 **(a)** 21 μT **(b)** 45 μT
20.24 **(a)** 0.33 mT **(b)** 0.060 mT
20.25 0.20 m
20.26 **(a)** 0.59 μT **(b)** two cables
20.27 **(b)** $B_{centre} = 2B_{ends}$
20.28 2.4 kA
20.32 4.5 mV
20.34 **(b)** 0.70 V
20.35 2.6 mV
20.36 **(b)** about 0.3 mV
(c) 5 mΩ
20.37 0.35 Wb or 0.35 T m^{-2}
20.38 **(a)** 15 mWb
(b) **(i)** 0
(ii) 7.5 mWb
(iii) 0
(iv) 15 mWb
(v) 30 mWb
(vi) 0
20.39 1.6 T
20.40 0.58 Wb
20.43 for both consider the effect of the induced current
20.45 **(a)** 17 μWb
(b) –17 μWb
(c) 12 μV
20.46 **(a)** 20 mV **(b)** 1.7 A
20.47 **(b)** 10 μWb s^{-1}
20.48 **(b)** **(i)** $\times\frac{1}{2}$
(ii) same
(iii) ×2
(c) **(i)** $\times\frac{1}{4}$
(ii) same
(iii) × 4
20.50 **(a)** 0.50 mV clockwise, zero, 0.50 mV anticlockwise
(b) 0.25 A clockwise, zero, 0.25 A anticlockwise
20.51 **(b)** 28 T s^{-1} **(c)** 14 V
20.53 **(a)** 240 V **(b)** 0.10 A
(c) 24 W
20.54 **(a)** 3.6 W
(b) 7.2 W
20.55 **(b)** 0.48 W
(c) 75 A
(d) 3.0 g
20.56 **(a)** 110
(b) 5.5 A
(c) 45 mA
20.58 **(b)** about 3 kV